AUTOMATING BUSINESS PROCESS REENGINEERING

PRENTICE HALL INTERNATIONAL SERIES IN INDUSTRIAL AND SYSTEMS ENGINEERING

W.J. Fabrycky and J. H. Mize, Editors

AMOS AND SARCHET *Management for Engineers*
AMRINE, RITCHEY, MOODIE AND KMEC *Manufacturing Organization and Management, 6/E*
ASFAHL *Industrial Safety and Health Management, 3/E*
BABCOCK *Managing Engineering and Technology, 2/E*
BADIRU *Comprehensive Project Management*
BADIRU *Expert Systems Applications in Engineering and Manufacturing*
BANKS, CARSON AND NELSON *Discrete Event System Simulation, 2/E*
BLANCHARD *Logistics Engineering and Management, 4/E*
BLANCHARD AND FABRYCKY *Systems Engineering and Analysis, 2/E*
BURTON AND MORAN *The Future Focused Organization*
BROWN *Technimanagement: The Human Side of the Technical Organization*
BUSSEY AND ESCHENBACH *The Economic Analysis of Industrial Projects, 2/E*
BUZACOTT AND SHANTHIKUMAR *Stochastic Models of Manufacturing Systems*
CANADA AND SULLIVAN *Economic and Multi-Attribute Evaluation of Advanced Manufacturing Systems*
CANADA, SULLIVAN AND WHITE *Capital Investment Analysis for Engineering and Management, 2/E*
CHANG AND WYSK *An Introduction to Automated Process Planning Systems*
CHANG, WYSK AND WANG *Computer Aided Manufacturing*
DELAVIGNE AND ROBERTSON *Deming's Profound Changes*
EAGER *The Information Payoff: The Manager's Concise Guide to Making PC Communications Work*
EBERTS *User Interface Design*
EBERTS AND EBERTS *Myths of Japanese Quality*
ELSAYED AND BOUCHER *Analysis and Control of Production Systems, 2/E*
FABRYCKY AND BLANCHARD *Life-Cycle Cost and Economic Analysis*
FABRYCKY AND THUESEN *Economic Decision Analysis*
FISHWICK *Simulation Model Design and Execution: Building Digital Worlds*
FRANCIS, MCGINNIS AND WHITE *Facility Layout and Location: An Analytical Approach, 2/E*
GIBSON *Modern Management of the High-Technology Enterprise*
GORDON *Systematic Training Program Design*
GRAEDEL AND ALLENBY *Industrial Ecology*
HALL *Queuing Methods: For Services and Manufacturing*
HANSEN *Automating Business Process Reengineering, 2/E*
HAMMER *Occupational Safety Management and Engineering, 4/E*
HAZELRIGG *Systems Engineering*
HUTCHINSON *An Integrated Approach to Logistics Management*
IGNIZIO *Linear Programming in Single- and Multiple-Objective Systems*
IGNIZIO AND CAVALIER *Linear Programming*
KROEMER, KROEMER AND KROEMER-ELBERT *Ergonomics: How to Design for Ease and Efficiency*
KUSIAK *Intelligent Manufacturing Systems*
LAMB *Availability Engineering and Management for Manufacturing Plant Performance*
LANDERS, BROWN, FANT, MALSTROM AND SCHMITT *Electronic Manufacturing Processes*
LEEMIS *Reliability: Probabilistic Models and Statistical Methods*
MICHAELS *Technical Risk Management*
MOODY, CHAPMAN, VAN VOORHEES AND BAHILL *Metrics and Case Studies for Evaluation Engineering Designs*
MUNDEL AND DANNER *Motion and Time Study: Improving Productivity, 7/E*
OSTWALD *Engineering Cost Estimating, 3/E*
PINEDO *Scheduling: Theory, Algorithms, and Systems*
PRASAD *Concurrent Engineering Fundamentals, Vol. I: Integrated Product and Process Organization*
PRASAD *Concurrent Engineering Fundamentals, Vol. II: Integrated Product Development*
PULAT *Fundamentals of Industrial Ergonomics*
SHTUB, BARD AND GLOBERSON *Project Management: Engineering Technology and Implementation*
TAHA *Simulation Modeling and SIMNET*
THUESEN AND FABRYCKY *Engineering Economy, 8/e*
TURNER, MIZE, CASE AND NAZEMETZ *Introduction to Industrial and Systems Engineering, 3/E*
TURTLE *Implementing Concurrent Project Management*
VON BRAUN *The Innovation War*
WALESH *Engineering Your Future*
WOLFF *Stochastic Modeling and the Theory of Queues*

AUTOMATING BUSINESS PROCESS REENGINEERING

Using the Power of Visual Simulation Strategies to Improve Performance and Profit

Second Edition

Gregory A. Hansen

To join a Prentice Hall PTR Internet mailing list, point to http://www.prenhall.com/register

PRENTICE HALL PTR
Upper Saddle River, New Jersey 07458
http://www.prenhall.com

Library of Congress Cataloging-in-Publication Data

Hansen, Gregory A.
Automating business process reengineering: using the power of visual simulation strategies to improve performance and profit/ Gregory A. Hansen.—2nd ed.
p. cm.—(Prentice-Hall international series in industrial and systems engineering)
Includes bibliographical references and index.
ISBN 0-13-576984-1
1. Organizational change—Computer simulation. 2. Organizational effectiveness—Computer simulation. 3. Quality control—Computer simulation. 4. Industrial management—Computer simulation.
I. Title. II. Series
HD58.8.H3633 1997
658.4'06'0113—dc20

96-42332
CIP

Production Editor: *Kerry Reardon*
Acquisitions Editor: *Bernard Goodwin*
Cover Designer: *Lori Obermeyer*
Cover Design Director: *Jerry Votta*
Marketing Manager: *Dan Rush*
Manufacturing Manager: *Alexis R. Heydt*

The publisher offers discounts on this book when ordered in bulk quantities. For more information contact: Corporate Sales Department, Prentice Hall PTR, One Lake Street, Upper Saddle River, NJ 07458. Phone: 800-382-3419, FAX: 201-236-7141, E-mail: corpsales@prenhall.com

Printed in the United States of America

10 9 8 7 6 5 4 3 2 1

ISBN 0-13-576984-1

Prentice-Hall International (UK) Limited, *London*
Prentice-Hall of Australia Pty. Limited, *Sydney*
Prentice-Hall Canada Inc., *Toronto*
Prentice-Hall Hispanoamericana, S.A., *Mexico*
Prentice-Hall of India Private Limited, *New Delhi*
Prentice-Hall of Japan, *Tokyo*
Simon & Schuster Asia Pte. Ltd., *Singapore*
Editora Prentice-Hall do Brasil, Ltda., *Rio de Janeiro*

Contents

Preface

As the global economy becomes more competitive, business is moving away from a focus on customer satisfaction and quality and back to a focus on the bottom line. Total Quality Management (TQM) and Continuous Process Improvement (CPI), both of which emphasize customer satisfaction, are giving way to Business Process Reengineering (BPR) and an emphasis on the bottom line. We can see this change in focus in just about every business. For example,

- Customer service centers are moving to automated attendants, requiring customers to wade through a maze of selections to get the answer they want, if they get it at all.
- Computer hardware vendors are placing software updates on the Internet, requiring their customers to first find the software and then download it.
- A major software developer released a new operating system with known defects in order to meet artificial deadlines established by its marketing department.

These are all examples of decisions made by strategic, financial, and operational managers, and they are aimed at cutting costs. Trade-offs are being made between customer satisfaction and profit. This is a radical departure from the 1980s philosophies that quality and customer satisfaction come first. In fact, these examples may prove that the commitment to quality never really existed, and that business managers simply paid lip service to those concepts.

Whatever the case, many of these decisions have been made as a result of Business Process Reengineering efforts undertaken within the companies.

Unfortunately, many decisions of the types described above are not providing the returns that the businesses expected. Unanticipated issues arise that affect the cost of the changes that have been implemented, and not enough analysis has been done to deal with those issues. For example,

- Some companies have found resistance to automated phone attendants and many people simply wait to talk to a human. This results in an increase in staff over and above the staff level projected in the BPR analysis. Therefore, anticipated costs savings from staff reductions do not materialize.
- Many hardware vendors have found the number of people who actually use the Internet for serious business purposes are fewer than imagined. Moreover, executive-level personnel are accustomed to a level of service that does not include downloading software for several hours. Therefore, these vendors have not been able to decrease their customer service staff levels and their costs are as high as ever.
- Sales of the software vendor's new operating system were not what it had expected, even though there were brisk sales upon its release. Many customers were simply waiting for the next release and, in fact, many corporations are refusing to update their operating systems until the software becomes stable.

When problems such as these arise, it is Business Process Reengineering that is blamed. BPR is starting to be called a failure. The advocates of BPR, not willing to accept the blame for its failure, point the finger back at management. They claim that management has not made enough of a commitment to change.

This theory may be true; however, the failure of BPR is more deeply rooted. The simple fact is this—business and government organizations continue to attempt to improve their processes by using time-worn techniques that have not worked in the past, are not working in the present, and will not work in the future. Compounding this problem is the plethora of books pretending to present methods to accomplish the radical change BPR promises, but these books are no more than presentations of the failed techniques wrapped in different words.

It is time to move away from nebulous goal setting and "warm fuzzy" process analysis techniques and move toward empirical measures of success and cold, hard data.

Compare these two goal statements:

1. "Implement a strategic planning and prioritization process based on customers' needs."[1]
2. "Implement a strategic planning prioritization process that increases profits by 25 percent."

[1]Andrews and Stalick, *Business Reengineering: the Survival Guide,* (Englewood Cliffs, NJ: Yourdon Press, 1995), p. 44.

Or these two goal statements:

3 "Implement an application development process that involves customers upfront and assures quality."[2]

4 "Implement an application development process that reduces cycle time by 30 percent by eliminating rework."

Which of these goal statements makes more sense to you? Which of these goal statements defines goals you can measure? Which of these goal statements can be used to determine the success of an effort? The first and third goal statements are typical of an approach to BPR that is similar to the approaches that have been in TQM and CPI, that is, those approaches that have failed to deliver long-lasting effective change. The second and fourth statements are the type that will be discussed in this book—the type business leaders should be making.

The focus of BPR *is* correct. BPR views business as a whole—as a *system*—and BPR advocates suggest that business must use new techniques to measure its performance and to make changes. Unfortunately, many BPR advocates are simply recycling old approaches to change and, if this trend continues, BPR will be declared a failure and the gains that have been made will be lost.

Fortunately, the move to empirical analysis of business processes is gaining momentum. In addition, there has been movement toward technological assistance for BPR. The use of modeling and simulation, although not yet widespread, seems to be increasing exponentially. There is a wide variety of BPR simulation tools available—this book will help management personnel and BPR practitioners decide not only when to invest in BPR modeling and simulation technology but also how to most effectively use that technology.

[2]Ibid.

About the Author

Gregory A. Hansen is President of Computer Aided Process Improvement (CAPI), a firm specializing in the application of modeling and simulation to business process definition, analysis and reengineering. Mr. Hansen has more than twenty years experience applying emerging software technology to organizational analysis and process improvement. He is an entrepreneurial, quality-oriented professional with a reputation for achieving optimum results and customer satisfaction, and a proven leader with a record of building highly motivated and productive teams.

Mr. Hansen developed the simulation models contained on the CD that accompanies this book. The following describes the simulation models and their references in the book:

1. IBMASIS and IBMTOBE are both referenced in Chapter 1 (So You Don't Think You Need Simulation)
2. DOCREV is referenced in Chapter 15 (Process Reengineering Case Studies)
3. PILOT is referenced in Chapter 9 (Modeling and Simulation Terminology and Techniques)
4. THERMO is not referenced in the book, but is an example of a "balancing loop." It is here for your enjoyment.
5. The models in the CREDIT Folder are referenced in Chapter 13 (Developing Simulations: Step by Step Examples)

6 The models in the SUPPORT folder are referenced in Chapter 16 (Applications of Computer Aided Process Reengineering)

If you have questions about these models, please contact CAPI at:

Computer Aided Process Improvement
830-13 A1A North
Suite 327
Ponte Vedra Beach, FL 32082 USA
PH: 904-285-2126 Fax: 904-285-3272
Email: 73024.542@compuserve.com
Web Site: www.capi.net

About the CD-ROM

The CD contains Extend RunTime along with several of the models discussed in the book. The RunTime program is a limited version of the Extend application which allows you to run simulation models, make changes to model parameters, and save and print the results. Unlike the full version of Extend, the RunTime version does not allow you to build your own models or build your own blocks (the components of models). Using the RunTime application, you can run the example models as they are configured in the book. Alternatively, you can make changes to the models, run the simulation, and view the results of those changes. You can save your changes and you can print the picture of the model, the contents of block dialogs, and any graphs that the simulation run generates.

Instructions for installing and using the Extend RunTime application and example models:

Macintosh

- Requirements: PowerMacintosh or Macintosh 68020+ computer running System 6.0.7 or better, with 8MB of RAM and 10MB of hard disk space.
- Installation: The Extend RunTime application and example files must run on your hard drive, not on the CD. To install the files onto your hard drive, first close the folder "Automating BPR" and then copy it from the CD to your hard drive.
- Use: To open a model, double-click the "Automating BPR" folder to open it, then either double-click the Extend application and "Open" a model

from the File menu, or double-click a model file directly. To change values, double-click a block icon on the model worksheet and make changes in the block's dialog. Run (simulate) the model using the commands in the Run menu. Print and Save using commands in the File menu.

- For more information: See the ReadMe file in the "Automating BPR" folder.

Windows

- Requirements: 386, 486, Pentium, or Pentium Pro computer running Windows 3.1, Windows 95, or Windows NT 3.5+, with 8MB of RAM and 12MB of hard disk space.
- Installation: The Extend RunTime application and example files must run on your hard drive, not on the CD. To install the files onto your hard drive:

1. Launch the installer:
   - For Windows 3.1 and Windows NT 3.5, select "Run" from the File menu in the Program Manager.
   - For Windows 95 and Windows NT 4.0, select "Run" from the Start menu in the Task Bar.
2. In the dialog that appears, type the letter of the CD drive followed by "Install" (for example, if the CD is drive D, type "D:Install".) Click "OK" and the Installer will begin loading files.
3. When the "wINSTALL Message" dialog appears, select "OK" or press the Enter key to begin the installation.
4. When prompted, choose a hard drive location and the name of a directory for installation ("autobpr" is suggested).
5. As it runs, the Installer will create a short-cut (Windows 95 or Windows NT 4.0) or a program group (Windows 3.1 or Windows NT 3.5) on your hard drive.

Note: If you are using Windows 3.1, you will also need Win32s version 1.25. This Microsoft operating system extension provides the performance advantages of 32-bit processing. If Win32s is not present on your computer, or if the version you have is older than 1.25, the Extend RunTime Installer application will automatically install it for you. The Installer also gives you the option of reinstalling Win32s if you choose. Choose this option if you are not sure if your computer's current installation of Win32s is complete.

- Use: To open a model, access the Programs submenu in the Start menu and click the Extend application (Windows 95 or Windows NT 4.0), or double-click the Extend group in the Program Manager window, then double-click the Extend icon (Windows 3.1 or Windows NT 3.5). Once the application has been launched, "Open" a model from the File menu. To change values, double-click a block icon on the model worksheet and make changes in the block's dialog. Run (simulate) the model using the commands in the Run menu. Print and Save using commands in the File menu.

- For more information: See the ReadMe file in the "autobpr" folder.

AUTOMATING BUSINESS PROCESS REENGINEERING

1

So You Don't Think You Need Simulation

All politics is local.
Tip O'Neill, former Speaker of the House of Representatives

All BPR is local.
Sandra Yin, Internal Revenue Service

Both statements are elegant in their simplicity. The comment by Speaker O'Neill explains why national politics never seems to accomplish anything—politicians deal with local problems that affect their constituents before addressing larger problems. Similarly, Ms. Yin's comment explains why Business Process Reengineering (BPR) is not providing the revolutionary changes we have expected. Because the tools and techniques used in BPR efforts are the same tools and techniques used in Total Quality Management (TQM) and Continuous Process Improvement (CPI), BPR can only address a limited set of process parameters. Therefore, BPR, as most commonly practiced, deals only with change in the small, that is, it is local.

Think about the methods used by TQM and CPI practitioners: interdisciplinary teams, cross-functional teams, quality circles, and so on. These terms all mean the same thing—create a team of people to discuss ways of improving a process. Usually the team's goal is to recommend process changes quickly, and for the changes to be implemented quickly. Unfortunately, the emphasis on quick change eliminates the possibility of making meaningful, long-lasting changes.

The problem is this—business processes are too complex to be analyzed and improved simply by bringing people together to talk about them. Process parameters, such as time to complete a task, material flow, staff size, and so on, determine to some degree how a process behaves. As the number of process parameters increases, the complexity of the process increases as well.

For example, a process with two parameters can have four possible change scenarios: You can change parameter 1, you can change parameter 2, you can change parameters 1 and 2, or you can change nothing. When you have a process with four parameters, you will have 16 possible change scenarios. With ten parameters, you will have 1024 change scenarios! In fact, for every "*n*" change parameters, there are 2^n possible change scenarios, including the scenario in which nothing is changed. The tools and techniques currently being used in most BPR efforts cannot deal with this level of complexity.

In order to address large-scale change, business processes must be treated as systems, and a systems analysis approach must be used in process improvement efforts. That approach must utilize modeling and simulation, or what is called Computer Aided Process Reengineering (CAPRE™) technology. CAPRE technology is highly graphical, rule-based modeling and simulation software specifically designed to be used in BPR activities.

Systems Analysis Approach

To demonstrate the need for a systems analysis approach, I will use an example from Hammer and Champy's book *Reengineering the Corporation* (HarperCollins, 1993). This is the IBM credit example described on pages 36–39. On those pages, the process is described as follows:

1. Field sales personnel call in requests for financing to a group of 14 people. (Step 1 begins.)
2. The person taking the call logs information on a piece of paper. (Step 1 ends.)
3. The paper is taken upstairs to the Credit Department. (Step 2 begins.)
4. A specialist enters the information into a computer system and does a credit check. (Step 2 continues.)
5. The result of the credit check is written on a piece of paper and sent to the Business Practices Department. (Step 2 ends.)

6 Standard loan contracts are modified to meet customer requirements. (Step 3.)

7 The request is sent to a "pricer," where an interest rate is determined. (Step 4.)

8 The interest rate is written on a piece of paper and sent to a clerical group. (Step 5 begins.)

9 A quote is developed. (Step 5 ends.)

10 The quote is sent to Field Sales via Federal Express. (Step 6.)

The process, according to Hammer and Champy, took an average of six days, and sometimes as long as two weeks. Because of this delay, customers occasionally canceled deals. IBM management decided to correct this problem.

Two IBM managers had a "brainstorm." They decided to walk through the process and determine exactly how much time was required to process a loan request. They discovered that the process took only 90 minutes! They determined (apparently) that the delays were caused by "hand-offs," or transfers between departments. IBM management decided to "reengineer" the process by replacing the specialists involved in the current process with "deal structurers," or generalists.

In the reengineered process, the generalists would handle the process from beginning to end. To facilitate processing, the generalists would receive assistance from a sophisticated computer system and, when there was a problem, would seek help from one of the remaining specialists.

BPR at Its Best?

When this new process was implemented, cycle time was down to four hours, and a "small head count reduction" was achieved. The reduction of cycle time from six days to four hours seems to be very significant, and this example seems to be an example of BPR at its best. But is it?

There are a number of questions that must be asked about this example, not the least of which is the following: If the application processing time is 90 minutes, why does the new process have an average cycle time of four hours?

To demonstrate the power of a systems analysis approach to BPR, and the value of CAPRE technology, I examined the process in more detail to determine if a better solution could be found. I used the dynamic modeling and simulation tool Extend+BPR™ to develop a model of the process and to test scenarios.

Consider the information about the process that we have. Processing time is 90 minutes, and cycle time is six days. Without any other information, and based on the fact that IBM created a new process to eliminate hand-offs, we can assume that the six-day cycle time is caused by the hand-off delays. But, there is a great deal of information missing, such as,

1 At what rate do phone calls arrive?

2 How many people (total) are involved in the process?

3 Is processing credit applications the only task the people in the process perform, or, if not, is it their top priority task?

To develop a model of the process, I had to make certain assumptions, such as,

1 The rate of phone calls will not overwhelm the phone call takers.
2 There is one worker for each task in the process for every phone call taker. In other words, for each of the 14 phone call takers, there is one person in each of the other four tasks, or a total of 70 people in the process. This is not out of line considering Hammer and Champy's description of the process.
3 Total labor, or processing, time is 90 minutes, divided up more or less evenly among the five tasks.
4 Hand-offs between tasks range between one and two days and take an average of one and one-half days.
5 Each worker in the process makes credit application processing his or her top priority task. This is a very important assumption because, if processing credit applications is not the top priority task, IBM's basis for changing the process is seriously flawed.

Using Extend+BPR™ [1], I developed the model shown in Figure 1.1. The model depicts one set of five application processors, since my assumption was that there are 14 identical sets of processing teams. Using the times discussed earlier and shown in the model, average processing time turns out to be seven days, including a one-half day delay for Federal Express. (It is not clear from the description in the book if the six-day

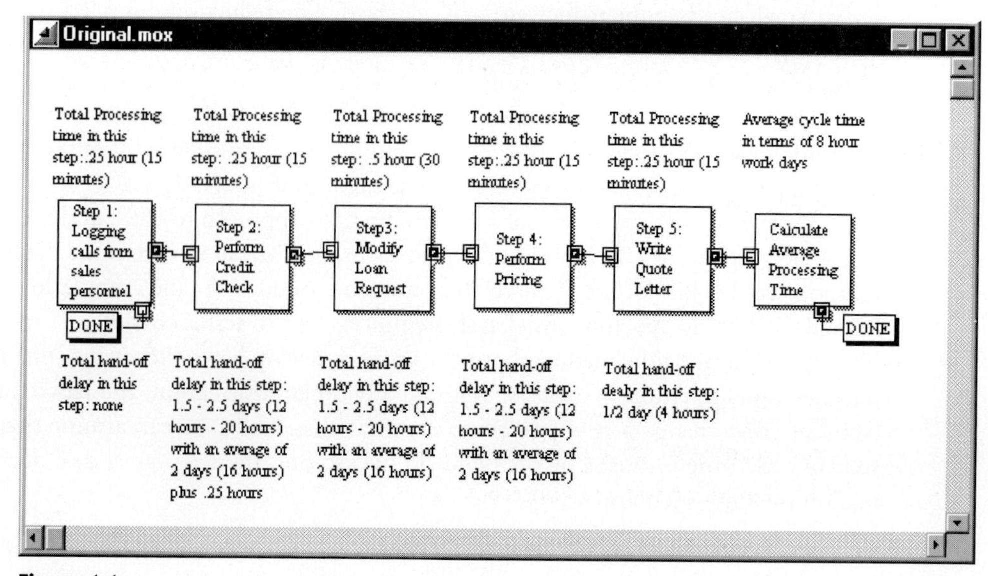

Figure 1.1

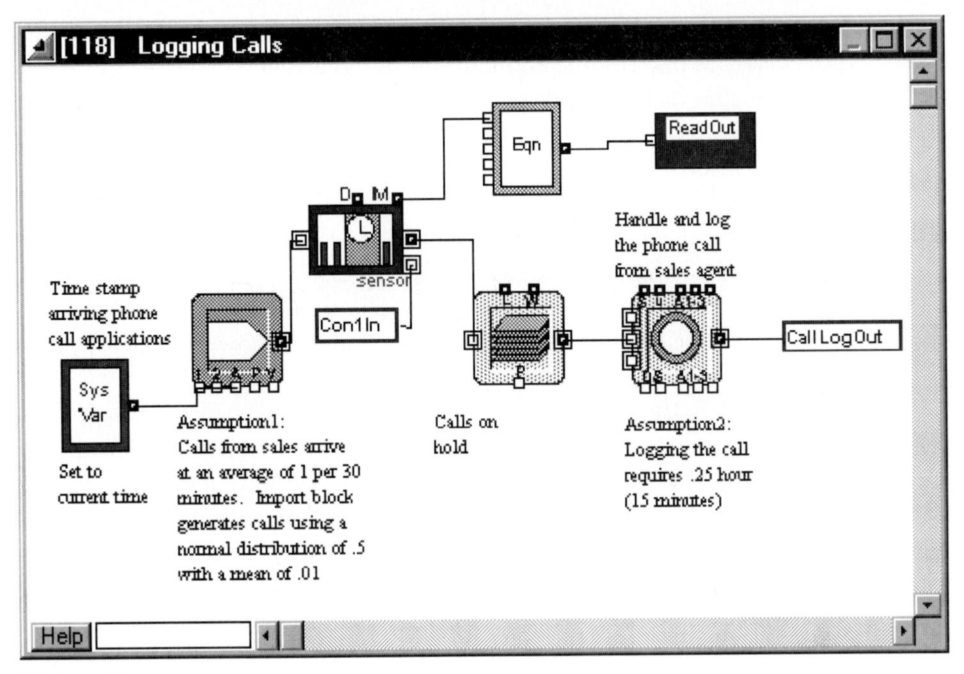

Figure 1.2

average cycle time included the Federal Express delay.) Cycle time is expressed in terms of 8-hour working days, rather than 24-hour days. Since we are dealing in terms of days and measuring cycle time, not productivity, this distinction is not important.

Each of the blocks shown in Figure 1.1 is a hierarchical block, containing more detail. For example, an exploded view of Step 1 is shown in Figure1.2.

Process Parameters

Before proceeding, let's stop and consider all the parameters for the process that have already been defined.

1 Processing time for each step
2 Hand-off delays between each step
3 Federal Express delivery time
4 Priority of tasks at each step in the process
5 Number of workers in the process
6 Rate of arrival of phone calls

[1]Extend and Extend+BPR are trademarks of Imagine That, Inc., San Jose, California.

Hammer and Champy suggest that IBM focused on hand-off delays and viewed those delays as one parameter. Hand-off delays actually represent five separate parameters, one for each task. For example, the hand-off between Task 1 and Task 2 could be one day, while the hand-off between Task 3 and Task 4 could be three days, and so on. Hand-offs, therefore, provide 32 change scenarios themselves!

Similarly, the priority each processor applies to credit applications is another important parameter and, like hand-offs, provides another 32 possible change scenarios. It is possible, for example, that one specialist treats credit application processing as a low priority task and processes applications only once per week. This situation would cause large cycle time delays, regardless of how the other specialists prioritized the processing of credit applications.

Whatever the case, the Hammer and Champy example suggests that IBM decided to eliminate all hand-off delays and to prioritize credit application processing by assigning personnel to handle the process from beginning to end. The example further suggests that significantly fewer personnel were involved in the reengineered process but does not provide any details of staff levels. Therefore, a question comes to mind: If there are, in fact, 14 sets of five personnel involved in the process, how many generalists are required to do the work of those people? Can one generalist replace all five of the people in a sequence of tasks? Remember I assumed that there are four other personnel for each phone call taker.

To determine if one generalist can replace the phone call taker and the four specialists, I built a simple Extend+BPR model, shown in Figure 1.3.

Figure 1.3 depicts an *Import* block, which is used to introduce work into a model; a *Stack* block, which is used to provide storage for items of work; an *Operation* block,

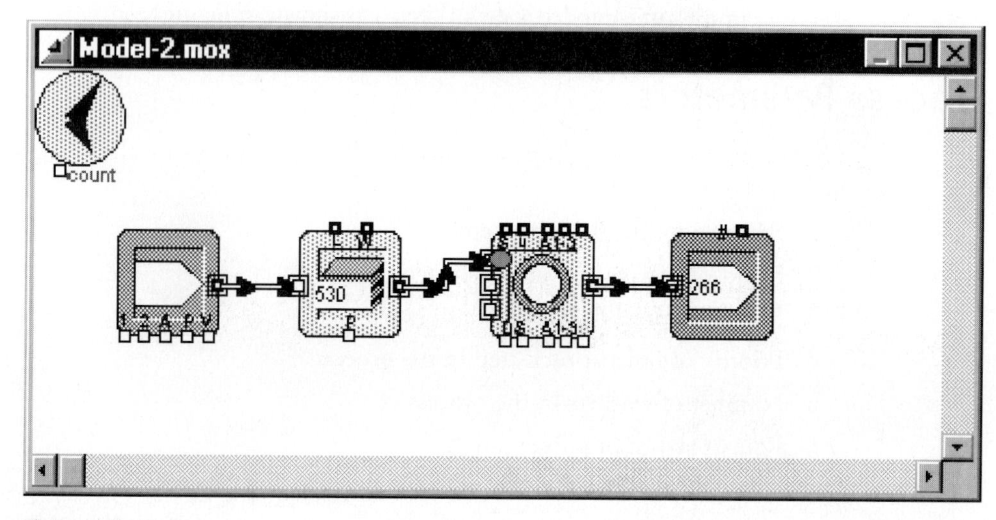

Figure 1.3

which is used to represent work in progress; and an *Export* block, which is used to represent items of work leaving the model. In this model, phone calls are introduced at the same rate as in the original model, that is, at an average rate of one call every 30 minutes.

Based on the information contained in Hammer and Champy's book, I chose 15 minutes as the time required to perform the credit application review process. When this model is executed for a simulated 400 work hours, or ten weeks, we can see that 266 applications are processed, but more than 500 remain to be processed. Clearly, one individual cannot replace the five individuals in the original process. Now the question is, how many individuals are required?

Using the sensitivity analysis feature of Extend, I ran the model using two, three, and four workers. The model with three workers provides the results closest to those described by Hammer and Champy. Graphically, cycle time for credit applications is shown in Figure 1.4.

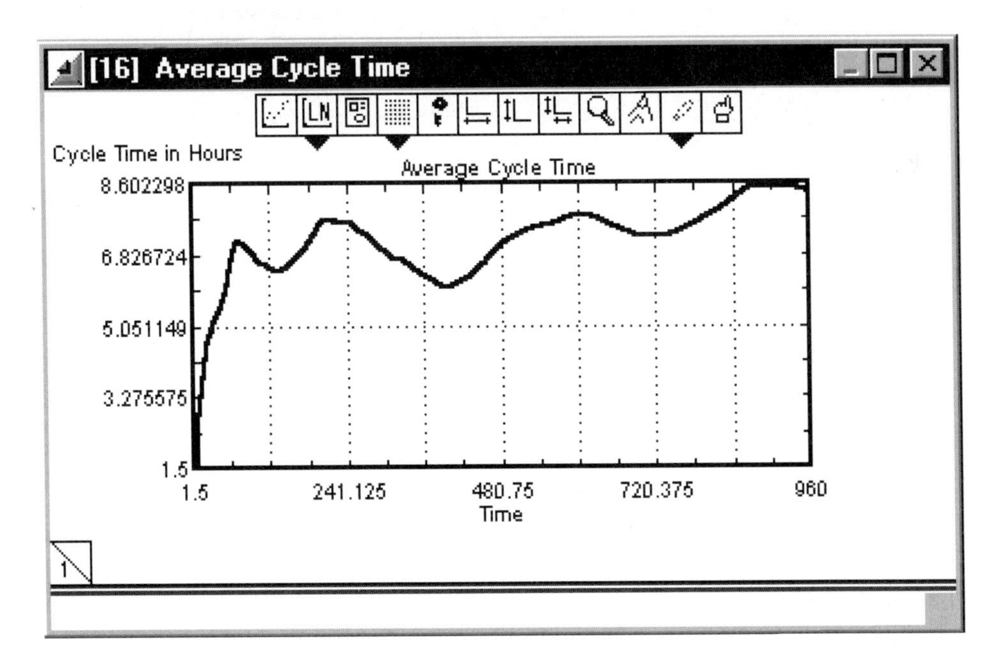

Figure 1.4

Disturbing Plot

This plot is somewhat disturbing, however. First, it shows that cycle time is as high as 9.6 hours, not the 4-hour average mentioned by Hammer and Champy. Also, cycle time is on an increasing slope, implying that, perhaps, three workers are not quite enough. However, let us assume for the time being that three workers can handle the work that was previously being performed by five workers.

This would certainly seem like a success story and demonstrate the power of management by walking around. However, there is one factor that has not yet been addressed: Hammer and Champy stated that, on occasion, the generalists would have to seek help from a "small pool" of specialists. One must ask—how often does this occur, how long does it take to find a specialist, how long does it take to resolve the problem, and will the three generalists still be sufficient to handle all the work required?

To answer these questions, we must make some more assumptions. First, we must assume that the generalist will make every effort to resolve any issues associated with the credit application before seeking help from a specialist. Therefore, the generalist will work on the application for the normal average of 90 minutes. Second, we must assume that the specialists are not readily available; therefore, some time is spent tracking down a specialist to assist in the problem. We must then assume that the specialist spends some time reviewing the credit application and solving the problem and that the generalist stays with the problem until it is resolved. After all, the key to the process is the elimination of hand-off delays, so it is unlikely that the credit application will simply be put into interoffice mail or laid in an in-box.

The problem with determining what happens in these situations is that we have no hard data to work from; therefore, we have to make some logical guesses. Note that we have now added three possible change parameters to be analyzed: the time required to find a specialist, the percentage of credit applications that need a specialist's help, and the time required for the specialist to resolve the problem. Analysis of the process now becomes very difficult.

To determine what happens to cycle time when generalists require help, I developed a second model of the process in which there are *four* generalists, and the generalists request help from the specialists about a certain percent of the time. Using the sensitivity analysis capability of Extend+BPR, I executed four change scenarios, varying the percentage of requests for help from 5 percent to 20 percent, increasing the amount in increments of 5 percent for each run. Figure 1.5 shows the results of this simulation.

This graph shows that as the percent of problems the cycle time increases. Even at the lowest level of percent and time, the cycle time has increased and, more importantly, the graphs reveal an increasing slope. An increasing slope of cycle time indicates that applications are being stacked up awaiting processing, and that cycle time will increase ad infinitum.

Changing Scenarios

This model also demonstrates that four generalists are insufficient to handle the flow of applications. Now the question is, How many workers are required? Note that I have not yet considered the availability of specialists or the number of specialists available. I have worked with only one parameter—the percentage of problems—and

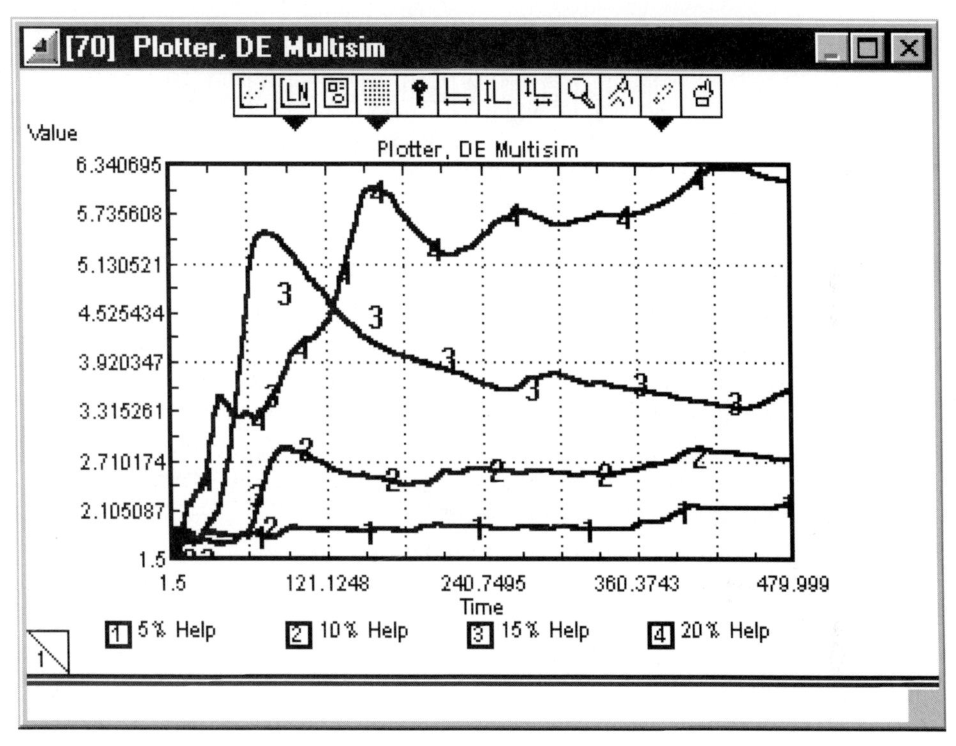

Figure 1.5

the change scenarios have already become very complex. Remember, for *n* process parameters, there are 2^n change scenarios.

Figure 1.6 shows the results of changing scenarios when five generalists are available.

This graph reveals that with five generalists, cycle times are more in line with those presented by Hammer and Champy. We know now that five generalists are required to replace the five original workers. This seems like a great deal of effort simply to get back to where you were in the first place.

Engineered First

Is this a good example of Business Process Reengineering? Regardless of the fact that it is published in what is considered the seminal work on the subject of BPR, the answer is *no*. This is a good example of the type of results one might expect if the techniques of TQM and CPI are applied to a problem. This is an excellent example of process reimplementation and of change in the small, that is, change accomplished by analyzing a small number of parameters in isolation.

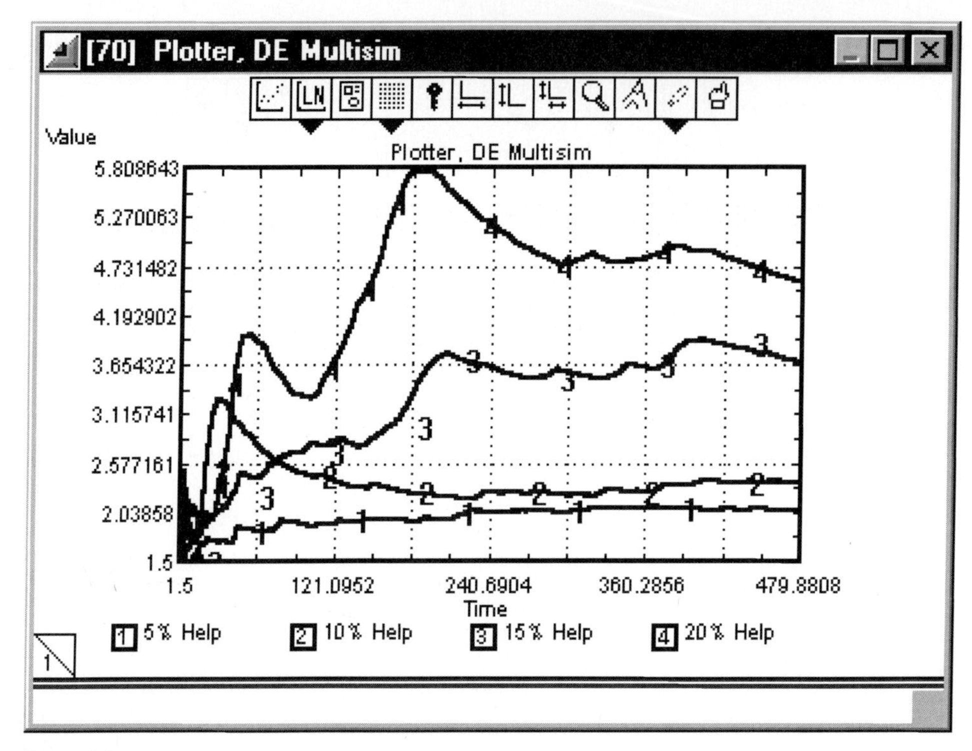

Figure1.6

Now ask yourself: Can we do better? Is there some way to look at a process and its parameters and predict the results of changes? If we return to the original process, there are two important parameters that can drive cycle time: the delays associated with hand-offs and the priority the workers in the process give to credit application processing.

What if we made no changes to the process at all other than eliminating hand-off delays (including the Federal Express delay) and ensuring that a credit application was handled as soon as it was received? What would cycle time be under those conditions? When an Extend+BPR model of the process is run under these conditions, the cycle time is revealed to be 0.5 day, or four hours, exactly the same result achieved by IBM after restructuring the process.

Now consider what would happen if there were some way to reduce the amount of time required to perform each step in the process. What if phone calls were eliminated and field sales personnel submitted applications by electronic data interface? That would take the phone call handling time down to zero. It might also reduce processing time at each step slightly, since processors would no longer be retyping vital information.

Figure 1.7 is a model of the original process with hand-off delays removed, with the time for phone call processing set to zero, and with a 10 percent decrease in processing time at each step in the process. Note that cycle time is now only slightly more than one hour, a vastly superior (400 percent better) result than was achieved by IBM.

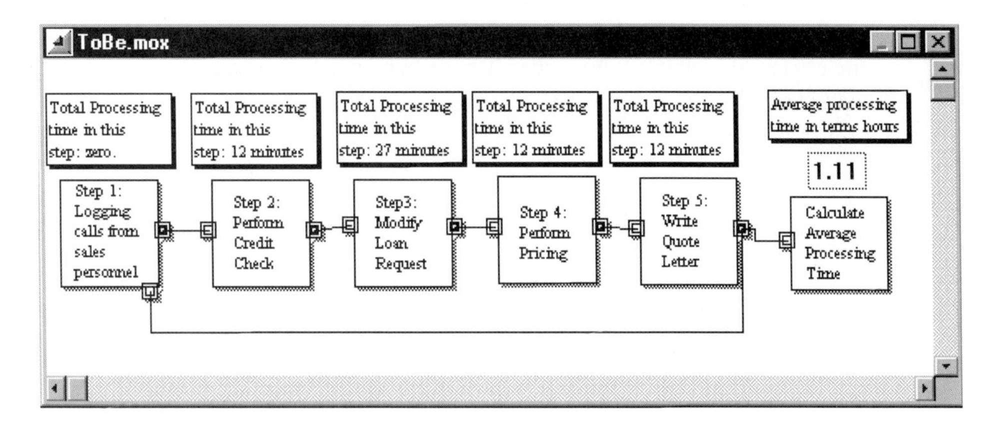

Figure 1.7

Is this a good example of BPR? Hammer and Champy define reengineering as "the fundamental rethinking and redesign of business processes to achieve dramatic improvements in critical contemporary measures of performance." This definition implies that business processes have been thought out or designed somewhere in the past. I contend that business processes evolve over time and are not designed; therefore, before a process can be reengineered, it must be engineered.

I contend that by establishing the process parameters and experimenting with those parameters, we have engineered the process just described. Our experimentation allowed us to find an optimum solution to the cycle time problem, better than the solution found by IBM. Therefore, the example I presented is a good example of business process reengineering.

Is this the best result that can be achieved? Probably not. If each field sales office had a system that could do all the work required to process an application, then the need for the home office process would be eliminated altogether. However, that might require an investment in technology that would be cost prohibitive at this time. It is a process that should remain under consideration, since technology is increasing in sophistication and decreasing in cost every day.

This leads to the concept of process management. Modeling and simulation allow a company to continually test change scenarios for a process, a capability that is missing from current popular BPR practices.

This example demonstrates conclusively that business processes are too complex to be addressed by simplistic methods such as *management by walking around,* and that as process parameters increase in number, process complexity and the number of change scenarios increase exponentially. This example also demonstrates the need for automated assistance in BPR efforts.

Industry in the United States is facing increasing challenges of cost reduction, competition, and so on. It is more important than ever to take advantage of technology in business processes, not only for implementing changes, but also for predicting the effects of changes before they are implemented. Without doing so, industry (and government) may not be able to uncover the most effective methods of reengineering their processes.

2

A New Perspective on Change in Business

Theories of business and management change frequently. Many companies have discovered that the idea of "just-in-time" deliveries of raw material does not always work. They are experiencing cynicism on the part of employees who suggest that company policy is not "drive out fear" but "drive in fear." The idea of lifetime employment as espoused in Theory Z is all but gone in the wake of corporate downsizing. Business is ready for new approaches, and one of those approaches is Business Process Reengineering, or BPR.

According to Hammer and Champy, BPR is the "fundamental rethinking and redesign of business processes to achieve dramatic improvements in critical contemporary measures of performance such as cost, quality, service and speed."[1] A dramatic statement and an interesting concept, but note the abundance of the *re-* prefix in the sentence—*re*thinking, *re*design, *re*engineering. The *re*-prefix means "again"; therefore, this definition of BPR implies that business processes have been previously thought out, designed, or engineered. I contend that the following is true: Business and government processes in the United States (and in most of the rest of the world) have never been engineered; rather they have evolved.

A business process cannot be reengineered until it has first been engineered.

[1]Hammer and Champy, *Reengineering the Corporation*, p. 32.

Engineering is the "application of scientific and mathematical principles to practical ends such as the design, construction and operation of efficient...systems."[2] These principles must also be applied to process reengineering. Unfortunately, most BPR approaches, although claiming to represent radical change, are no more than the continuation of the evolution that has led to the processes that exist today. Such approaches to BPR emphasize increasing communications about processes. The only differences in the many BPR approaches being popularized are the differences in their approach to increasing communications. Whereas communications may be important, talking about business processes is only part of the BPR effort.

Before a business considers reengineering any process, it should first consider engineering the process. Process engineering is the application of engineering disciplines to the analysis and improvement of processes. Although a process cannot be reengineered if it has never been engineered, a process can be engineered and reengineered at the same time by applying process engineering methods. The application of scientific methods to business process reengineering is a radical, revolutionary departure from the comfortable, philosophical process reengineering approaches we continue to hear about.

In the late 1970s and early 1980s, the field of study called operations research (OR) was booming. In most large industries one could find entire OR departments dedicated to analyzing and modeling manufacturing systems, banking systems, and so on. The need for this type of systems analysis was driven primarily by the emergence of computers as readily available resources. Not only were tasks being performed in minutes instead of hours, but technology was also changing so quickly that minutes of operation were being changed to seconds of operation. The effect of these radical changes had to be analyzed to determine their overall impact on a business.

The theory of operations research is that "systems are sufficiently complex that carefully developed models are necessary for understanding (their) performance. The behavior of complex systems needs to be understood in order to be able to design new systems or to improve the operation of existing systems."[3] Since the words *system* and *process* are synonymous, the theories of operations research are as applicable to business processes as they are to any other types of systems. It does not matter whether a system (or process) under study is comprised solely of human or mechanical objects.

For some reason, however, the practices of operations research are not being applied to the analysis of human processes. Perhaps this is because business leaders continue to see "quick fixes" to problems, and they believe that "the process of modeling a system is not an easy task... is very expensive to construct and often requires a great amount of effort to maintain and use."[4] Maybe we have been sensitized so

[2]*American Heritage Dictionary*, Atlanta, GA, Houghton Mifflin Company.

[3]MacNair and Sauer, "Elements of Practical Performance Modeling" (Englewood Cliffs, NJ: Prentice Hall, 1985)

[4]Ibid

much that we don't want to think of people as objects. Maybe modeling and simulation tools have been too complicated and expensive to appeal to business managers.

Whatever the reason, almost all popular process reengineering approaches are based on the practices of Total Quality Management (TQM) and Continuous Improvement (CPI). TQM and CPI are philosophical approaches to business process improvement that emphasize corporate cultural change. Being philosophical, the approaches to CPI and TQM change with regularity. Unfortunately, given the tremendous investment that has been made in Continuous Improvement and Total Quality Management, the payback to industry and government has been quite small.

The success stories of TQM and CPI tend to be based on anecdotal data. Consider the following. A speaker at a TQM meeting mentioned that Nissan had saved 50 seconds on an assembly-line task by standardizing screw heads. The time was saved by eliminating the need to change the attachments on power drills. The speaker went on to say, "Until the United States gets to this level of detail, we will continue to be behind 'the curve.'"

This story seems impressive. However, one must ask: 'What does this mean in terms of the overall process? Were cars produced 50 seconds faster, or did the worker stand around idly for 50 seconds?' Depending on the answers to these questions, this story can be a success story, a failure story, or a story not worth mentioning. Unfortunately, this type of anecdotal story permeates the TQM/CPI world. Managers hear these anecdotes and demand similar solutions from their employees. More often than not, process reengineering efforts that use TQM and CPI result in little improvement at all.

Clearly, an engineering approach to process reengineering that incorporates modeling and simulation is necessary. Fortunately, the emergence of low-cost software known as Computer Aided Process Reengineering (CAPRE) software has significantly reduced the difficulty of modeling and simulation. CAPRE is a highly-graphical, object-oriented technology that provides the capability of predicting the effect of process changes in cost, time to completion, quality, and so on, *before* an organization implements them. This capability is not available with traditional CPI, TQM, or popular process reengineering concepts.

To help put the relationship of CPI, TQM, and computer aided process reengineering in perspective, I reference the Software Process Maturity Model developed at the Software Engineering Institute, a federally funded research and development center located at Carnegie-Mellon University in Pittsburgh, Pennsylvania. This model classifies processes into five levels of maturity and, although designed to analyze software development processes, is applicable to business processes in general.

The tools used in CPI and TQM efforts can be grouped into similar levels of sophistication, and I will map them to the process maturity model to demonstrate their use in process improvement efforts. Using examples, I demonstrate how organizations can ascend through process maturity levels using process reengineering techniques, and how CAPRE technology can be used to optimize processes.

I also reference the work of Dr. W. Edwards Deming and Peter M. Senge. Dr. Deming is probably the most widely known and quoted advocate of Continuous Improvement, and he has put forth theories of management that are the foundation of many CPI philosophies. Unfortunately, many attempts at CPI begin and end with Dr. Deming's theories, and this is part of the reason for the lack of success of CPI. The use of CAPRE technology does not invalidate any of the theories put forth by Deming; in fact, the importance of those theories is reinforced in process modeling and simulation development activities. In this book, I suggest how Dr. Deming's theories relate to process engineering.

Dr. Senge has taken the cause-and-effect analysis suggested by Deming to a new level of sophistication and understanding. He has promoted a diagramming technique that facilitates such analysis and suggests that there are recurring types of cause-and-effect archetypes we should be aware of.[5] Causal analysis is essential to determine the type of cause-and-effect analysis one wants to perform using simulation features of CAPRE technology.

I believe that industry and government in the United States are frozen in place with Continuous Improvement and Total Quality Management. As I promote the idea of computer aided process reengineering and the power of CAPRE technology, I sometimes encounter people who suggest that CPI and TQM are the cure-all for American industry. I don't believe that and, frankly, I don't think any CPI or TQM theorist would suggest that his or her approach is the *only* acceptable approach. My goal in writing this book is to convince management and workers that, although CPI and TQM are necessary and fundamental elements of process improvement, they must look beyond these theories.

The approach to process reengineering I just described requires diligence, patience, and a commitment to change from management. It is not sexy, it is not necessarily fun, and there is not a lot you can do to make it exciting. This approach is not a replacement for TQM and CPI but rather an augmentation of these practices. It *is*, however, effective. It will save corporations money by providing a mechanism to predict the effects of reengineering changes before they are implemented. As more organizations begin to practice computer aided process reengineering, it will generate its own form of excitement.

[5]Peter M. Senge, *The Fifth Discipline*, (NY, NY: Doubleday, 1990).

3

The State of Business Process Reengineering

Some of the most commonly used buzzwords in industry today are *Continuous Process Improvement, Total Quality Management,* and *Business Process Reengineering*. Continuous Process Improvement (CPI) and Total Quality Management (TQM) represent philosophies of change that embrace open communications and the elimination of barriers between management and nonmanagement personnel. Business Process Reengineering (BPR) is the activity of implementing changes to a process that result from the application of these philosophies.

The goal of CPI and TQM is to improve the quality of products by improving the quality of the processes that produce those products. Terms such as the *boundaryless organization, empowerment, drive fear out,* and so on, have become closely associated with CPI and TQM. Since TQM and CPI are essentially the same, the terms will be used interchangeably throughout this book.

Continuous Process Improvement is often described as

- A way of thinking (about customers, coaching versus judging, etc.).
- An integrated approach to doing work (empowerment, total involvement, teamwork, etc.).
- A set of tools and techniques (workout, quality circles, process mapping, etc.).
- A set of beliefs about people (cultural diversity, trust, etc.).
- A body of managerial and business knowledge (best practices, etc.).

There is nothing wrong with these ideas; however, although U.S. industry has invested heavily in CPI and TQM training, the improvements realized to date have fallen short of expectations. The primary reasons for this limited success are as follows.

1. There are no formal and systematic approaches to CPI and TQM; instead, there are many variations on one theme all professing to be correct.
2. The tools and techniques taught in CPI and TQM classes are mechanical and limited in their ability to present an overall view of a process.
3. CPI and TQM training stress analysis of the *results* of existing processes, that is, the quality of products produced by a process. There is little emphasis on predictive analysis of proposed changes to processes.
4. Management in U.S. industry looks for quick solutions to problems, and the failure of immediate success has lessened the enthusiasm of management for both CPI and TQM.

The acceptance of the Continuous Process Improvement philosophies is related to the state of the economy. When times are good, management is more likely to accept the "warm, fuzzy" aspects of CPI. When the economy is turning downward, CPI tends to be replaced by more hard-line, bottom-line approaches to business. Participative change is replaced by dictatorial change. When workers perceive that management is abandoning CPI, they tend to become cynical and abandon it as well.

Suddenly, companies find that their investments in CPI and TQM have been wasted, and they are faced with reintroducing the concepts and retraining their staff.

An excellent example of the cyclical, economy-related nature of TQM is *Theory Z*.[1] Theory Z was a widely accepted and discussed theory of management popularized in the early to mid-1980s. One of the basic principles of Theory Z is that companies should try to provide permanent (lifetime) employment for their employees, as is the tendency in Japanese companies. One company regularly mentioned as a practitioner of Theory Z was IBM, a company that was proud of the fact it almost never laid off employees. Economics and changing business dynamics have changed that dramatically, and IBM is no longer considered a lifetime employer. Not surprisingly, references to Theory Z have all but disappeared.

It is also not surprising that the failures of philosophies such as Theory Z have led to the falling popularity of CPI and TQM, and to the rise of BPR. BPR was initially

[1]William Ouchi, *Theory Z* (Santa Monica, CA Addison-Wesley, 1981).

greeted with great anticipation, since it seemed to offer a new approach to change; however, BPR has not introduced any new techniques or tools to effect change and it, too, is starting to decline in popularity.

Computer Aided Process Reengineering

None of this suggests that CPI and TQM are invalid; rather, this book suggests that the philosophical approaches of CPI and TQM are not adequate to meet industrial and governmental needs of increasing productivity and quality. To meet those needs, a more pragmatic approach that takes the tools and techniques of CPI and TQM to a new level of sophistication is required.

This book introduces such an approach, called *Computer Aided Process Reengineering*, which is the science of treating processes as systems and using systems analysis techniques to manage and improve the performance of those processes.

The concept of applying computer-based systems analysis has been limited mainly to manufacturing processes and has not effectively been applied to more general business processes. Moreover, such analysis, until recently, has required computer expertise that was not generally widespread; however, technology is emerging that will

- Dramatically change the manner in which we approach process reengineering.
- Change Process Reengineering from an evolutionary process to a revolutionary process.
- Make automated process analysis available to anyone possessing basic personal computer skills.

This technology consists of computer-based process modeling and simulation system tools, hereafter referred to as *Computer Aided Process Reengineering* (CAPRE) tools.

CAPRE tools have the following characteristics:

- They utilize a graphical representation approach to define a process flow.
- They provide an analytical capability through the use of an underlying rule base.
- They provide predictive capabilities through simulation.

CAPRE tools are not a replacement for the techniques currently taught in CPI training. Those techniques are an essential and fundamental part of a formal approach to Process Reengineering, defined in this book as the *Rules of Process Reengineering.* The Rules of Process Reengineering emphasize the following points:

- The tools and techniques of TQM and CPI apply only to the beginning stages of process reengineering efforts.
- Adherence to the Rules of Process Reengineering and utilization of CAPRE technology are required to achieve optimum process performance and to avoid counterproductive process changes.
- Attempts at process reengineering that do not apply the Rules of Process Reengineering can lead to unwanted and counterproductive results.
- CAPRE technology, when used in conjunction with the Rules of Process Reengineering, can accelerate process reengineering efforts.

The SEI Process Maturity Model

The SEI Process Maturity Model describes "the extent to which a software organization has adopted and institutionalized a continuous improvement focus"[2] and is depicted in Figure 3.1.

Level	Characteristics
5 Optimizing	Improvements Fed Back into the process
4 Managed	Process Defined and Measured
3 Defined	Process Defined with Standardized Results
2 Repeatable	Process Informally Defined with Predictable Results
1 Initial	Ad Hoc/Chaotic

Figure 3.1 Software Engineering Institute Process Maturity Model.

Although created primarily to measure the effectiveness of software development processes of government contractors, the maturity model represents a general theory of process evolution that can be applied to any process under investigation. The SEI uses this model to assess the level of process maturity of an organization; the model is used in this book to discuss how organizations can ascend through maturity levels.

[2]"An Analysis of SEI Software Process Assessment Results: 1987–1991," Technical Report, CMU/SEI-92-TR-24.

I have previously stated that the application of Continuous Process Improvement and Total Quality Management philosophies have failed to generate the results expected by industry and government in the United States. This statement was based partially on observation and partially on the results of extensive process analysis activities conducted by the Software Engineering Institute. The SEI analysis reveals that, of the organizations studied, 81 percent have Level 1 software development processes and 12 percent have Level 2 software development processes![3] In other words, a full 93 percent of the organizations studied have rudimentary software development processes. This statistic alone reinforces the need for automated process reengineering tools.

A theory associated with the maturity model is that productivity and quality of products produced increase as the maturity level of a process increases, and the risk of errors introduced into the process decreases as the Level increases. Case studies performed by the SEI support this assertion; however, there is another aspect of risk that must be considered, and that is the risk of introducing counter-productive changes.

As processes migrate to higher maturity levels, the probability of implementing counterproductive changes decreases, but the *risk* (potential negative impact on cost, productivity, or quality) of those changes, increases. Figure 3.2 depicts this relationship.

Level	Characteristics
5 Optimizing	Improvements Fed Back into the Process
4 Managed	Process Defined and Measured
3 Defined	Process Defined with Standardized Results
2 Repeatable	Process Informally Defined with Predictable Results
1 Initial	Ad Hoc / Chaotic

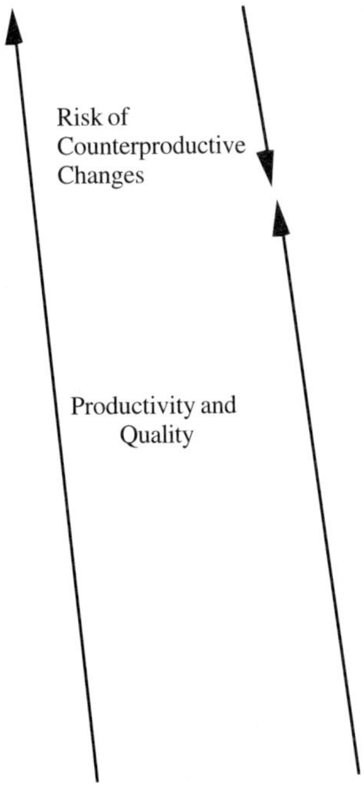

Risk of Counterproductive Changes

Productivity and Quality

Figure 3.2 Productivity, Quality, and Risk Related to Process Maturity Level.

[3]Ibid.

This requires more explanation. When changes are made to a Level 1 or Level 2 process, the positive or negative effect of those changes is usually quite minimal, since those processes, by definition, are not as productive as more mature processes. On the other hand, a Level 4 process is a productive process and one in which changes are welcomed as part of a Continuous Process Improvement effort. Since Level 4 processes are well defined and measured and are executed as efficiently as possible, a change to such a process that negatively affects it can be *measurably* counterproductive.

When a process reaches Level 5, however, the *risk* of counterproductive changes decreases. Assumptions about productivity gains, cycle time reduction, cost improvement, and so on, are tested in a Level 5 process *before* changes are implemented. Suggested methods of implementation can also be "tweaked" or fine-tuned to assure maximum payback. It is the Level 5 process that takes advantage of computer aided process reengineering techniques and the Rules of Process Reengineering that are presented in this book.

Counterproductive changes to mature processes can have disastrous effects. For example, a company with a mature Level 4 process decides to invest in technology to reduce cycle time. Since the process is well documented and measured, management has made some assumptions about productivity gains and has concluded that its investment will yield large cost reductions through increased productivity. When the technology is in place, management is surprised to learn that productivity has changed very little; in fact, an increase in productivity would require an additional investment in more expensive technology. This is not an atypical situation, and examples of such process reengineering attempts are included as case studies later in the book.

In the example just cited, the simulation capabilities of a CAPRE tool would have provided management more information about the overall effect of the technology insertion and would have allowed it to make a more informed decision. These capabilities are essential for problem prevention and without them, even the most well-intended change can fail.

Deming's Theories of Continuous Improvement

Although not stated in any of the referenced Software Engineering Institute documents, the process maturity model neatly encapsulates some of the theories put forth by Dr. W. Edwards Deming. Dr. Deming is an advocate of using statistical methods to achieve process improvement and has said, "you have to know *what* to do, and then do your best."[4] The following statements imply a process maturity migration: Look at a process, determine how to improve the process, implement changes, and then look at it again.

[4]Mary Walton, *The Deming Management Method,* (NY, NY: Putnam Publishing Group, 1986).

Dr. Deming is also known for the 14 points that sum up his philosophy of management.[5] Deming's 14 points follow.

1 Create constancy of purpose for improvement of product and service.
2 Adopt the new philosophy (of rejecting poor workmanship).
3 Cease dependence on mass inspection.
4 End the practice of awarding business on price alone.
5 Constantly improve the system of production and service.
6 Institute training.
7 Institute leadership.
8 Drive out fear.
9 Break down barriers between work areas.
10 Eliminate slogans, exhortations, and targets for the work force.
11 Eliminate numerical quotas.
12 Remove barriers to pride of workmanship.
13 Institute a vigorous program of education and retraining.
14 Take action to accomplish the transformation.

Many of these points are validated through Continuous Process Improvement efforts, and the applicability of these points will be discussed through the use of examples.

In addition, Dr. Deming implies that

- Planning requires *prediction* of how things will perform. Tests and experiments of past performance can be useful, but *not definitive*.
- Workers work in a *system* that is beyond their control. It is the *system*, not individual skills, that determines how they perform.

These two statements sum up the main focus of this book. Although Dr. Deming claims that a system (or process) determines how workers perform, almost all of the TQM and CPI philosophies that are currently being practiced focus on the "people aspect" of processes. There seems to be some philosophical belief that process improvements can be accomplished by changing the attitudes of workers. Secondly, although Dr. Deming states that planning requires prediction of the effects of process changes, there is virtually no emphasis on prediction in popular process reengineering approaches.

The need to predict changes to processes before an effort is made to improve them is clear. Deming has suggested the use of several techniques to perform such predictive analysis: causal chains, flow diagrams, histograms, run charts, Pareto dia-

[5]W. Edwards Deming, *Out of the Crisis,* (MIT Press, 1986).

grams, control diagrams, and scatter diagrams. Unfortunately, most of these are statistical analysis techniques that analyze process results *after* a process is in place. This book supports an approach of predicting the results of a process change *before* a process is in place.

Prior to the advent of CAPRE tools, predictive process analysis was difficult to perform and required the development of computer programs. Since the programs themselves were difficult to analyze, it was often difficult to determine if errors in predictions were the result of faulty analysis or faulty programs. CAPRE tools offer simple yet powerful mechanisms for performing predictive analysis of process changes.

Senge's Theories of Causal Analysis

One of the process analysis tools suggested by Dr. Deming is cause-and-effect analysis. A popular diagramming technique, called *fishbone diagramming*, is used to graphically represent the factors that may lead to certain results of processes. An example fishbone diagram is depicted in Figure 3.3.

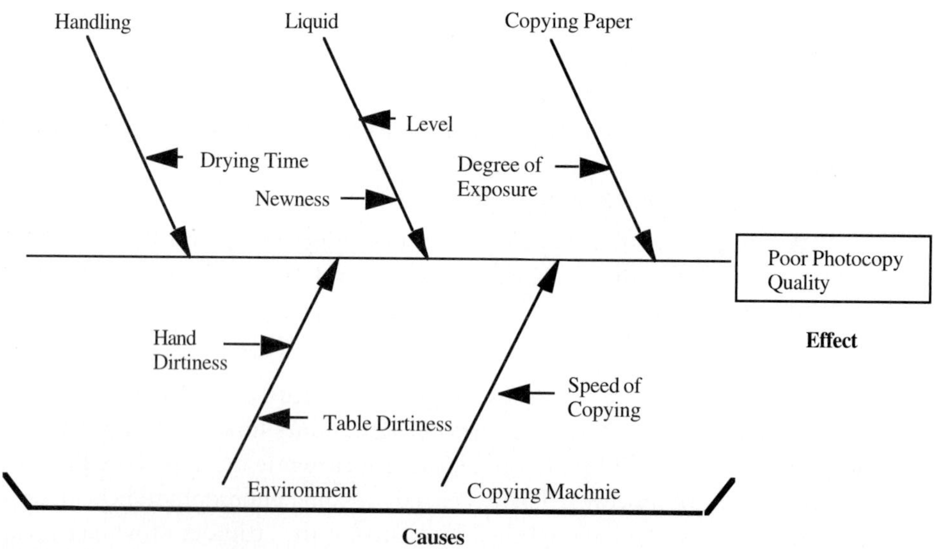

Figure 3.3 Example Fishbone Diagram.

The diagram in Figure 3.3 lists the possible causes of poor copying quality of a photocopy machine. Although fishbone diagrams are useful in conveying information about process elements, these diagrams are cumbersome, difficult to understand, and do not examine the *relationships* between process elements. Moreover, cause-and-

[6]Peter M. Senge, The Fifth Discipline, (NY, NY: Doubleday, 1990).

effect analysis is typically applied when a problem occurs and is rarely applied as a means to determine the effect of process changes.

Recently, interesting work has been done in the area of causal relations by Peter M. Senge of the Sloan School of Management at MIT.[6] Dr. Senge has developed a concept of causal loop archetypes which can be useful in understanding why certain process patterns develop. The basic structure of a causal relation is in Figure 3.4.

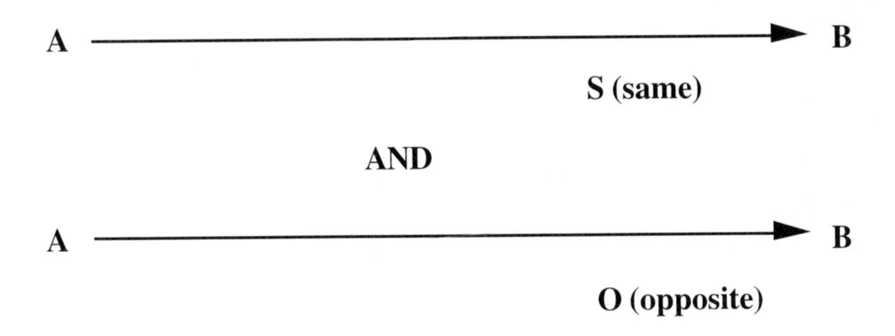

Figure 3.4 Causal Relations.

The first relation can be read as "a change in A implies the same change (S) in B" and the second relation can be read as "a change in A implies an opposite change (O) in B."

Senge theorizes there are patterns of causal behavior, or archetypes, that can explain why events happen in certain ways. For example, one archetype defined by Senge is the *vicious circle*, represented in Figure 3.5.

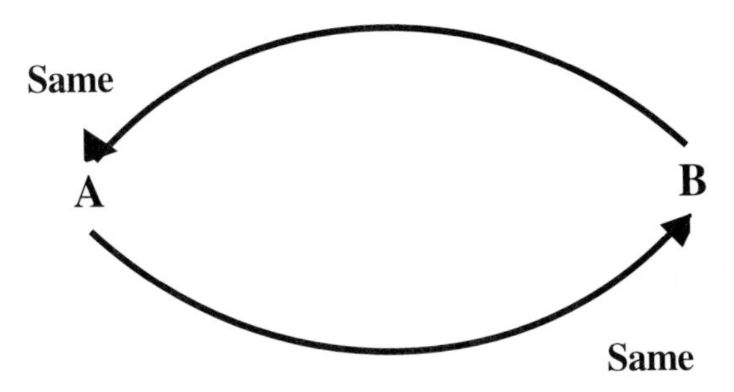

Figure 3.5 Vicious Circle.

This is interpreted as "A implies an increase in B which implies an increase in A which implies an increase in B...", and so on. Senge uses the gasoline crisis of the late 1970s as a prime example of a vicious loop. Gasoline was expected to be in short supply, so people rushed off to fill their tanks. When lines developed at the gas pumps, the idea that gasoline was in short supply was reinforced, so more people rushed to the pumps. Eventually, people were trying to fill their cars with gas even though they had almost full tanks.

When analyzing processes in a cause-and-effect manner, the archetypes presented by Senge can help add insight into the process. The diagramming technique used by Senge is useful in determining possible effects of changes to processes before they are implemented and can be used in conjunction with CAPRE tools as part of what-if analysis.

Process Examples

Two examples are used throughout the book to describe levels of processes and migration between levels. The first example is the Origami Process,[7] an exercise in team building that is very popular and often used in Continuous Process Improvement training. The Origami process is used in this book to demonstrate

- Evolutionary migration through process maturity levels.
- The use of process improvement tools and techniques applicable to each level.
- The limitations of conventional Continuous Process Improvement methods.
- The process optimization capabilities of CAPRE technology.

The second example is an actual process used by a major manufacturer to process purchase orders for personal computing equipment. The purchase order process is used to demonstrate that

- Attempts at process reengineering that are made without proper analysis can be counterproductive.
- Migration from a Level 1 process to a Level 5 process can be accomplished quickly by using CAPRE technology.

Origami Process Description

The Origami Process example is a fictitious assembly line process and is a commonly used training exercise, the goal of which is to emphasize Deming's philosophies. (For those unfamiliar with the exercise, a description of the process is included in the appendix.)

[7]The name *origami* is used for brevity and represents a paraphrased version of a Continuous Improvement training exercise called *The Flying Star Ship Factory.*® The Flying Star Ship Factory is a product offered by Block Petrella Weisbord, Inc, 1009 Park Avenue, Plainfield, NJ 07060.

The Origami Process is ideal for presenting the concepts of computer aided process reengineering for the following reasons:

- It is, on the surface, a simple, straightforward process that is easy to understand.
- Attempts to improve the process reveal this simple process is actually quite complex, as are most business processes.
- Process reengineering requires cooperation between participants and open communication, so it is a good exercise to emphasize the value of TQM and CPI philosophies.
- It represents an ideal process for learning modeling and simulation.
- The complexities of the process allow for the development of equally complex computer aided process reengineering models.

The product produced in the Origami Process is a star-shaped paper toy. Students are assigned tasks and, under the guidance of a facilitator, given several opportunities to improve the process. One student is selected to act as a manager, and the others are designated as workers. The rules that define the manner in which the process will proceed initially are as follows:

- Workers are paid on a piecework basis—the more they produce, the more they earn. This is done knowing that, even though it is a training exercise, students will tend to compete and create pockets of suboptimization.
- A manager assigns tasks to workers, and workers can perform only the tasks to which they are assigned. They are strictly prohibited from performing other tasks.
- The manager is paid based on the number of final products that pass inspection. The manager's goals can confuse other workers, since he stresses quality and quantity simultaneously.
- Workers may not communicate among each other—all communication is handled by the manager; therefore, if some workers are idle due to the lack of raw material, they must get the attention of the manager and request the materials.
- A material handler moves items from one process step to the next as directed by the manager. If not instructed to deliver anything, the material handler remains idle. The material handler is paid a flat rate, so he is concerned neither about quality nor throughput.
- An inspector reviews the final products and separates them into passing and failing categories. The inspector is focused solely on quality without regard for production or order fulfillment; therefore, there can be conflict between the manager and the inspector.

Note that a process is not defined; instead, tasks and operating rules are defined. This may seem like an extreme example, but it is used in CPI and TQM training because it reflects the nature of actual manufacturing processes in place in U.S. industry.

Purchase Order Process Description

The purchase order process, like most other business processes in place today, is one that has evolved and expanded over time. It is a process that was most likely suitable for ordering equipment a number of years ago, but given the explosion of the number of personal computing systems available, the number of requests flowing through the system increased dramatically, and the process began to break down.

Basically, the process behaved as follows:

- A user sent information regarding computer equipment requests to a contact in a particular department.
- The information was used to generate a purchase order, which was then passed among various people for validity checks, budget checks, and approvals.
- The purchase order was passed on to the Purchasing Department, where a purchasing agent contacted a vendor and placed an order.

The product of this process is a completed purchase order, and it would appear to be as simple a process as possible; however, there were numerous problems associated with the process and numerous attempts were made to correct those problems. Again, this process is one in which the participants know their tasks and rules have been established, but a process has not been defined.

Summary

These two examples are easy to understand. The Origami Process will show how application of the Rules of Process Reengineering facilitates the migration of a business process from Level 1 to Level 5. The purchase order process will demonstrate the dangers of attempting changes without following process reengineering rules.

4

Level 1 (Initial) Processes

Initial, or Level 1, processes are the most common processes in business today. These are the processes that have evolved and grown over time. As new functions in a process are required, more tasks are created and more workers become involved in the process. The complexity of the process increases until no one is absolutely certain why the process behaves as it does. Eventually, management looks at the process and wonders, how did this ever grow so large and become so inefficient? At this point, management wants to reengineer the process.

A Level 1 process is one in which results are variable and the mechanisms used to create a product or provide a service change frequently. In a Level 1 process, the quality of the end product, the time to completion, and the costs of making the product are all likely to change as business conditions change. Since a Level 1 process evolves through the addition of components, it is not documented and the participants tend to lack an overall understanding of the process. Success in a Level 1 process usually comes from the efforts of employees rather than any management plan. Accordingly, Level 1 processes typically suffer from suboptimization.

Level 1 processes, then, have the following characteristics.

1. A process has evolved and is known to the participants, but it is not documented.
2. The process participants know their specific tasks and how to perform those tasks but lack understanding of the overall process.
3. There are usually one or more product inspections, but typically inspection occurs at the end of the process.

4 Although workers may perform well and meet quotas, the quality of the product produced is unpredictable.

5 There is usually strong management present.

6 Since participants in the process lack an understanding of the overall process, there is little communication among workers about possible improvements.

An undocumented process with a high degree of division of labor is a Level 1 process. If the end results of the process are erratic, then it is most likely a Level 1 process. Most start-up companies rely on Level 1 processes, which may explain the high failure rate of those companies. Level 1 processes are those processes most likely to improve from the application of Deming's 14 points. The primary mechanism for improvement is communication among workers, among workers and management, and among management.

The Origami Process At Level 1

Before the Origami team-building exercise begins, one student is designated as the manager, and the others are designated as workers. Workers are assigned tasks, given training in the tasks they will perform and provided an opportunity to practice those tasks. As mentioned in the previous chapter, workers are paid based on what they produce, and the manager is paid based on what is accepted.

When the exercise begins in earnest, it is not surprising that there is chaos and the quality of the product is poor. Since the goal of the exercise is to promote team building, the Continuous Process Improvement instructor will, at some point, stop the process and convene a brainstorming session. The purpose of the brainstorming session is to discuss the students' observations of the process and to promote communication about possible process improvements.

Since both management and workers are included in the session, the instructor will typically point out that he or she is invoking the following of the Deming's 14 points:

1 Drive out fear (point 8)

2 Break down barriers between staff areas (point 9)

In addition, the session will be used to demonstrate the value of *team building, participative management, total involvement*, and several other TQM/CPI philosophies.

After a discussion of limited duration, the students will be instructed to vote on the process improvements that have been suggested. The following improvements are almost always approved and implemented, although not in the order presented.

1 The students determine their goal should not be to produce as many products as possible, but to produce as many acceptable products as possible.

This reinforces Deming's first point, that is, create constancy of purpose.

2 This first suggestion invariably leads to the elimination of piecework pay, reinforcing Deming's point 11—eliminate numerical quotas. Not every student always agrees with this suggestion. The paper cutters, for example, produce as much as they can regardless of whether or not the cut paper is used; however, this suggestion is usually adopted.

3 The next suggestion revolves around workers' tasks. Work areas are arranged so material can easily be passed between tasks. As a result, the function of material handler is eliminated. The student performing that function is then trained in all aspects of the process and asked to move from area to area depending on the needs of other workers.

   The manager is instructed to determine where the critical needs for assistance are and to assist the material handler in his decision process. This suggestion reinforces two of Deming's 14 points:

   - Institute leadership (point 7).
   - Institute a vigorous program of education and retraining (point 13).

4 Occasionally, other workers, particularly the paper cutters, will recognize that they either have idle periods or have produced more material than can be used in the near term. In keeping with the team building concept, these individuals volunteer to help out in other tasks when needed; therefore, they are trained in one or more other tasks, thus reinforcing Deming's point 6—institute training.

5 Inspections are not eliminated, but the emphasis is changed from one of determining the quality of the workers' efforts to determining the effectiveness of process changes. This reinforces point 3 (cease dependence on mass inspection) and also Deming's philosophy of statistical control.

Finally, the suggestions are implemented with everyone accepting the responsibility for success or failure of the new process. This reinforces point 2 (adopt the new philosophy), and point 14 (take action to accomplish the transformation). The group agrees to meet again to determine if other improvements can be made to the process, thus reinforcing Deming's point 5 (constantly improve the system of production and service).

When the improved process is implemented, quality is increased. The number of products produced is reduced, but the ratio of rejects to accepted products is reduced as well. Quick mathematical analysis invariably proves that the cost of acceptable products has been reduced. That cost can be calculated as

Cost per Acceptable Product = Total Cost of Products / (Total Products – Rejected Products)

Therefore, the value of applying Deming's theories to a Level 1 process has been demonstrated.

The Origami Process also demonstrates the value of causal reasoning. Although it is not the purpose of the exercise to promote causal reasoning, it does occur. Whenever anyone says "if we do A, then B will occur," causal reasoning is being used.

What techniques were used to improve the Level 1 Origami Process? The primary mechanism used to improve the process was *communication* among the participants. In the brainstorming session, students discussed problems, proposed options for solving the problems, and recommended actions. Process Reengineering Rule 1 is, therefore, *Talk about the process!* There are many theories about how communication should take place, and all are applicable at this level.

A point of interest: While trying to improve the Origami process, a minimum of 11 of Deming's 14 points are referenced. This demonstrates that Deming's philosophy is best suited for Level 1 processes and, in fact, CPI and TQM as currently taught in the United States are best applied to Level 1 processes. This can be stated as *CPI and TQM are designed to get industry and government out of chaos and into repeatable processes.* Perhaps this is why Deming's famous book was entitled *Out of the Crisis.* Until industry and government move beyond these basic concepts, organizational processes will continue to function at suboptimal levels.

The Purchase Order Process: First Attempt at Reengineering

The purchase order process was one that evolved over time, adapting to changes in the needs of the user community in a very informal manner. As government regulations became more stringent, the process became more formalized, but it did not change. The purchase order process, therefore, is representative of many of today's business processes—they have evolved; they have not been planned.

The purchase order process consisted of the following participants: people who wrote purchase orders (referred to as either users or customers), people who processed purchase orders, people who approved the purchases, people who actually issued the purchase orders, and vendors of computer equipment. All of the participants had some complaint about the process, such as

1 A high number of purchase orders were being rejected because the information contained in them was inaccurate or incomplete. Product names,

models, configurations, and so on, changed frequently, but because there was limited communication with users of the process, those changes were not being reflected in the purchase orders.

2 The amount of time required to process purchase orders was inordinately long.

3 Vendors were delivering equipment prior to receiving the purchase orders based on "good faith" letters from users. This created legal and insurance problems as well as warranty problems.

4 Budgets were being overrun.

5 Ordered equipment was being returned to the vendors because it had either been obtained without proper signatures or no budget existed. This imposed an expense on the vendors and made them less likely to cooperate in the future.

The first attempt to reengineer the purchase order process was approached in a completely different manner than the first attempt at reengineering the origami process. In this case, management decided

1 It had to reengineer the process quickly to reduce complaints, and

2 It would reengineer the process based on a limited set of information.

Management concluded that the best way to reengineer the process was to reorganize. In other words, it decided that there was nothing wrong with the process, but there was something wrong with the way the process was being managed. Reorganization, when it occurs in response to a problem, is the antithesis to Deming's point 8 (drive out fear). In fact, reorganization as punishment can be paraphrased as "drive *in* fear."

After organizational changes were made, problems associated with the process were initially reduced, but then they returned. In fact, the problems in the purchase order process eventually increased. This result leads to some questions.

- Why did management decide to reorganize without investigating the process thoroughly?
- Why did problems initially decrease?
- Why did problems eventually return and ultimately increase?

The first question can be answered by considering Senge's theory that humans tend to develop *mental models* about processes. Mental models are "deeply ingrained assumptions, generalizations, or even pictures that influence how we understand the world and how we take action."[1] In the case of the purchase order process, management had concluded that the group responsible for handling the orders was doing a poor job, and that the solution to the process problems was to put some other group in charge.

[1] Peter M. Senge, *The Fifth Discipline* (NY, NY: Doubleday Publishing, 1980), pp. 8–9.

The second and third questions can be answered by the fact that management was trying to improve the process by means of a quick fix and was ignoring systemic problems in the process. What resulted is a phenomenon called a "fix that backfires." This phenomenon is depicted in Figure 4.1.

Figure 4.1 A Fix That Backfires.

The diagram in Figure 4.1 shows that

1 As management's perception of the effectiveness of an organization or process decreases, the threat (or likelihood) of a reorganization increases.
2 When a reorganization is threatened or actually occurs, there is a good probability that productivity will increase. When faced with a reorganization, workers will put forth an additional effort, either to prevent the reorganization or to impress new managers.
3 When productivity is increased, management's perception of the organization responsible increases. Its belief that reorganization is the answer to process problems is also reinforced, increasing the likelihood of future reorganizations.

Typically, however, productivity changes due to the efforts of the personnel involved do not last for any appreciable period of time. Figure 4.1 also shows that

1 Reorganizations, particularly when used as an assessment of blame, have a side effect—they produce a fear of failure and fear of management.
2 When there is a fear of failure, there is a reluctance to take any risk associated with process changes, no matter how small or how reasonable the risk may be.
3 Without *systemic* changes, a process will not improve.

4 The lack of systemic changes results in a loss of the initial improvements and the process degrades instead of improves.

5 The process repeats; that is, as management's perception of the effectiveness of an organization or process decreases, the threat (or likelihood) of a reorganization increases.

This is a classic example of Deming's theory that workers work in a *system* that is beyond their control. It is the *system*, not individual skills, that determine how they perform.

Process improvements based on mental models

1 address symptoms, not problems and

2 quick fixes make it harder to determine the real nature of process problems.

The way to avoid improvements based on mental models is Process Reengineering Rule 1—*communicate and increase awareness of the process.*

Summary

To improve a Level 1 (initial) process, the individuals who participate in the process, regardless of their role, must communicate their views of the process and their roles in the process to the other participants. From these discussions, understanding of the process will increase, suggestions for improvements will be made, and some changes will be implemented. Developing verbal communication skills is essential for an organization to be able to migrate to a Level 2 process.

An organization, however, cannot rely on verbal communication alone to either develop a good process or to implement meaningful change. Verbal communication has the following deficiencies:

- It is subject to interpretation.
- Information can be lost when communicated verbally.
- Information can be incomplete or inaccurate.

To avoid these problems, an organization must take the next step to effectively communicate and define its processes—the organization must document those processes. The need to document processes is discussed in the next chapter.

TQM/Continuous Improvement Philosophies: How They Apply to Level 1 Processes

Think for a minute about all the TQM and CPI philosophies and techniques you know. For the most part, these philosophies can be grouped into two categories: (1) methods of increasing communication between workers and between workers and management, and (2) methods of making workers feel important and part of a team. The second category is really an extension of the first, since, if workers feel better about their jobs and themselves, they are more likely to communicate about their jobs.

"Six buzzwords make a power sentence."—Jim Maki, artist and philosopher

One problem with TQM and CPI is that they are very much like fads—whatever theory is popular at the time is the theory being promoted in industry. TQM and CPI suffer from an overuse of buzzwords. My colleague Jim Maki coined the above phrase—it is a clever way of saying that the more someone uses buzzwords, the more expert he or she sounds.

Why do the buzzwords associated with TQM, CPI, and process reengineering change so frequently? Probably because the theories they represent have failed to deliver on their promises. Rather than change the theories, experts in these areas reintroduce them under different names. Unfortunately, the results are always the same.

In the following sections, TQM and CPI buzzwords, present and past, will be discussed. For the most part, these buzzwords relate directly to increasing verbal communication in the work force. In other words, most TQM and CPI philosophies apply to Process Reengineering Rule 1. The buzzwords are not presented in any particular order.

Quality Circles. Quality control circles were supposedly formalized in Japan by Dr. K. Ishikawa in the 1960s. A quality circle is a gathering of workers, with or without managers present, in which ideas for improving the quality of their products are discussed. The goal of a quality circle is to bring to the attention of management "the full use of the successes of small groups of workers in the elimination of special causes of variability of product,...improvement in the system..., " and so on. Quality circles were popular for many years, but recently, the concept has become less popular and has been replaced by similar, newer philosophies.

One must wonder how a concept that was so popular for so long could simply fall from favor. The answer is simple—the concept did not work. Quality circles focus on small changes in small pieces of larger processes. Such changes, as will be demonstrated later in this book, rarely have any meaningful effect on a process and are often subject to failure. Since quality circles were not the problem-solving mechanism either management or workers thought they would be, quality circles began to be met with cynicism and a general lack of enthusiasm.

Workout. Workout is a phrase supposedly coined by Jack Welch of General Electric. Workout appears to be an updated version of quality circles, but it also represents a more focused approach to problem solving. A workout session is a gathering of employees and management in which methods of finding "high payback" fixes to systemic problems are discussed and sometimes mapped (with flow charts). Once again, workout is primarily a method of talking about problems, and the results will be similar to the results of quality circles.

Cross-Functional Teams. A cross-functional team is a collection of people from different organizations that are involved in a process. The point of a cross-functional team meeting is to encourage discussion of how different subgroups can work together more effectively. Cross-functional teams are usually associated with white-collar processes, but they have also been used in manufacturing environments. Cross-functional teams will meet to discuss mechanisms for sharing information more effectively, providing services more effectively, and so on. Cross-functional teams simply define the make-up of quality circles and workout sessions.

Interdisciplinary Teams. This is another buzzword for cross-functional teams.

Customer-Supplier Relations. Customer-supplier relations is a concept taught in basic sales training. TQM theorists have now discovered that dealing with people and organizations who use what you produce should be of interest to everyone. Customer-supplier relations is another theory of communication that suggests we should ask our customers (those in a process who receive our products) if our products meet their needs. If not, then the customers and the suppliers should work together to determine what can be done to fix the problem. (In sales, this is called *negotiations.*)

Participatory Management. This popular TQM theory (also known as management by walking around) encourages managers to participate in discussions with their employees, rather than simply listen to their suggestions after a quality circle or similar meeting. The theory is if workers believe that management is truly interested in what they have to say and in making improvements, they will be more willing to talk about their jobs and make suggestions.

Suggestion Boxes. One of the oldest forms of worker-management communication mechanisms known is the suggestion box. The theory behind suggestion boxes is that, assured anonymity, workers will be unafraid to present solutions to management. Perhaps because suggestion boxes have been around for so long, they have become the brunt of many jokes. Suggestion boxes have all but disappeared from the workplace in the United States; however, they remain popular in service industries, such as hotels and restaurants. In these cases, they are being used to get the opinions of customers, not of workers.

Coaching versus Judging. This philosophy is aimed mostly at managers, but it is also being taught in interpersonal skills training. The theory is that, when listening to an argument or suggestion by an employee or peer, one should avoid the temptation to make counterpoints or countersuggestions. Instead, one should try to elicit more information from the speaker and "lead" him or her to the "correct" conclusion through coaching. Supposedly, a coach provides guidance and help when requested but avoids all appearances of being dictatorial. A coach never says, "I don't think that will work," but might say, "Tell me again how that will help." Coaching versus judging is another philosophy of communication.

Empowerment. This is currently one of the most popular buzzwords in industry and government. It is a theory of management that gives employees the power to institute change and promote ideas. Unfortunately, that power rarely, if ever, comes with the additional power of setting schedules, obtaining personnel and funding resources, and so on. If employees were truly empowered, the need for management would be eliminated. Without being given control of budgets and staff, the most empowered employees can do is talk about changes and convince others to implement the changes.

Total Involvement. This philosophy suggests that employees can become involved in all aspects of business operations. It is similar to cross-functional teams, except that employees are encouraged to make suggestions about subjects outside their area of expertise.

Self-Directed Teams. These teams are a combination of quality circles (and derivatives) and empowerment. This is a theory that not only individuals, but teams of people can also be empowered to effect change in an organization.

Cultural Diversity. This philosophy recognizes that people are different and is an effort, being promoted in industry and government alike, to bring people of different backgrounds, races, religions, and so on, together to communicate. It is a theory aimed mostly at white males under the assumption that industrial power in the United States is held by that group of people. Again, this is a theory of how communication among people can be enhanced and encouraged.

Quick Market Intelligence. This concept, supposedly started by Wal-Mart, brings together business leaders from a company to discuss the required responses to changing market conditions. It is not very different from other group communication concepts, except that the focus is on marketing and sales.

Brainstorming. This is another group communication concept, although there are some accepted rules for brainstorming sessions. In a brainstorming session, all ideas are written on scraps of paper and taped to a wall. They are then categorized by application or type and discussed over and over until some consensus is found. This is another method of opening up and encouraging communications. There are some other buzzwords associated with brainstorming.

- *Storyboarding.* This term started in the film industry and refers to brainstorming sessions in which plots for films or television programs were developed. The act of pinning or taping ideas to a wall is storyboarding.
- *Off the Wall.* This term was supposedly invented in Disney studios. It refers to ideas that were so outlandish that the group decided to take them "off the wall."
- *Blackboarding.* This term started in the computer industry with the advent of electronic communications. It refers to a shared file into which people can place ideas. It is a method of communicating electronically.

Boundaryless Organizations. This philosophy indicates that all employees' ideas are valid, and that no employee should be afraid of approaching management with suggestions for change.

Team-Based Manufacturing. This theory has found its major application in the apparel construction industry. Team-based manufacturing suggests that, if a worker is not busy, he or she will help out a worker who is busy. In other words, workers will look at the backlog at various workstations and then help out the person or people who have the largest backlogs. The problem with this concept is that it is based on the fact that some workers are less busy than others and it avoids the real problem in the system—the lack of a consistent and steady flow through the manufacturing floor. Moreover, it is self-defeating, since, as soon as a worker leaves a workstation, the backlog at that station will begin to build.

The Seamless Organization. The seamless organization is one that is supposedly free from the constraints that hinder most organizations, although those constraints are not well defined. The main approach to creating a seamless organization is the collapsing of vertical functions into cross-functional teams. In other words, this is an old concept given a new name.

Benchmarking. Benchmarking is the measurement of organizational effectiveness through the use of empirical measures, such as productivity and output. Unfortunately, it is a measurement of the results of a process, not the benchmarking of the individual processes that make up the process. It is a step in the right direction but does not deal with the intricacies of processes and the reasons why measures change in response to changing conditions.

Vertical Partnering. This is an idea that works, but it is not really a new idea. Basically, the concept is that two organizations will create a strategic partnership in which the purchasing organization will guarantee business to the selling organization in exchange for a number of considerations. Typically, these considerations include the vendor supplying the shelves for the purchaser, determining the inventory requirements of the vendor, and so on. In other words, the seller manages the supply chain for the purchaser and eliminates the purchaser's need for warehousing space for high-volume products. While this may be a new idea for retail stores, it is a practice that has been in place in supermarkets for many years.

Value-Added Mapping. This is a technique in which processes are analyzed from a functional perspective, with the following question being asked of each function: Does this function add value to my customer? Again, this is a type of analysis that is being couched in terms of the customer, when, in reality, it should not be. For example, a favorite target of advocates of value-added mapping is a purchasing department. The case can be made that purchasing adds no value to a finished product; therefore, there is no value to the customer. This, of course, is ridiculous. Without purchasing departments, there would be no raw material to create a product and, therefore, no product to be delivered.

The real question *should be*: Can the company justify the expense of this function, and are there ways in which we can reduce the expense of this function? By changing the emphasis from being customer oriented to profit oriented, we can focus on the real issue—the costs of processes. If the cost of purchasing, for example, can be reduced, then profits might be increased, even if very slightly, or a company may choose to pass these savings along to the client to increase its competitive position.

Communication: The Foundation of Process Reengineering

Not surprisingly, when it is suggested that TQM is mostly differing theories of how to talk and how to encourage talking, many TQM experts claim that it is more—that it is also a commitment. Perhaps what they are saying is that TQM and CPI are also *listening*. Along with listening comes *commitment*.

Take the example of quality circles. Besides the fact that they did not provide the results expected of them, it is possible that they are no longer popular because, although workers made the required commitment to communication and change, management did not. Management has supported quality circles in terms of money, facilities, training, and so on, but has usually ignored the suggestions that resulted from quality circles.

Some managers have claimed that the suggestions for process improvements that came from quality circles could not, accurately or with certainty, be phrased in terms of cost reduction or in cycle time reduction; therefore, there was a high degree of risk associated with implementing the changes. Workers who have participated in quality circles will claim that management has simply paid "lip service" to quality circles and other similar concepts. Management encourages communication but has not made the commitment to listen.

As this continues to happen, workers lose their enthusiasm for the currently popular TQM philosophies and buzzwords and begin to feel cynical. Eventually, a new philosophy comes about, new buzzwords are invented, and the process begins all over again.

This discussion of TQM buzzwords may seem like an attack on TQM, but it is not. Communication is the essence of process reengineering and process improvement, and without effective communication, computer-aided process improvement is an impossibility. Since the theories of how to communicate change with regularity, organizations should pick one and move on.

Question: Why Do These Approaches to Change Persist?

One has to wonder not only why new techniques come and go with such regularity, but also why, when a new technique appears, it is accepted so rapidly by management. I think the answer is that trying to create change at level 1 is easy. What could be more simple and straightforward than bringing people together in a room and letting them talk about a process? There is no investment in technology and a minimal investment in time. There is also little prospect for meaningful change so no one is disappointed with the results of the change effort.

BPR is no different than any other effort toward improvement. The old adage "You get what you pay for," is directly applicable to BPR efforts. If you don't commit time and resources to change, you won't get change. What is about to be described in the next few chapters is a rigorous approach to change that requires commitment from both management and employees.

Summary

A Level 1 process that is changed using nothing but Level 1 change techniques will remain a Level 1 process. To get out of the chaotic process state, an organization must migrate through the five levels of change. The next chapter begins a description of that migration path.

5

Level 2 (Repeatable) Processes

A Level 2 process is exactly what its name indicates—repeatable. The quality of products produced by a Level 2 process is predictable and falls within established parameters.

What differentiates a Level 2 process from a Level 1 process?

> As projects grow in size and complexity, attention shifts from technical issues to organizational and managerial issues.

A Level 2 process enables people to work more effectively by incorporating lessons learned in *documented procedures* and to continually improve.

There are two key issues addressed by this statement: (1) A Level 2 process is a *documented* process, and (2) there is communication among process participants. The last chapter focused on the need to communicate about processes, and this chapter focuses on the need to document processes.

Documented procedures provide the foundation for consistent processes that can be reused across an organization. In other words, the documentation of process tasks (processes in the small) begins to provide some overall understanding of processes in the large.

The preceding statements can be encapsulated into a description of the characteristics of a Level 2 process.

1 The results of a Level 2 process are predictable (although not necessarily of high quality).
2 Communication exists among process participants.
3 Individual process tasks are defined and documented.
4 There is an *awareness* of an overall process by the participants in individual process tasks; however, the overall process itself is not defined or documented.

To migrate from a Level 1 process to a Level 2 process, an organization must

1 Institute some TQM or CPI philosophy that encourages communication among workers and management.
2 Make the commitment to listen to workers.
3 Create a repository for information provided by workers.

Process Reengineering Rule 2, therefore, is *document the process*. It is the documentation of the process that is the key to making a process repeatable. The reasons for this are as follows.

1 The success of Level 1 processes is determined by the abilities of the participants in the process. If a participant leaves an organization or is no longer available, there can be tremendous impact on the quality of products produced and productivity if the tasks performed by that person are not well documented.
2 Typically, process reengineering suggestions that result from traditional TQM efforts, such as quality circles, are focused on individual process tasks. Therefore, a firm understanding of those tasks is essential.
3 The documentation of process tasks allows them to be reused (repeated) by individuals who are not expert in the tasks themselves.

In addition, the documentation of a task must include the specifications of the inputs required and the products produced. This information is essential for building an overall process definition as the migration through process levels continues.

Documentation of process tasks begins to provide the functional view of a process. That view depicts what tasks are performed and what products are produced. In the remainder of this chapter, the Origami Process will be used to show how documentation can enhance the understanding of the process and make it repeatable, and the purchase order process will be used to show how a lack of documentation can lead to counterproductive attempts at reengineering. Common problems associated with written documentation will then be discussed, accompanied by a suggested template for the documentation of tasks.

Migration of the Origami Process to Level 2

The Origami Process is an exercise focused on the team building and communication aspects of TQM and CPI. Although the training exercises do result in process improvements, the process is a Level 1 process since it is undocumented. The success of this process is clearly dependent on the abilities of the participants and their willingness to work as a team.

The Origami Process can be used as a training mechanism in developing documentation and, therefore, as an exercise in migrating from Level 1 to Level 2. For example, the instructor can ask the paper cutters to document the paper cutting task. This is the simplest task of the process, and it may take some time to convince a student that it can be documented at all; however, a student will typically write something like this:

1 Get a stack of paper.
2 Measure the paper (to specifications).
3 Cut the paper.

This documentation can then be given to students who are unfamiliar with the cutting task. When these students are asked to "repeat" the cutting task, the results will vary, since each student will perform the task in a slightly different way. Even though the cutting task is very basic, the fact that the results vary demonstrates the need for accurate documentation.

At this point in the exercise, the instructor will ask one of the paper cutters to add detail to the written documentation. The result will read something like this:

1 Get a stack of yellow paper.
2 Place the stack on the cutting machine at the 6-inch marker.
3 Cut the paper.
4 Place the cut stack into a pile to be picked up by the material handler.
5 Put the waste into the recycling bin.

This is better documentation but, again, repetition of the task by people unfamiliar with it will yield different results. Reasons for this are

- The size of the stack of paper (the number of individual pieces of paper in the stack) referenced in the documentation is undefined.
- It is not clear whether the long or short side of the uncut paper is placed at the 6-inch marker on the cutter.

At this point, the instructor will encourage the experienced paper cutter and the inexperienced paper cutters to discuss the task. This discussion will provide some more information about the task, such as, the stack of paper to be cut consists of ten pieces. The size of the stack, we will also learn, was based on experimentation by the experienced cutter and represents a "best practice."

What this demonstrates is that documentation of a task is an iterative process. Process Reengineering Rule 2 makes use of the communication aspects of Process Reengineering Rule 1. After discussing the task and listening to the inexperienced cutters, the cutting task documentation can be enhanced to read

1. Get a stack of *ten pieces* of yellow paper.
2. Place the *long side* of the stack on the cutting machine at the 6-inch marker.
3. Cut the paper.
4. Place the cut stack into a pile to be picked up by the material handler.
5. Put the waste into the recycling bin.

Although this is an extremely simple task, the documentation contains some very important information.

1. The input requirements of the task are defined; that is, uncut yellow paper.
2. The tool required for the task is defined; that is, a cutting machine.
3. The activity that represents the task is defined. In fact, several activities are defined: getting the paper, placing the paper on the machine, cutting the paper, placing the cut paper in a repository, and placing the waste in a repository.
4. A specification for the cutting activity is defined.
5. A product is defined.
6. The participation of another functional group (the material handler) is defined.

While this may all seem trivial, it is extremely important when the reengineering of a process is being considered. For example, if the paper cutter being used is manually operated, what will be the effect on the process if it is replaced with an automatic cutter? This type of detailed information is required for effectively modeling a process to determine the effect of changes on the process.

The Origami Process is an ideal exercise for learning how to develop documentation. The tasks involved in the process are very simple but, as shown earlier, can require detailed documentation. The documentation of the process tasks also reiterates the necessity of effective communication and reinforces Process Reengineering Rule 1.

Purchase Order Process: Second Attempt at Reengineering

In the previous chapter, it was demonstrated that decisions based on limited information or mental models can lead to counterproductive results, and that quick fixes should be avoided. The purchase order process, at this point in time, remains a

Level 1 process, even though management had attempted to improve it by reorganizing. In fact, the results of the process had gotten worse. The number of purchase orders processed had decreased and the number of policy violations had increased.

Recognizing the increasing nature of its problem, management decided to apply TQM theory and organized a purchase order process reengineering team. The members of the team consisted of the management responsible for the process and the workers who actually processed the purchase orders. The first order of business of the team was to discuss the process itself.

Through these discussions, it was determined that there were three mechanisms of obtaining equipment:

- Outright purchases of equipment.
- Equipment leased for fixed periods of time.
- Equipment rented for undetermined periods of time.

It was also discovered that a slightly different process was required for each type of order; therefore, there were actually three *different* processes being discussed. In addition, management learned that each processor had differing levels of expertise in these processes and, among them, they had an informal agreement as to who would handle each type of purchase order.

Management decided to make the informal processes formal and assigned each processor to the task of handling either purchases, leases, or rentals. Management then informed the user community of the change in procedures and identified the individuals and the types of orders each would handle. Management had implemented another quick fix, but was certain that this time the fix would yield favorable results.

Not surprisingly, the process problems grew worse. The number of purchase orders processed decreased and the number of policy violations increased. Management had attempted to utilize process reengineering rule 1 (talk about the process) but failed to understand that *all* the participants in the process had to be represented. In this case, the users of the process were not involved and were unable to provide input. The changes management made to simplify the process of handling purchase orders actually increased the complexity of writing purchase orders. The result of the changes to the process can be captured in the causal loop diagram shown in Figure 5.1.

This causal loop demonstrates that

1. Management assumed that having separate processes for each type of order would increase the speed of processing.
2. An increase in the speed of processing would increase customer satisfaction.
3. An increase in customer satisfaction would result in a decrease in attempts to bypass the process.

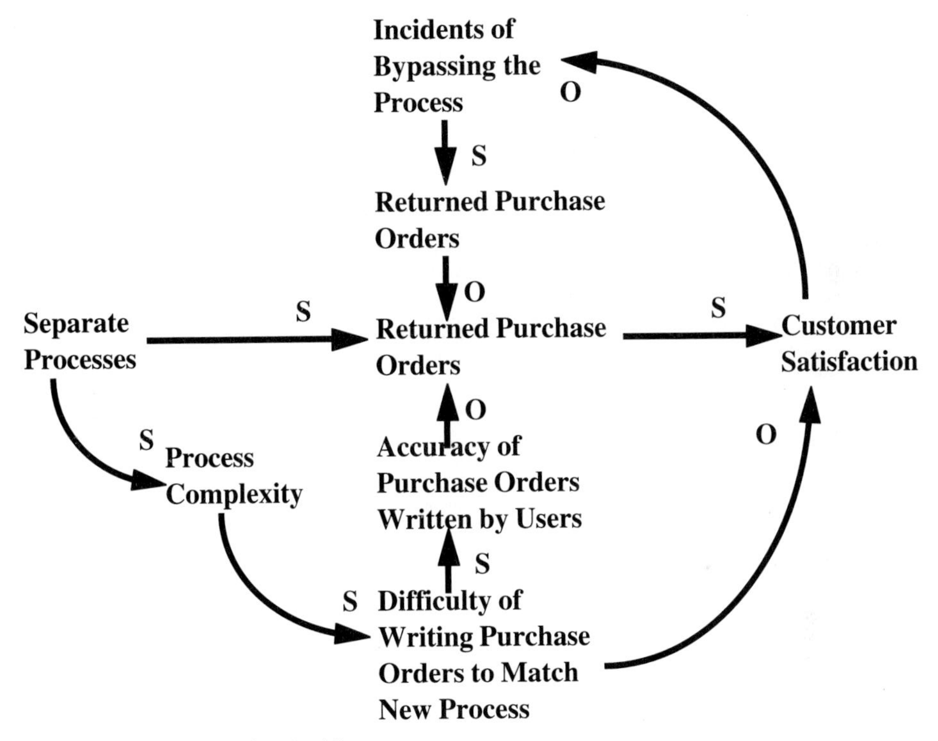

Figure 5.1 Another Fix that Backfires.

4. A decrease in the number of times the process was bypassed would result in a decrease in the number of purchase orders being returned by vendors.
5. A decrease in the number of returned purchase orders would result in an increase in the speed of processing, since processors would not be doing rework.

These all seemed to be valid assumptions; however, by failing to talk to the users of the process (the customers), management was denied some very important information.

- Users typically did not know whether equipment would be purchased, leased, or rented. Moreover, they did not *care* how the equipment obtained, only that it be obtained. It was actually the purchase order processors who made the determination of how equipment would be obtained, a function that still existed in the improved process. The process had become more complex in the view of the users.
- When users attempted to work within the new process and determine the method by which equipment would be obtained, they became confused. They found that writing purchase orders, once a simple process, had now become quite complex and difficult.

This had two results: (1) Users either wrote inaccurate purchase orders, or (2) they bypassed the system. In either case, their satisfaction with the system decreased and the number of system bypasses increased.

Management had taken the first step toward process reengineering by opening a dialog about the process; however, they did not understand that both customers and suppliers are participants in a process, a concept that is fundamental to TQM and CPI philosophies. Moreover, they had not documented the process and had missed important details, such as the nature of the input to the processors and the details of the tasks they performed.

We will look at management's further attempts to improve this process in the next chapter.

The Computer Aided Process Reengineering Method of Documentation

The migration from a Level 1 process to a Level 2 process requires that process participants communicate their understanding of the process and their roles in the process. The information obtained through the verbal communication of the process forms the foundation for Process Reengineering Rule 2—*document the process*. The act of documenting a process not only makes tasks repeatable but also serves to clarify and refine the information obtained from communicating the process. In the remainder of this chapter we will discuss methods of developing useful documentation so that the processes described can be repeated.

There is an inverse relation between the amount of effort put forth by TQM and CPI experts regarding verbal communication and written communication. Considering that written documentation is the primary method of communicating processes to workers, one would think that TQM and CPI philosophies would address the need for effective documentation skills.

Computer aided process reengineering requires that documentation be well written and contain specific, detailed information. Unfortunately, there is a tendency in industry and government to write documentation that overlooks and omits details and that combines the descriptions of several tasks into one. There are a number of possible explanations for this tendency.

- Companies are intent on creating processes to make products. Documentation of a process is usually done (if at all) after the process is in place, and often as an afterthought.
- Process documentation is typically written by an expert in the process. Experts don't feel the need to describe what they believe is obvious.

- The development of documentation is time-consuming and costs money. If a company cannot see the benefits of well-written documentation, it is likely to apply minimum resources to documentation efforts.
- Since processes tend to evolve over time, the need for documentation can simply be overlooked.

However, the simple truth of the matter is that it is ultimately more expensive not to produce good documentation than it is to produce it. A suggestion for developing documentation that is consistent with computer aided process reengineering is the following.

1. Start with the "whats" of the process; that is, what is needed to perform a task, what are the tasks performed, what is produced, and so on.
2. Specify the tools, inventory and staff required to perform a task.
3. Break down tasks into subtasks. For example, a task may consist of setup, actual work, breakdown, preparation for shipping, and so on.
4. Provide a description of the products that result from the task.

The template in Table 5.1 can be used to develop documentation in the manner just described.

Some of these points require some more explanation.

1. *Inputs to task.* Note the template suggests defining types, or classifications, of inputs. For example, purchase orders which are inputs to a purchasing task can be classified as "less than $5000," "more than $5000 but less than $100,000," and "more than $100,000." This classification scheme implies that different types of processing may take place depending on the value of the purchase order.
2. *Staff required.* Define the staff required to perform only the task being documented. If interim processing takes place and involves other staff, then the task being defined should be broken into subtasks, one which occurs before the interim processing and one which occurs after the interim processing. The interim processing should be defined as a task by itself.
3. *Conditional processing.* This part of the documentation would include a description of any specialized processing that might occur in the task. For example, in the case of the purchase orders described earlier, specify the special processing that occurs for each type of order.
4. *Products.* Include all products, including work sent back to the suppliers, waste, and so on.

Documentation written using this template provides information about what tasks are performed and also *how* the tasks are performed (specifications of a task). This type of documentation begins to define the behavior of the process. All this

Table 5.1

Task Name	Define the name of the task or function being performed
Inputs to Task	Define the raw material, inventory, tools, and so on, required to perform the task. Specify different types or classifications of inputs, if any.
Staff Required	Define the number of individuals required to perform the task. Specify the functional group to which those individuals belong.
Suppliers of Inputs	Specify who supplies input to the task being performed and how the input is introduced.
Subtasks	If the task consists of subtasks, define what those subtasks are. If different individuals are involved, then define the subtasks as separate tasks in their own right.
Description of Task	Describe how the task is performed, including details, no matter how insignificant.
Conditional Processing	Specify how paths through the task are determined.
Recipients of Task Output	Specify where task outputs are sent, if that information is known.
Products	Define what is produced when the task has been completed.

information is necessary, both to migrate to a level 3 process and to utilize computer-aided process reengineering technology.

Summary

Documentation of a process requires *thinking* about the process. The mere act of writing down one's thoughts about a process will yield important information about the process. It is not unusual to develop process documentation iteratively, with each iteration revealing more important details that may have been overlooked during the previous iteration. *Thinking*, unfortunately, is an activity that some managers frown upon—they want action! Detailed investigation and documentation of a process compromise a necessary step in the migration through process maturity levels.

6

Level 3 (Defined) Processes

A Level 3 process is a process in which tasks have been formally defined and documented, and for which an overall view of the process has been developed. A Level 2 business process can be described from the aspect of *what* (what tasks exist and what gets produced). A Level 3 business process adds the concept of *how* (how the process flows) and the concept of *when* (the conditions under which certain process paths are taken). When an overall view of a process is presented, the process has been *defined.*

While migrating from Level 1 to Level 2, verbal communication and written documentation have been used to describe processes. Clearly, benefits result from the use of these practices, but there are also deficiencies with both verbal and written process descriptions. For example,

- Verbal descriptions of a process are often incomplete. Important details are often overlooked or lost in discussions.
- Written descriptions of a task cannot provide an easily understood view of a large or complex process and are best applied to individual process tasks.
- There is a tendency to develop written documentation from the perspective of process experts, so important information may be left out.

There is one other limitation of written documentation that has not yet been discussed; that is, written documentation cannot effectively present an overall definition of a process. There is simply too much information to be absorbed. Global views of processes are more effectively defined graphically, using one of several different types of diagramming techniques. These diagramming techniques are collectively referred to as *activity diagramming*.

The most popular activity diagramming mechanism is the flowchart. Flowcharts consist of symbols that represent process steps, storage functions, and decision points. While flow chartsprovide global views of processes, they are limited in their ability to provide an actual view of how a process works.

Migration of the Origami Process to Level 3

The Origami Process is an ideal process to show both the value and limitations of activity mapping. After the exercises suggested in the previous chapters, the Origami Process has become a Level 2 process—it has been documented and the process steps are repeatable. To further increase understanding of the process, Continuous Improvement classes usually have students develop flowcharts of the process. An initial flowchart of the Origami Process is shown in Figure 6.1.

This flowchart is representative of flowcharts that are typically created to accompany written documentation. A review of this chart will reveal that it is not quite accurate. For example, the paper cutting task is actually two tasks: a white paper cutting task and a yellow paper cutting task. These tasks occur in parallel and can be represented by the flowchart in Figure 6.2.

This flowchart provides a more informative view of the process; for example, it demonstrates that certain tasks occur in parallel. It also assumes certain facts; for example, it assumes that there is an adequate supply of yellow and white paper available for the task labeled "Assemble Final Product." Although a decision symbol could be inserted prior to that step to ask if there is one piece of yellow paper and one piece of white paper available, this would begin to complicate the flowchart and render it incomprehensible; however, this is an important fact to be considered to make sure the process runs smoothly.

At this point, it is important to emphasize the limitations of activity diagramming. The primary function of an activity diagram is to provide an overall view of the process steps required to produce a single product. It is *not* a representation of the process as a whole. Some of the deficiencies of activity diagrams are as follows:

1 They do not accurately represent parallelism in a process. *All* the tasks in the Origami Process are being done at the same time—paper is being cut, folded, painted, and so on. Activity diagrams represent the tasks required to produce a single product, and this representation is mostly serial in nature.

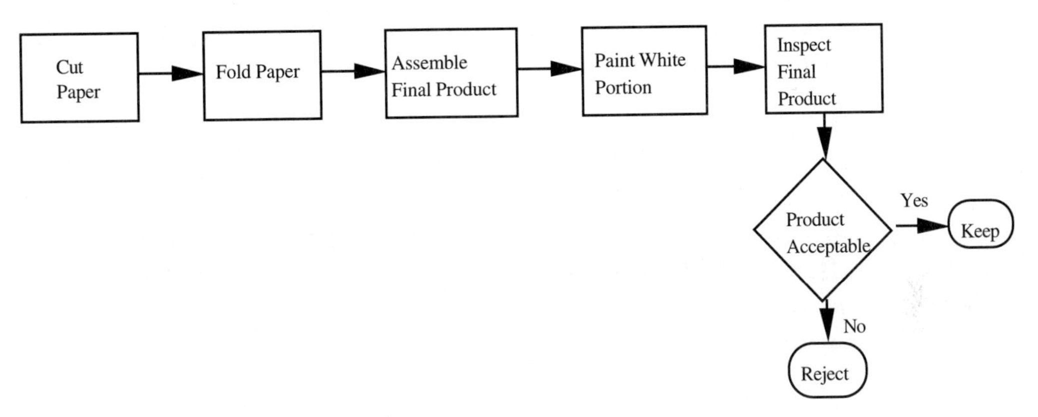

Figure 6.1 Initial Flowchart of Origami Process.

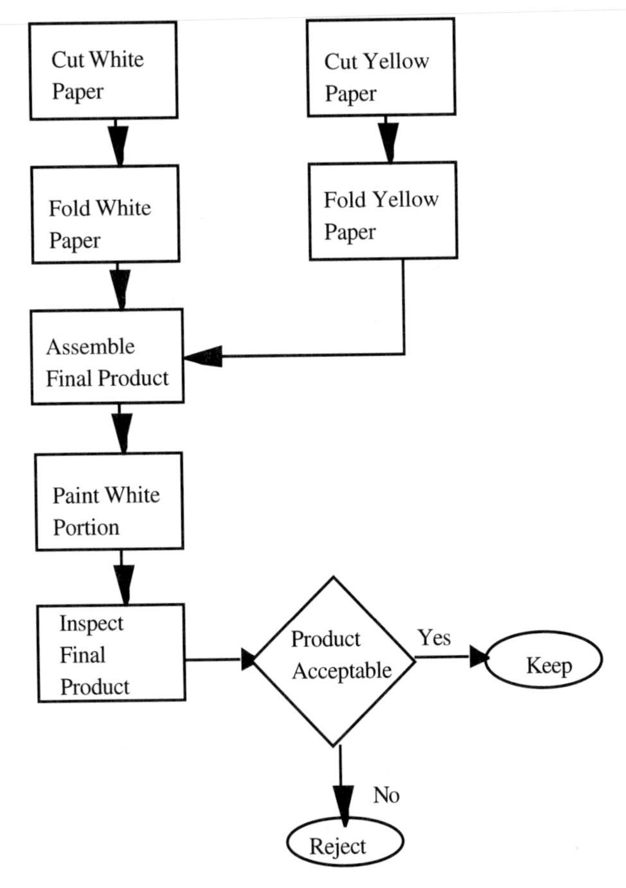

Figure 6.2 Second Origami Process Flowchart.

2 They do not provide any sense of timing or production flow. The flowchart in Figure 6.2 *implies* a sequence of events, but the process does not follow that sequence exactly.

3 They do not provide any mechanism for measuring process parameters, such as

- the number of personnel assigned to a process step
- the number of final products produced after a period of time
- the productivity of personnel involved in the process
- the total amount of time to produce one product

Activity diagrams also do not represent the *transactions* between steps or the existence of storage areas. For example, flowcharts, such as the one in Figure 6.3, can depict sequences in which material may or may not be transferred between tasks.

Figure 6.3 *implies* a transfer of cut paper from the cutting task to the folding task, but such a transfer cannot be accurately depicted without the addition of several more symbols. If every transaction in a process were graphically represented, an activity diagram would become overly complicated very quickly. However, transactions take time and must eventually be understood to provide a more accurate global view of a process.

For example, in the Origami Process as it is first practiced, a material handler moves inventory from process task to process task. The actions of the material handler represent the transactions between process steps. These transactions are scheduled, or occur, based on some decision-making process by the manager. Understanding the actions of the material handler is essential to understanding the overall process, but these actions are difficult to diagram. The material handler *interacts* with the production process but is not part of the process itself.

However, since the actions of the material handler are essential to the understanding of the overall process, they should be captured in a flowchart of the process. Figure 6.4 is an attempt to show those actions.

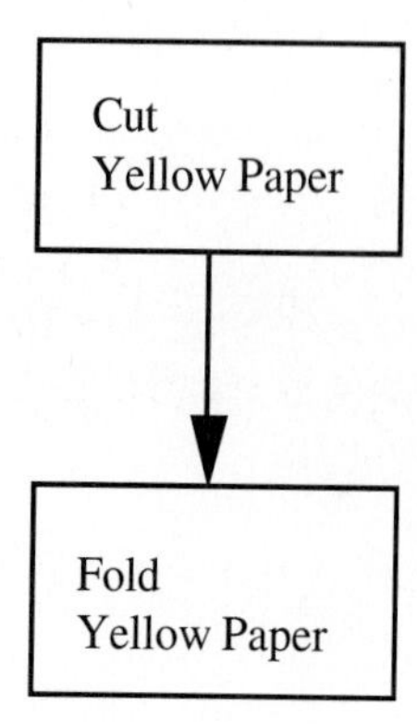

Figure 6.3 Flowchart Transaction.

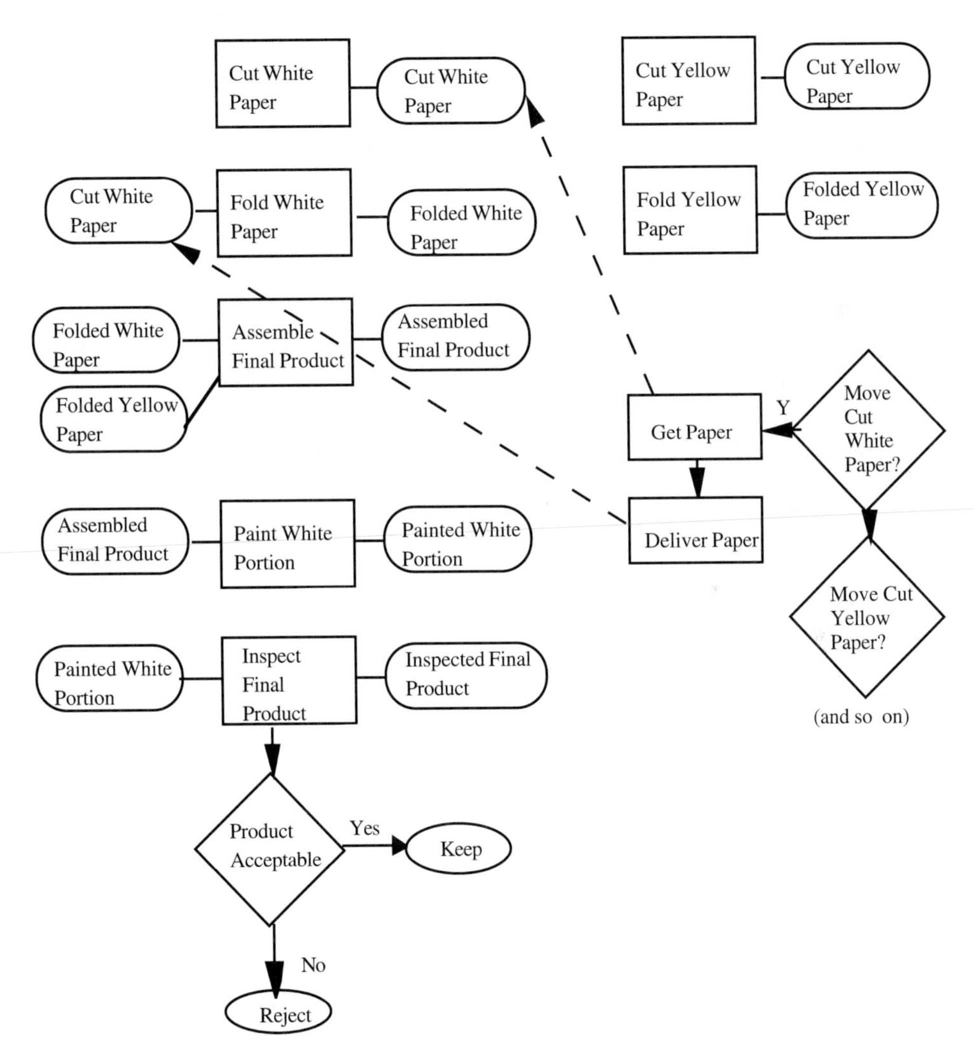

Figure 6.4 Flowchart of Origami Process Showing Material Handler.

The flowchart of the process is quickly becoming *extremely* complicated. Additional symbols have been added to represent the output of tasks and the input into tasks. Decision boxes represent the conditions that dictate the actions of the material handler, and dashed lines represent movement to and from storage areas. This flowchart of a relatively simple process demonstrates that as systems grow more complex, the ability to capture them in flow diagrams diminishes.

Activity diagrams are, however, very valuable tools. They present

1 An overall view of a process.

2. A sense of "how" the process operates. For example, the Origami flowchart at least shows that there is some parallelism, even though the total parallel nature of the process is not provided.
3. A limited sense of when things happen. For example, the decision symbol in the Origami flowchart represents a check of the condition of the product, in this case, acceptable or unacceptable. When the condition is acceptable, the product is kept; otherwise, it is rejected.
4. The true complexity of a system if developed correctly.

This last point is very important. One can look at the last flowchart of the Origami Process and ask if a material handler is really necessary. The system would be less complicated if the participants in each task simply moved their finished products to the input stacks of the next task. Then the rationale for the material handler can be explored to determine if changes could be made to improve the process.

Point three in the preceding list references the *when* aspect of flowcharts, and this concept requires more discussion. TQM and CPI experts often propose the use of *event diagrams* to depict the events or conditions that cause processes to flow. An event diagram of the Origami Process would appear as shown in Figure 6.5.

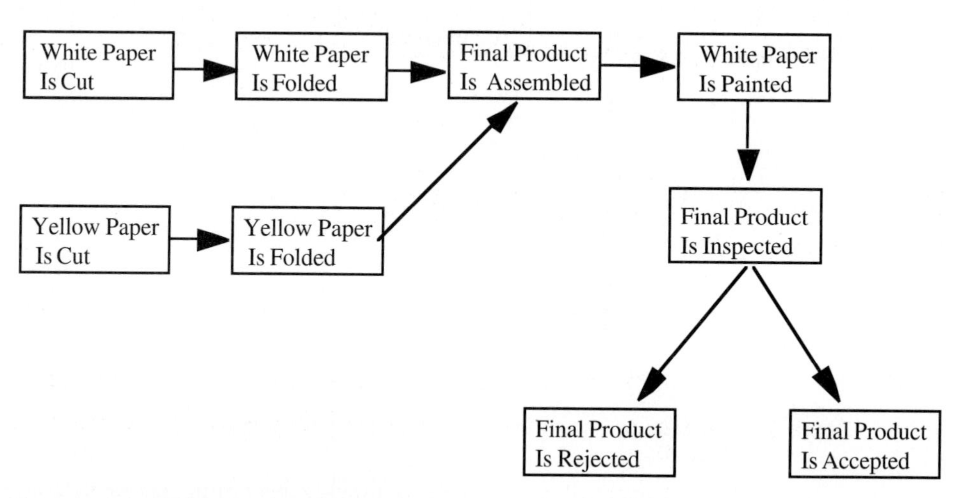

Figure 6.5 Origami Process Event Diagram.

Figure 6.5 shows the events or condition changes that occur as a product flows through a process. It can be interpreted as when the paper is cut, it can be folded; when it is folded, it can be painted; when it is painted... and so on. Event diagrams *assume* that some transformational activity occurs at each step or some measurement (in this case a measure of quality) occurs. The theory behind event diagramming is that depicting events is sufficient for understanding process flow, but this is a debatable point. This type of diagramming is becoming popularly known as *state transition diagramming*.

Event diagrams suffer from some of the same deficiencies as flowcharts, mainly, they represent a process in a serial nature even though the processes being diagrammed may be highly parallel.

Sophisticated process diagrams can be developed by combining representation methods. For example, activity diagramming and state transition diagramming have been successfully combined to represent organizational processes and to improve those processes.[1] This combination, however, requires two separate sets of diagrams and the communicative power of each diagram is diminished. In addition, state transition theory requires an object-oriented view of processes that is not universally known or accepted.

Whatever method is chosen, activity diagrams provide valuable information regarding processes. It is necessary to develop an overall view of a process and to define interactions with other processes before the parameters that define the behavior of a process can be understood. Meaningful improvement cannot be attempted until a global view of a process has been developed.

Process Reengineering Rule 3 is, therefore, *diagram the whole process.* Activity diagrams, event diagrams, process maps, or a combination of these can be used to represent the process, but the information that must be captured are the activities and conditions that affect process paths.

Purchase Order Process: Third Attempt at Reengineering

At this point, two attempts have been made to reengineer the purchase order process:

1. A reorganization changed the personnel involved in the process.
2. Three separate processes were created to handle the different types of purchase orders.

After these attempts at reengineering, the process remained a Level 1 process—chaotic. Management had been using two measures to determine the success of the process: (1) the time required to process a purchase order, and (2) the number of times the process had been bypassed. With each attempted improvement, the time required to process a purchase order increased and the number of attempts at bypassing the process increased.

[1] Hansen and Kellner, "Software Process Modeling," SEI Technical Report, CMU/SEI - 88 - TR - 9, May 1988.

Management then decided to investigate the process further by discussing it with people writing purchase orders (i.e., the users of the process) and reinterviewing the individuals who processed them. After a series of interviews was conducted, management learned that (1) the users were combining different types of purchases into one purchase order, and (2) the purchase order processors were rewriting purchase orders received from users to separate them into the categories of purchases, leases, and rentals. They were doing so to adhere to the procedures just implemented.

This resulted in increased process time and, therefore, increasing user frustration, so more users began to bypass the process. In addition, a new problem emerged: Because the system was being bypassed with increasing frequency, budgets for computing equipment were being exceeded. This really got management's attention!

To overcome the problems in the process, management came up with a unique approach. It developed a document entitled "How to Write Purchase Orders," which was distributed to users of the purchase order process. In other words, management was empowering the users of the process to improve the process. Figure 6.6 depicts management's reasoning.

Management reasoned that if the purchase order preparation process could be documented,

1 Users would write better and more informative purchase orders.
2 As the purchase orders received became better, the speed of the process would increase.
3 As the speed of the process increased, user satisfaction would increase and the number of incidents of bypassing the process would decrease.
4 As the number of incidents of bypassing the system decreased, the speed of the process would increase as well since purchase order processors would not be occupied chasing problems.

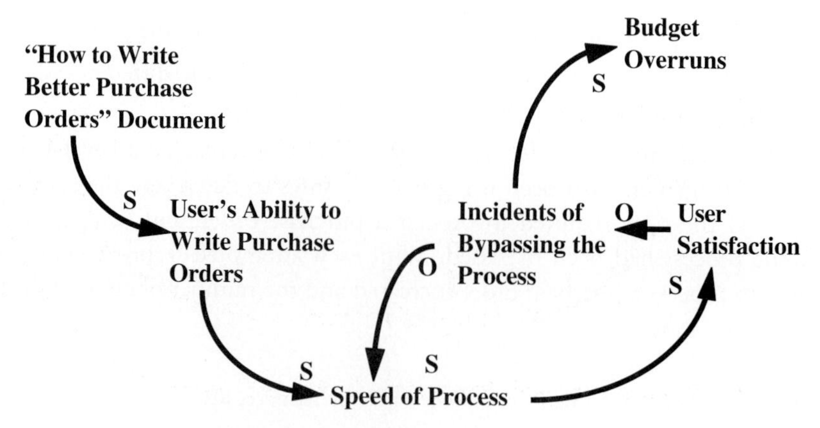

Figure 6.6 How the User Affects the Purchase Order Process.

This defines a *virtuous cycle*. A virtuous cycle is the same as a vicious cycle, except that the results are beneficial.

Management also made one other change: To deal with the budget problem, it decided to have the purchase order processors do budget checks for every purchase order received and to reject those that resulted in a budget being exceeded.

Once again, this quick fix backfired. Although management's reasoning was not entirely incorrect, there were some unexpected side effects.

1 The check points put in to prevent overruns decreased the speed of the process.
2 The decrease in speed frustrated users and as frustration increased, the incidents of system bypass increased.
3 The increase in the ability of users to write purchase orders increased their sense of independence and they became more likely to bypass the system than before.

There was also another, more fundamental problem. Users, to this point, had not been asked to participate in process reengineering discussions, and they did not like the fact that they were being held responsible for fixing a process they did not "break." Therefore, users saw "How to Write Purchase Orders" as an opportunity to bypass the process more efficiently.

The result: more budget overruns! The very problem management tried to eliminate became worse. The causal diagram in Figure 6.7 shows what actually happened in the process.

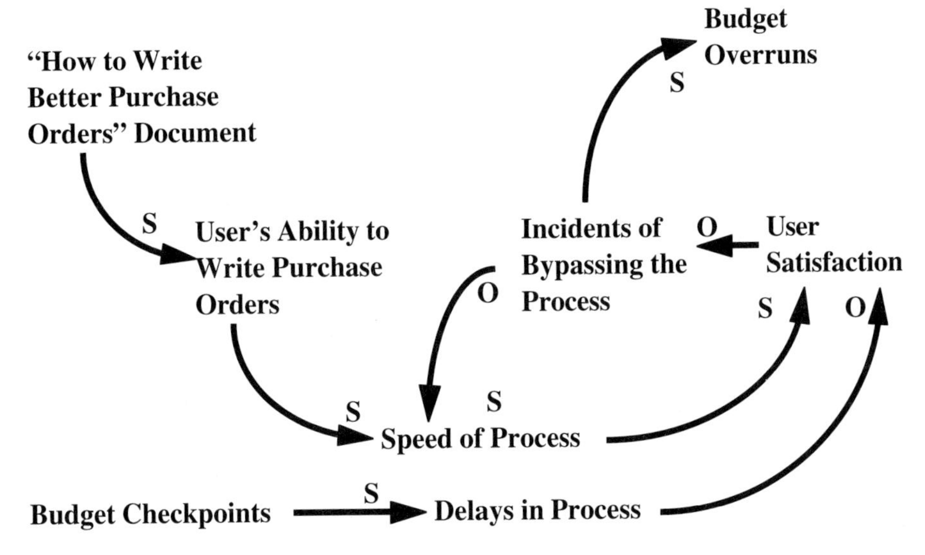

Figure 6.7 Virtuous Cycle Turns Vicious.

Once again, management had attempted to improve the process without taking the necessary analytical steps. It can be argued that it invoked Process Reengineering Rule 1—it opened up communication about the process. It can also be argued that it invoked Process Reengineering Rule number 2—it documented a part of the process. Management did not, however, document the whole process. Management also violated a fundamental principle of TQM and CPI by failing to talk to its customers.

Documenting the whole process might have given users insight into the problems created by bypassing the system and might have led to more focused improvements. Although this seems to be a story about continuing failures of attempts at process improvement, it is also a story about a learning process. Management was beginning to take the steps suggested by computer aided process reengineering, but it was also still looking for the quick fix. The two are incompatible and, in this case, management had more work to do.

Summary

Migration from a Level 2 process to a Level 3 process requires an understanding of the global process being investigated. The first step in creating this understanding is to apply Process Reengineering Rule 1—communicate about the process. The second step is to apply Process Reengineering Rule 2—document the process tasks. The third step is to apply Process Reengineering Rule 3—*diagram the overall process*. Any diagramming technique can be used, as long as the technique portrays both process activities and events (conditions).

Engineered Flowcharts

There are ways to build flowcharts so they provide more information than traditional methods; that is, they can be "engineered" to capture some of the behavioral aspects of a process. For example, the flowcharts of the Origami Process were done using a *data flow* diagramming technique that has been popularized by software developers. That particular diagramming method uses symbols, such as the symbol shown in Figure 6.8, to represent storage areas for information.

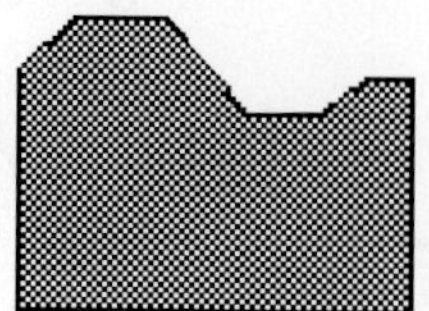

Figure 6.8 Data Flow Diagram Storage Symbol.

A data flow diagramming technique was used to develop the activity diagram of the Origami Process, and a special symbol was used to represent storage of physical objects. This symbol is very generic, and it may not be clear in other flow diagrams whether the storage area is being defined for tangible items or intangible items, such as data. A solution to that problem is to create and use a symbol that represents a storage area for tangible items, as shown in Figure 6.9.

Figure 6.9 Physical Storage Flowchart Symbol.

This can be used in a flowchart as shown in Figure 6.10.

Figure 6.10 provides more information than is normally found in a flowchart. It shows that spokes, rims, and tires must all be available for the activity of building wheels to take place. Figure 6.10 also shows that the results of the build wheels activity are, in fact, wheels.

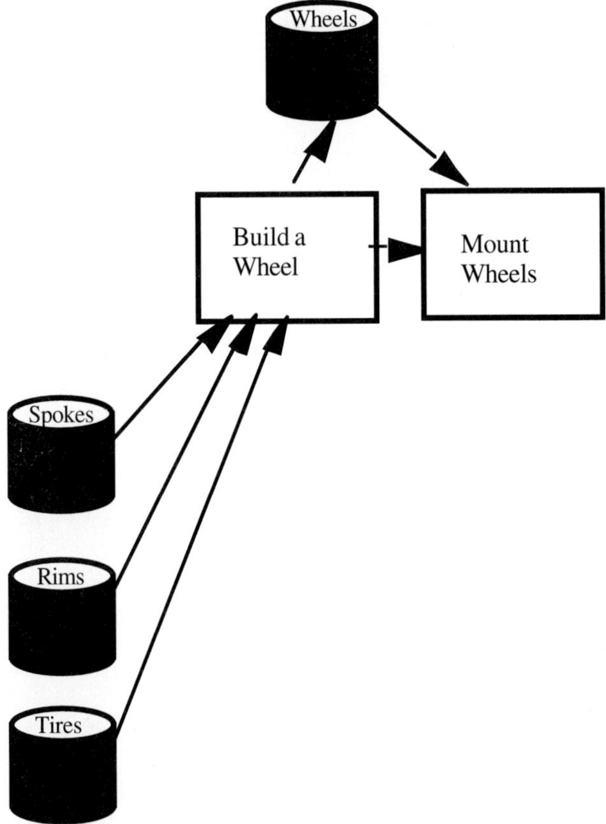

Figure 6.10 Flowchart with Symbol Storage Symbols.

In Figure 6.10, it is intuitively obvious that all the input items must be available for the activity to take place. That is not always the case, so a symbol representing an AND condition can be used to build more informative flowcharts. Consider the sequence of events shown in Figure 6.11.

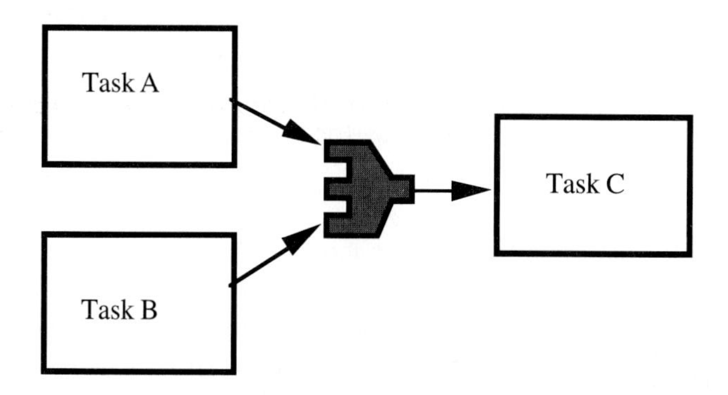

Figure 6.11 Flowchart with AND Condition Symbol.

The additional symbol, called a merge symbol, specifically denotes that Task A *AND* Task B must both be completed before Task C can begin. This represents a *condition* that dictates how the process will flow, and the conditions that affect a process must be known to create effective models.

Identifying people or groups responsible for activities helps make flowcharts more informative. Figure 6.12 shows how groups of individuals can be identified in flowcharts.

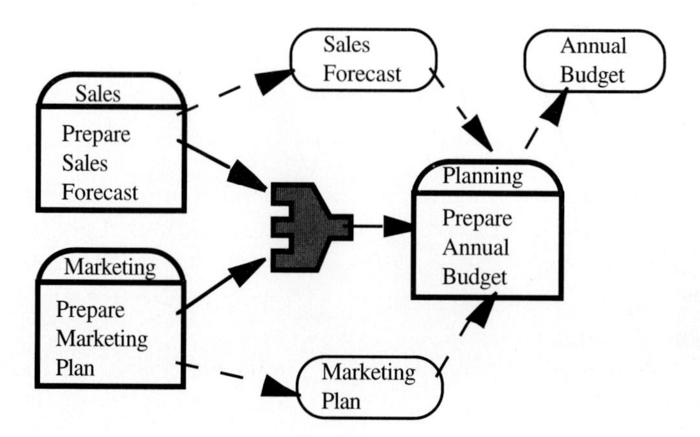

Figure 6.12 Flowchart Using AND Symbol.

This flowchart contains a great deal of information, such as,

- There are three organizations involved in the process.
- The activities Prepare Sales Forecast and Prepare Marketing Plan must both be completed before the activity Prepare Annual Budget can begin.
- The first two activities produce information that is used by the third activity.
- The third activity produces a data item called the Annual Budget.

Multiple paths based on condition checks in flowcharts can be problematical. For example, to depict the selection of three possible paths based on three possible conditions, a flowchart would normally be drawn as shown in Figure 6.13.

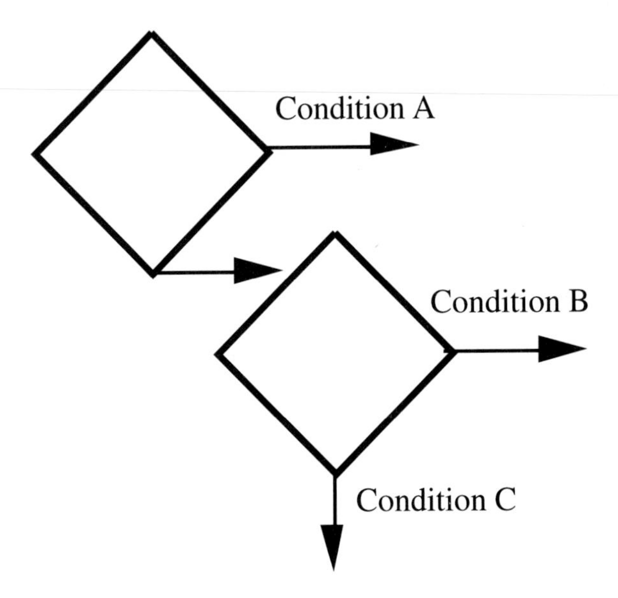

Figure 6.13 Traditional Representation of Multiple Paths.

Traditional diagramming techniques limit condition checks to two possible values. When more than two exist, the diagram can become crowded and complex. Figure 6.14 depicts a relatively simple way to show more than two conditions.

This simple diagramming technique makes the representation of multiple paths easy to depict. This is sometimes referred to as a *case* symbol.

There are often activities that occur outside the normal flow of a process that are triggered by events. For example, if sales projections are missed by a certain percent, a company may convene a meeting to determine the problems and find solutions. Figure 6.15 shows how events can be diagrammed in flowcharts.

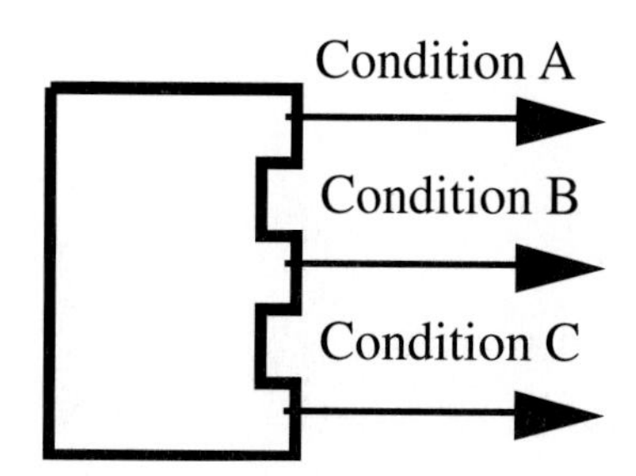

Figure 6.14 Alternative Method of Depicting Multiple Paths.

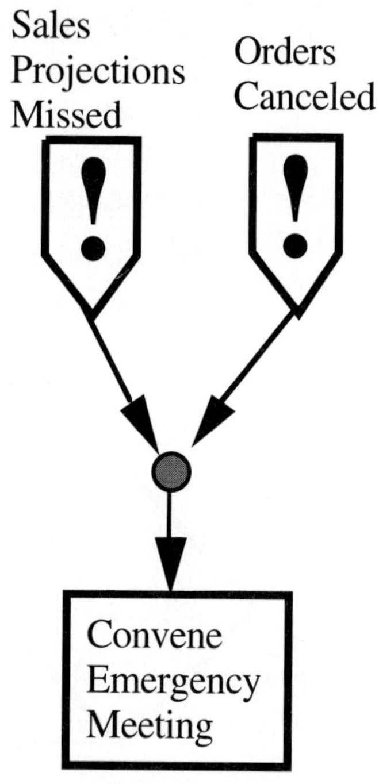

Figure 6.15 Flowchart Depicting Events and OR Condition.

Figure 6.15 introduces two symbols, one symbol representing an event, and one symbol representing an OR condition. The diagram in Figure 6.15 is interpreted as "*IF* sales projections are missed *OR* orders are being canceled, convene an emergency meeting."

There are also conditions that cause a process to stop. Such conditions should stand out in a flowchart and can be represented as shwon in Figure 6.16.

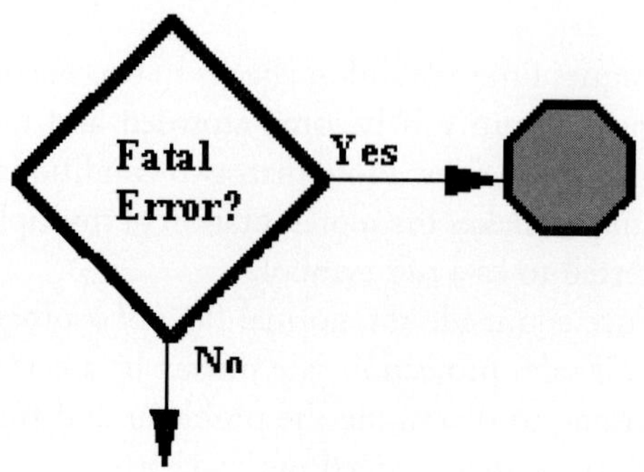

Figure 6.16 Flowchart Depicting Stop Condition.

This familiar symbol will immediately be recognized by anyone viewing the flow diagram.

Engineered flowcharts begin to graphically portray the IF… THEN…ELSE logic, as well as other logic, that determines the behavior of a flowchart. Spending the time adding such information to a flowchart will facilitate the development of process models and, therefore, the use of CAPRE technology.

7

Level 4 (Measured) Processes

So far, the migration of a process from Maturity Level 1 to Maturity Level 3 has required

- Application of Process Reengineering Rule 1—description of the process through the use of verbal communication.
- Application of Process Reengineering Rule 2—description of the process tasks through the use of written (textual) documentation.
- Application of Process Reengineering Rule 3—description of the overall process through the use of activity diagrams.

A Level 4 process is one in which workers communicate their views of the process, for which written documentation of process tasks has been developed, for which an overall view of the process has been developed, and for which *process parameters* that determine the behavior of the process have been defined. Therefore, to migrate from a Level 3 process to a Level 4 process, it is necessary to define process parameters and to measure the process.

A *process parameter* is anything of interest in a process that can be measured or for which a value exists. Process parameters include, but are not limited to

- Time to completion for development of the final product.
- Time to completion of each task in the process.
- Transaction times, that is, the time required to transfer the products of one task to the next task in the process.

- Input inventory levels at each step in the process.
- Quality of products produced by each task.
- Quality of final product.
- Productivity of workers, that is, how many parts or products each worker produces during a measured period of time.
- Factors affecting productivity. These can be many, such as difficulty of the task being performed, adequacy of machinery used in a task, setup time required, and so on.
- Conditions that determine the paths a process follows.
- Waiting or idle time of each worker.

Typically, industry and government in the United States focus solely on quality and productivity measures. This perspective of process measurement is reinforced by virtually all TQM, CPI, and popular process reengineering philosophies. This means that the number of products produced over a period of time is counted (productivity) or the percentage of items rejected is counted (quality). These measures provide only a very small view of overall process dynamics. There are many factors affecting quality and productivity, and these too must be measured. Those factors are the *rules* that define the behavior of a process.

A Level 4 process is a bridge between the defined and optimized process. Even after an organization has determined and measured the parameters associated with a process under investigation, attempts at process reengineering should not yet be made. The collection of process data is a prerequisite to the development of process models and for the simulation of processes.

Measuring the Origami Process

Once the Origami Process has been defined, students are instructed to look at the final products that have been rejected to determine the causes of the rejections. They are then asked to implement changes to the process to correct those problems. Because it is a training exercise, the Origami Process is "rigged" so that most of the rejects are caused by sloppiness in the painting task. An analysis of the rejected final products shows that there usually is some overlap from the painted white portions of the star to the yellow portions. Figure 7.1 depicts a rejected product.

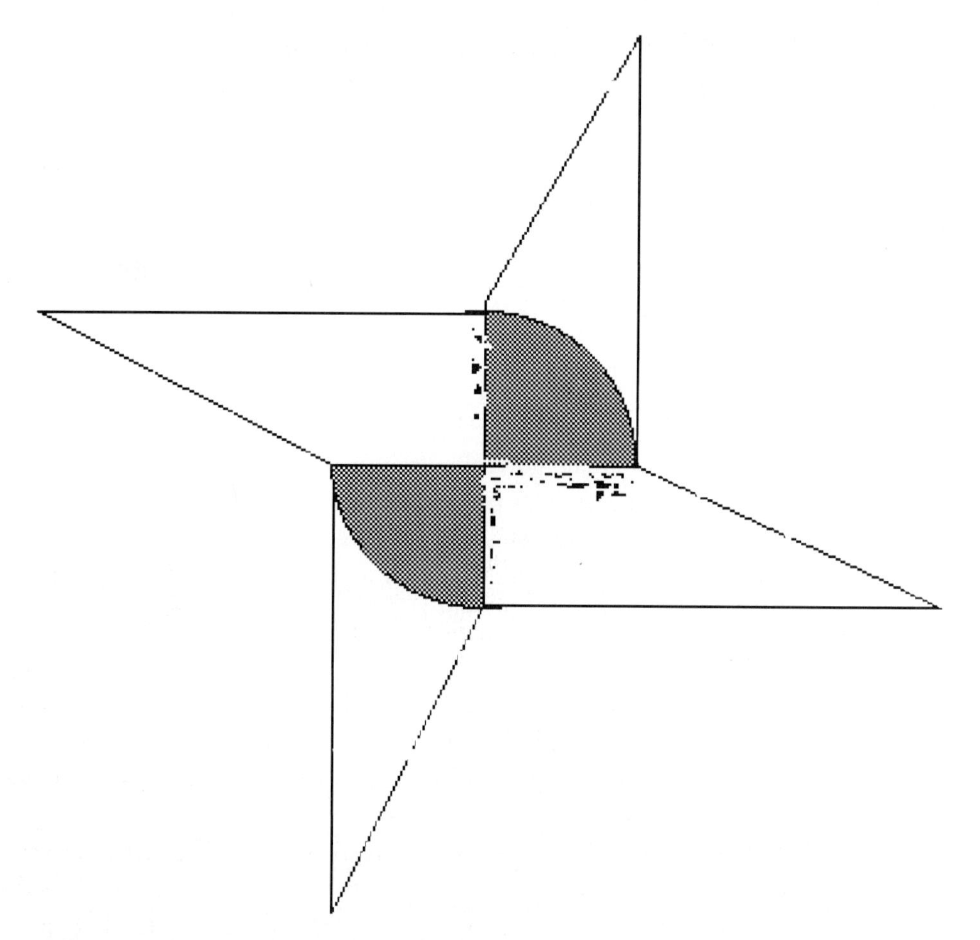

Figure 7.1 Rejected Origami Product.

When painters are asked about the problem, they indicate that painting something that has already been assembled requires patience, and that when they feel that they are under pressure, they tend to get sloppy. This explanation defines a causal loop depicted in Figure 7.2.

This loop represents a vicious cycle—the more pressure the painters felt, the faster they worked, the sloppier they got, and so on. Pressure on workers is a process parameter; however, it is the type of parameter that cannot be captured in written documentation or flowcharts. It can only be determined through discussions of the process, that is, by utilizing Process Reengineering Rule 1. Once the parameter is known to exist, it can be utilized in simulations of processes, as will be discussed in the next chapter.

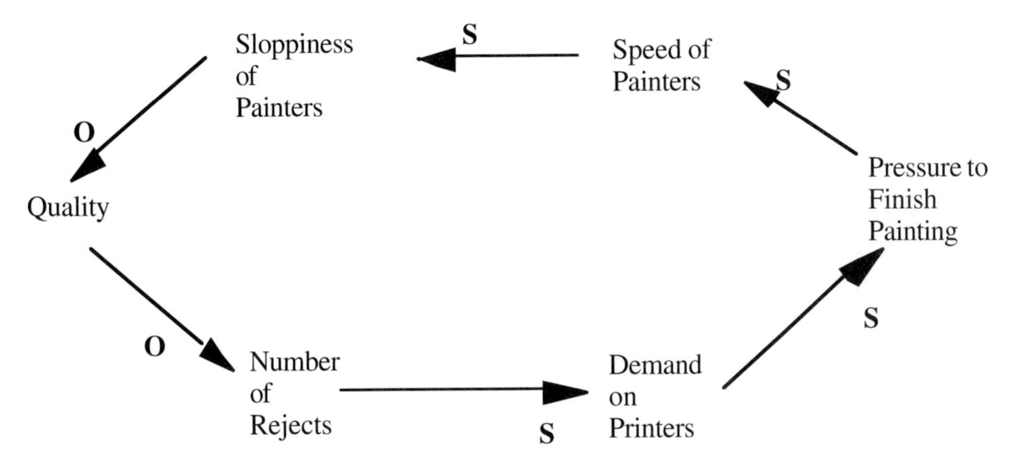

Figure 7.2 Effect of Pressure on Origami Painters.

When the effect of pressure on the painters is known, students are asked to find a solution to the problem. The solution must increase quality without decreasing productivity. After some discussion, the students agree on a solution: The white paper will be painted after it is folded, but before the final product is assembled. Because of the way the product is assembled, changing the process in this way actually allows the painters to be less exact in their jobs. Figure 7.3 uses flowchart symbols to depict the improved process.

When the change is implemented, quality of the final product increases dramatically. Sloppiness on the part of the painters is no longer an issue, so they can operate more quickly. Because any errors they make are hidden after the paper is folded, painting no longer contributes to rejected products. During the training, students have managed to increase quality while at the same time increasing productivity, primarily through the use of communication. Typically the training is complete at this point.

If the class were to continue the improved process for some period of time, however, problems would probably emerge. Initially, the whole process is adapted to handle the speed at which paper is being painted, since it is the most time-consuming task in the process. Earlier some of the standard improvements to the process typically made by students were mentioned. At this point, these improvements will be reviewed to determine if they are still valid.

- One of the first improvements made to the process was to make the material handler a "floating" worker, in which case the manager assigned that person to tasks whenever he or she thought a particular task was in need of

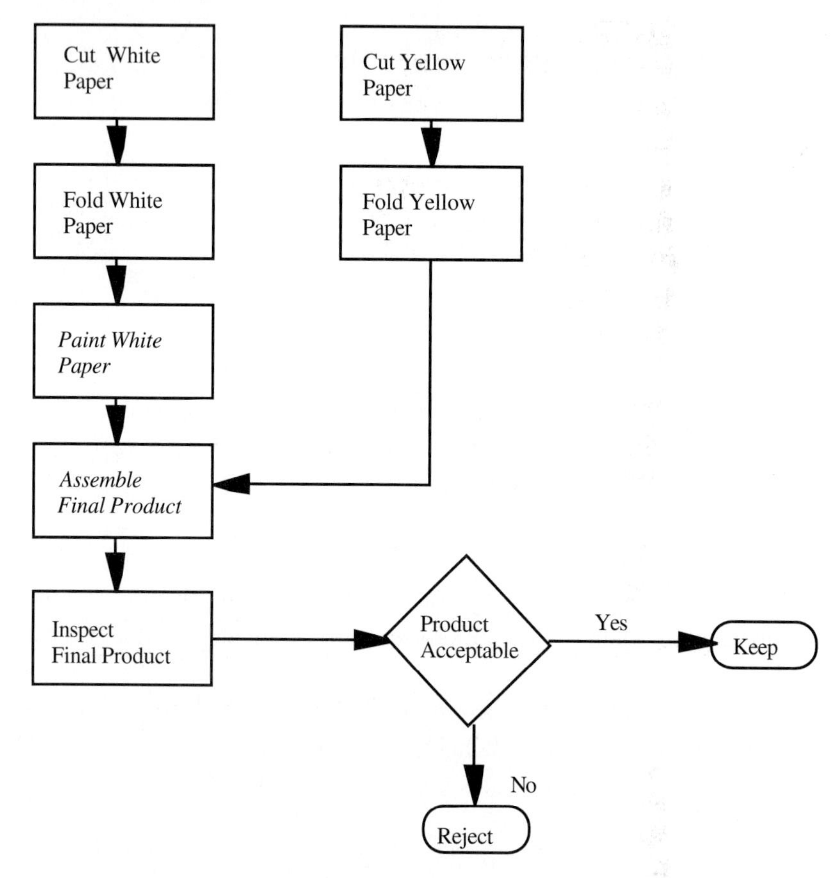

Figure 7.3 Revised Origami Flowchart.

help. If the material handler was being assigned to help the painters because they were slow, what would he or she do now?

- It was agreed in early discussions that the cutters produced paper faster than it could be used and often helped out in other tasks. Would the flow of paper established for the process now be adequate given that painting was proceeding at a faster pace?
- Paper cutting was also based on an assumed utilization of white and yellow at the same rate. With the change just implemented, utilization is now asynchronous because there is a step associated with white paper that is not associated with yellow paper.

Implementing the change described above represents a quick fix. There has not been enough information about the Origami Process parameters to determine how the change should be implemented, that is, how the overall process must be changed to account for the change in the timing of the painting task. Changes to a process that are made without considering *all* the possible effects resulting from the change can be counterproductive. Some of the questions that could be asked in relation to the quality problem described above are the following.

1. Under what *conditions* do the painters feel pressured? If the size of the "waiting to be painted" inventory is the factor, how can the process be altered to reduce the size of that inventory other than changing the manner in which products are painted?
2. What is the average productivity of painters? Under what *conditions* does the manager assign the material handler to the painting task? If an additional, full-time painter were added to the task, would the quality of the final product increase?
3. What other *conditions* cause products to be rejected? Are there other systemic problems that must be addressed?
4. If rejects could be eliminated completely, what would be an acceptable level of productivity for the process?
5. Raw materials are "pushed" to the painters. Would a process in which they "pulled" what they need be better?

All of the above define process *conditions.* The conditions that arise in a process and the decisions that affect the process when those conditions arise are the rules of behavior of a process. These rules can be neatly expressed in IF...THEN or IF...THEN...ELSE terminology.

Condition 1 can be stated as, "IF the inventory of products waiting to be painted is reduced, THEN the pressure on the painters will be decreased, THEN quality will increase. ELSE quality will remain poor." IF...THEN...ELSE expressions are oral or written representations of causal chains and closely resemble computer language representations of process logic. These expressions, combined with other process measures, form the foundations of computer simulations and are essential for predictive (what-if) analysis and are required to use CAPRE technology effectively.

Consider the logic used by the manager to dynamically assign personnel to tasks. The manager has based his decisions on conditions that he has noticed; in other words, he has observed events and developed rules associated with those events. If the process changes, are those rules still applicable? If not, will productivity be lost in some other task until new rules are developed? Answers to these questions can only be supplied by either

- Making process changes and waiting to see what happens, or

- Gathering information, performing what-if analysis, and predicting what will happen.

The first approach uses what are commonly called *pilot programs* by TQM and CPI advocates. A pilot program is an experimental implementation of a process or change to a process. Pilot programs require resources and are usually executed for short periods of time. If a pilot program were implemented to test the change to the Origami Process, it would most likely be successful, since it would be done under carefully controlled conditions. Pilot programs rarely, if ever, represent actual processes.

The most cost-effective way to test the effect of changes to a process is through the predictive capabilities of modeling and simulation. Determining and measuring process parameters are necessary components of predictive analysis. Therefore, Process Reengineering Rule 4 is *measure the process.*

To measure a process, an organization must look at process parameters, gather information about that process, and assign values to those parameters. The parameters can be anything of interest but should include timing analysis of tasks, productivity analysis, and quality analysis. In addition, parameters are the rules that determine the behavior of the process, and the values of those rules are the IF...THEN...ELSE statements that represent them.

It is within a Level 4 process that the statistical analysis tools promoted by TQM and CPI practitioners can be used. Bar charts, for example, are useful for isolating the analysis of quality inspections and can help determine where the focus of a process reengineering effort should be. Figure 7.4 is a bar chart of the initial inspections of the Origami products.

The bar chart demonstrates that a vast majority of rejects were caused by painting errors, so it is important to analyze the causes of those problems. It also shows, however, that other problems exist and they should not be overlooked. Scatter charts or diagrams can be helpful in looking at the causes of problems as well. We know that the amount of to-be-painted inventory caused the painters to work faster and, therefore, sloppier. A scatter chart of the painting process might have looked like the chart in Figure 7.5.

There are very few fundamental differences between a scatter chart, histogram, and trend chart. Any of these can be used to plot or determine the relations between process variables. These tools are useful for helping to determine where in the process problems might exist and serve as a guide for process analysis activities.

Once we have determined the process parameters associated with the Origami Process, taken measurements, and determined the conditions that affect the way the process behaves, we can develop simulations to test the effects of process reengineering suggestions. This will be discussed in detail in the next chapter.

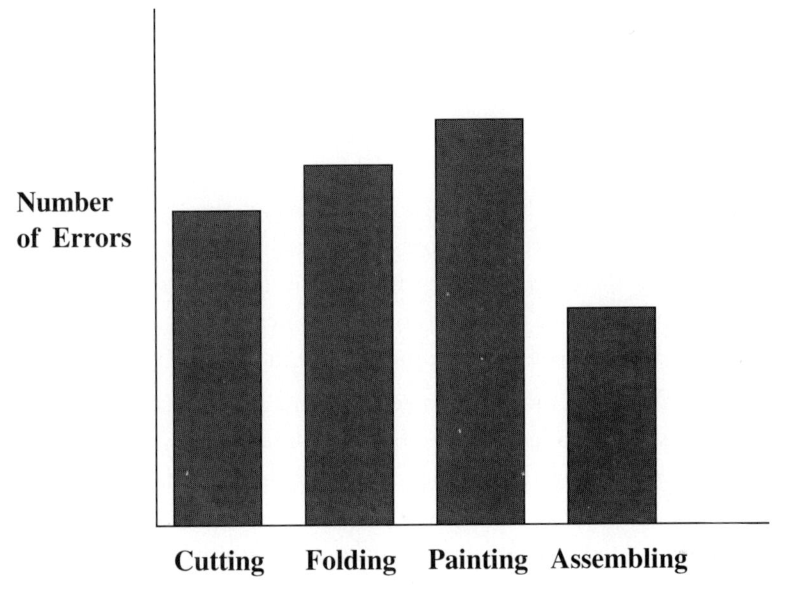

Figure 7.4 Bar Chart of Origami Rejects.

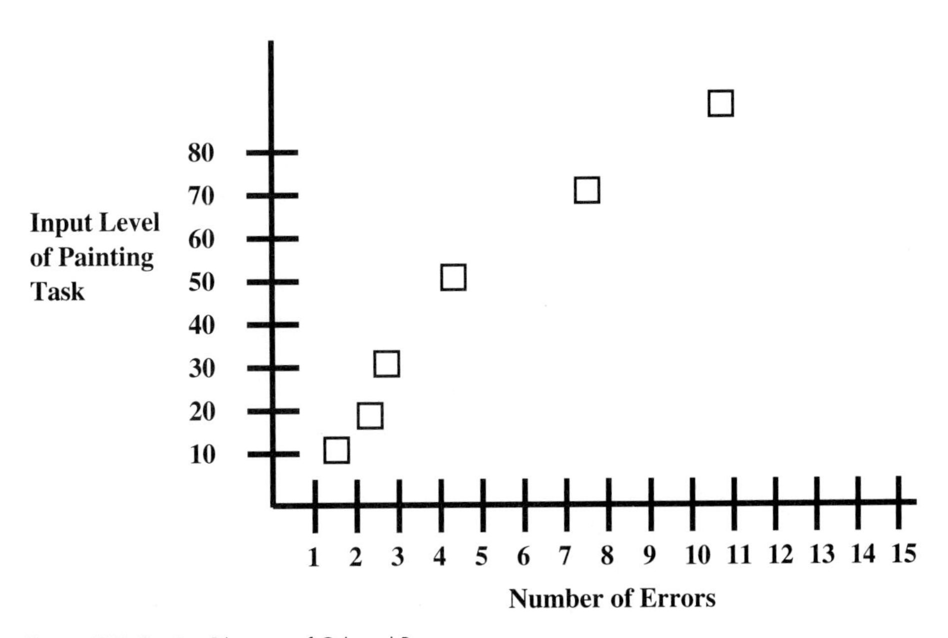

Figure 7.5 Scatter Diagram of Origami Process.

Purchase Order Process: Fourth Attempt at Reengineering

So far, three attempts have been made to reengineer the purchase order process, and each attempt has resulted in a deterioration of the process. After the third attempt, management decided to ask a *process reengineering facilitator* to determine what could be done to correct the process. The facilitator was an individual who had some knowledge of the process but was not part of the process and, therefore, had no preconceived notions about it or "territory" to protect.

The facilitator approached the process reengineering in the following manner:

1. The facilitator convened a meeting of the process participants for no other purpose than to learn more about it. Unfortunately, some managers failed to attend the meeting, but all other participants did. The facilitator learned that everyone was frustrated and upset and all had suggestions for reengineering. After a while, an informal description of the process emerged. The facilitator had used Process Reengineering Rule 1—get everyone talking about the process and agreeing to an informal description.
2. The facilitator had each participant write down a description of his or her task in the process. The facilitator instructed them to include a description of their interfaces with the other participants, that is, how they received a purchase order and how they sent it on to the next participant in the chain. Once all the descriptions were completed, the facilitator convened another meeting at which he read the descriptions to validate them. Some changes were made, but consensus was obtained. The facilitator had used Process Reengineering Rule 2—document the tasks in the process and validate those descriptions.
3. The facilitator then took the descriptions and developed flowcharts of the process, shown in Figure 7.6.

   The flowchart revealed a tremendous division of labor. Each task in the process was handled by a different individual, and the individuals changed depending upon the manner in which a resource was being obtained (purchased, leased, or rented). The flowchart also revealed a potential bottleneck in that there were delays getting approval signatures.

   The participants made numerous suggestions about ways to reengineer the process. The facilitator, being an advocate of computer aided process reengineering, pointed out that the overall process had been defined (Process Reengineering Rule 3) but had not yet been measured.

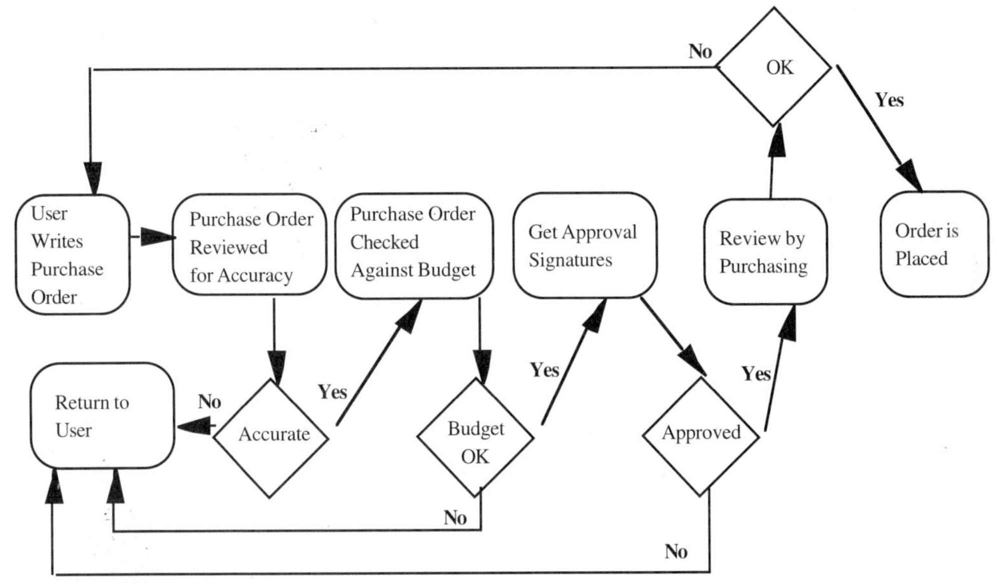

Figure 7.6 Purchase Order Process Flowchart.

4 The process reengineering facilitator asked the participants to determine the goals of the process and the factors (process parameters) that affected those goals. The group decided that its goal was to process purchase orders in the minimum amount of time required and to reduce, preferably to zero, the number of purchase orders that were returned to the users. In addition, they wanted to take steps to ensure that the system was not bypassed and that budgets were not overrun. Therefore, the process parameters of interest were

- Conditions under which purchase orders were rejected
- The time required to complete each task
- The time required to pass the purchase order to the next person in the chain after each task
- Conditions affecting the time to complete each task
- The labor costs of the process

5 The process reengineering facilitator began to measure the process by collecting information about the process parameters that had been defined. One interesting fact was that, on some occasions, purchase orders were passed along the chain by hand and sometimes by interoffice mail. The time for the transactions, therefore, ranged from several minutes to several days.

The delay caused by waiting to get approval signatures was also occasionally several days long. By looking at these parameters and determining the values for them, the facilitator and the participants were invoking Process Reengineering Rule 4—measure the process.

6. The facilitator then asked the process participants to make suggestions for process reengineering. They suggested
   - One person could handle all the tasks up to the point of passing the purchase order to the Purchasing Department.
   - Approval signatures were no more than a rubber stamp.
   - The person processing the purchase order should be delegated the responsibility of approving the purchase order. After all, a processor would not attempt to get signatures if the purchase order had passed all reviews.

To prevent bypassing the system, the participants suggested that the Purchasing Department accept only purchase orders with certain signatures (namely, the processors' signatures). This would force users into the system and might initially upset them. The processors recognized this possibility but concluded that if good service were provided, the system would eventually be accepted. They also agreed that instead of returning a purchase order to a user when mistakes were found, they would correct the errors over the phone. This decision added a new dimension of service to the process and reflected more customer sensitivity.

The participants were ready to present the findings to management, but the facilitator wanted the opportunity to predict the effects of the proposed changes. The facilitator also pointed out that by eliminating the need for approval signatures, the participants had taken management *out of the loop*! This might create nonlogical (or emotional) problems because management was being asked to give up control. The facilitator suggested that he would use causal diagramming and process predictions to influence management and to implement one of the corner stones of Continuous Improvement—*empowerment.*

What has just been described is a very fast evolution of a process. No changes had yet been made, but all the steps that make up an evolutionary process had been taken. It is important to note that the process had been rigorously defined and measured *before* suggestions for reengineering were considered. This rigorous analysis of the process provided the foundation for migration to a Level 5 (optimized) process. The analysis also provided the information required to develop a simulation of the process using computer aided process reengineering (CAPRE) technology. The use of CAPRE technology to improve the purchase order process will be discussed in the next chapter.

Process Drivers and Process Metrics

In the chapter on Repeatable (Level 2) Processes, I presented a template that could be used to assist in the development of written documentation of a process. Similarly, a template can be used to help define process parameters and to specify the measures associated with those parameters. Process parameters should be broken into two categories—those required to analyze a process, and those used to determine the effectiveness of a process. The first category is called *process drivers* and the second category is called *process measures.* Table 7.1 contains definitions of process drivers and measures associated with *process drivers.* A discussion of each process driver will follow the table.

Table 7.1 Definitions of Process Drivers and Measures

PROCESS DRIVERS	PROCESS DRIVER MEASURES
Inputs to Task	■ Rate of input to the task (a) By time period (b) By time of day ■ Types of input to the task
Staff Required	■ Staff specific to the task ■ Staff available part time for the task ■ Staffing levels by time of day
Time Required to Perform the Task	■ Average amount of time required for the task ■ Time required based on a distribution ■ Time required based on type of item being operated on
Conditional Processing	■ Types of conditions that might occur ■ Frequency with which conditions occur
Task Triggers	■ Conditions that initiate the task
Task Terminators	■ Conditions that terminate a task
Rework	■ Percentage of items that are returned for rework
Costs	■ Costs of labor per hour ■ Costs of raw materials ■ Amortized Costs of Tools and Equipment

These are not the only parameters associated with a task, but Table 7.1 describes the parameters that, at a minimum, are required to begin to establish process baselines effectively. Each is given more explanation in the following sections.

Inputs to Tasks

I have defined two parameters in the table: rate of input and types of input. It is very important to define rate of input as realistically as possible. For example, assume that you are trying to reduce turnaround time in a loan application process, and you have decided to measure cycle time and backlog of work. Your personnel may handle, collectively, 1000 applications per day. If you define your rate of input as 1000 per day, and your basis of time for your measurement is expressed in terms of days, then you will most likely have a backlog at some point of 1000, and you will have a loan that will have been in the backlog for one day.

If, however, you receive applications by mail twice per day and also receive applications that have been delivered in person and others by fax, then your rate of input must be defined accordingly. Assume that the 1000 applications arrive at a constant rate of 60 every one-half hour. Your backlog, in this case, may never exceed 60 and cycle time may never exceed one-half hour. Even though the number of applications received is the same, the measures associated with rate of input and the results of your analysis will be completely different.

It is also important to specify inputs by type. For example, assume you are analyzing a customer service call center. Your customer service representatives may receive phone calls from customers who are requesting pricing information, customers who are placing orders, customers who have a complaint, and so on. These are "types" of phone calls, and each type will require a different amount of time to be handled. When defining types of inputs, you must define the percentage of each type that is likely to occur.

A good way of defining both rate of input and type of input is through tables. For example, for the loan application process, we might define rate of input by time of day using Table 7.2.

Similarly, we could define types of phone calls for the customer service call center as follows:

1. A Type 1 call is a request for information;
2. A Type 2 call is an order being placed;
3. A Type 3 call is a complaint, and so on.

Once the types of inputs are defined, a table can be constructed to represent the measures, type, and percentage of type that are important to our process analysis. Such a table might look like Table 7.3.

Table 7.2

Time of Day	Number of Applications
8:00 A.M.	20
9:00 A.M.	60
10:00 A.M.	120
11:00 A.M.	200
and so on.	

Table 7.3

Type of Call	Percent Probability
1	0.25
2	0.5
3	0.25
and so on.	

If the rate of phone calls has been defined by time of day, as in the loan application example, we can then deduce the number of each type of call that is going to be received by time of day. When we have defined the time required to handle each call, we are well on our way to determining the number of customer service representatives who will be required.

Staff Required for the Task

The most popular and common way of defining staff requirements is through the use of the term *FTE*, which stands for Full Time Equivalents. When we say that we need 10 FTE for a task, we might be saying that, of our staff of 20 who can perform the task, 10 are available or necessary at any time. This implies that the staff may be doing other work. Availability of staff is a very important measure when analyzing a process.

Another important measure associated with staff is staff availability by time of day. Take the example of a customer service center. Phone call traffic is typically light in the beginning and end of a business day and heavy during the middle of a day. Again, tables can be constructed to represent staff availability by time of day, as shown in Table 7.4.

This is more succinct and accurate than simply stating that an average of 10 FTE is required to perform this task.

Table 7.4

Time of Day	Staff Available
8:00 A.M.	5
9:00 A.M.	10
10:00 A.M.	10
11:00 A.M.	15
and so on.	

Time Required to Perform the Task

Many data collection templates used in BPR activities suggest that you define the average time required to perform a task. While this is useful as a starting point, it is more useful to define a distribution of time associated with the task, or, when possible, define a distribution associated with the type of item being worked on in the task.

Assume that you have stated that it takes, on average, 30 minutes for a person to perform a particular task. Now assume that you state the time required to perform the task as "a normal distribution with a mean of 30 minutes, and a standard deviation of 5 minutes." This distribution is shown in Figure 7.7.

The specification of an average time of 30 minutes implies that a person can perform two tasks per hour. The use of a normal distribution implies that a person can perform as many as six tasks per hour (at a rate of 10 minutes) and as few as one task per hour (at a rate of 43 minutes). This type of distributional definition is important for determining the true availability of staff.

Other distributions can be used to define task time. Distributions can be applied to types of items being worked on as well. Using the example of a customer service call center again, assume that each type of call requires a different amount of

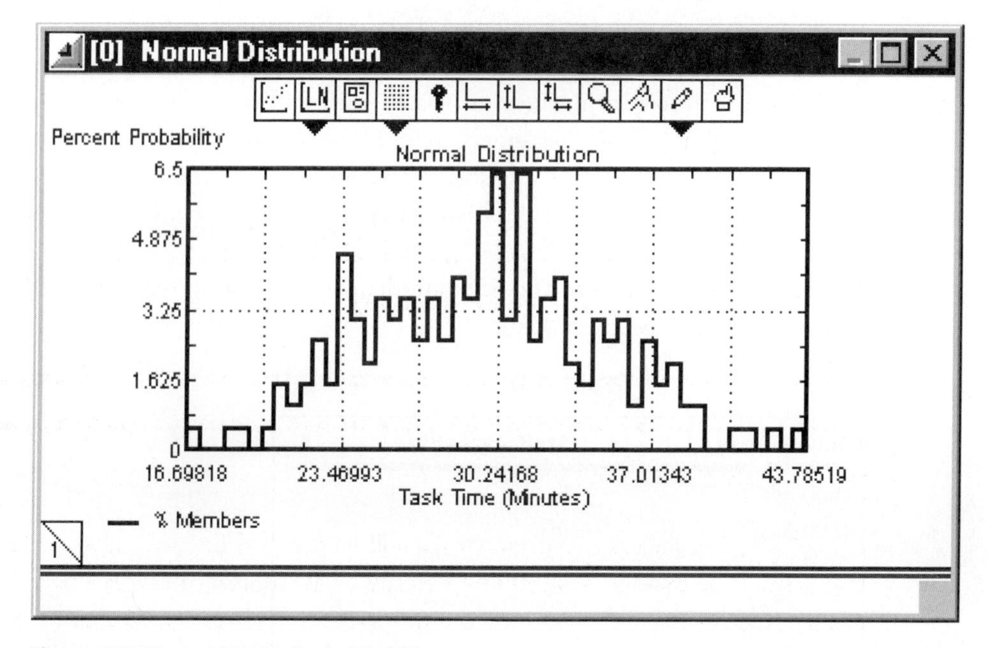

Figure 7.7 Normal Distribution of Task Time.

time, and the time is defined as an Erlang distribution (a type of distribution associated with phone call traffic). Erlang distributions are defined by two parameters, a *mean* and a *K parameter*, which defines the shape of the distribution. Again, we can construct tables that will dynamically create distributions to define task time, as shown in Table 7.5.

Table 7.5

Type of Call	Mean	*K*
1	10	2
2	15	5
3	8	8
and so on		

The distribution curve for a Type 1 call would then appear as shown in Figure 7.8.

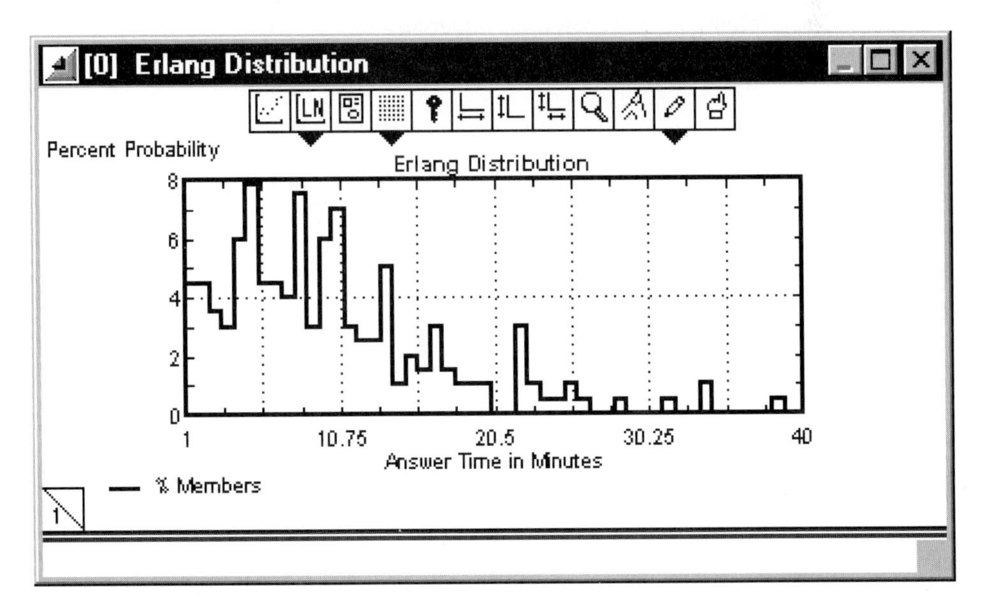

Figure 7.8 Plot of Erlang Distribution.

This curve suggests that most calls will be handled in 20 minutes or less, but that a small percentage will require as many as 40 minutes to be handled. This is greatly different than an average time and, when times are associated with items being worked on, more insight to the process and changes to the process can be realized.

Conditional Processing

No process is a straightforward left-to-right sequence of steps. There are usually conditions, such as exception processing, that require deviations fom normal processing, or paths taken based upon the type of item being worked on. For proper process analysis, it is important to determine what those conditions are and the frequency with which they occur. Whenever possible, tables should be constructed that show the frequency of conditions as related to other aspects of the process. For example, in the customer service call center example, we can specify what types of phone calls require additional work, and that information can be used to determine the total time required to respond to the phone call.

Assume, for example, that a certain percentage of each type of phone call defined earlier requires some type of follow-up analysis. We can construct a table that specifies the probability that follow-up will be required as in Table 7.6.

Table 7.6

Type of Call	Percent Probability of Follow-up Work
1	0
2	0.5
3	0.75
and so on.	

If different types of follow-up work are required, then the frequency of each type can be defined in the same table, along with the parameters describing the amount of time required to perform the follow-up work.

Task Initiation/Task Termination

Some tasks simply begin when an item of work arrives, and some terminate when a finished product is produced; however, it is possible that tasks begin based on conditions and terminate based on conditions. For example, it is possible that a task will be triggered only when a backlog of work has grown beyond a certain level, or it may terminate when a certain number of iterations on a particular item of interest have occurred. Defining these conditions helps constrain the length of time that a task will take for completion and provide valuable insight to the dynamics of the task.

Rework

The amount of rework, or iterations, in a task is perhaps one of the most overlooked but most important process drivers. If an item is worked on numerous times, the cycle time for that item will be greater than the cycle time of an item that is worked on only once. Rework can be defined as a probability; that is, you can state that there is a 10 percent probability that an item will undergo some type of rework in a process. If rework is not considered when defining process drivers, invalid results of analysis may occur.

Costs

Cost drivers should not be confused with cost metrics. Cost drivers simply define the cost of the resources involved in a process. These are used to determine the cost of a process, and the cost of a process can be defined in a number of different ways.

Process Metrics

Process metrics define the effectiveness of a process from a particular perspective. Process metrics can consist of

- Cycle Time—the total amount of time an item of interest is in the process.
- Throughput—the rate of, or interval between, finished products.
- Productivity—the number of finished products created per hour (or day, month, etc.).
- Cost—the cost of a finished product.
- Utilization of Staff—the percentage of time that any employee is actually engaged in his or her primary task.
- Queue, or Waiting, Times—the amount of time items spend waiting to be worked on.

Process metrics should be used for comparison between existing processes and changed processes. They should not be used as inputs to analysis. For example, some BPR experts suggest that cycle time should be an input to process analysis. Cycle time is a result of, or measure of, the effectiveness of a process. It should not be used as an input to the analysis of a process, but only as a metric.

Summary

Migrating from a Level 3 process to a Level 4 process requires that process parameters be defined and measured. The migration to a Level 4 process provides the information necessary to develop simulations of the process. Changes to a Level 4 process should be avoided until the effect of those changes has been tested, that is, until the migration to Level 5 is completed.

Migrating from Level 3 to Level 4 can also be time-consuming and difficult. It not only requires thinking about the process and diagramming the process, but understanding the process as well. In other words, at Level 4, it is not enough to simply understand what is happening in a process. It is also necessary to understand what is driving the process and what measures are important in the process. After establishing process parameters and collecting data about those measures, an organization is ready to proceed with meaningful change analysis.

8

Optimized (Level 5) Processes

To this point, the migration of a process form Maturity Level 1 to Maturity Level 4 has been described, and the migration has proceeded by following the Rules of Process Reengineering. After applying these rules,

- The process has been discussed and communicated.
- The process tasks have been documented.
- An overall view of the process has been developed.
- Process parameters have been defined and measured.

It is after these steps have been taken that reengineering changes to a process can be considered. We have seen that attempts at reengineering the purchase order process have failed, primarily because the Rules of Process Reengineering had not been applied prior to the implementation of changes to the process. The Origami Process has been reengineered to a degree, but there is no guarantee that the changes that have been implemented represent the best reengineering of the process.

A Level 5 process, according to the SEI Capability Maturity Model, is one in which improvements are constantly fed back into the process. Given the manner in which process reengineering is currently being practiced, most organizational processes will not achieve a Level 5 status.

The reason for this is simple: Process reengineering that utilizes TQM and CPI techniques represents a one-shot attempt at improving a process. Because of the time and expense associated with reengineering a process, multiple scenarios of improvement are not attempted. When any improvement is achieved, success is declared, and no further investigation takes place. When an improvement does not work as planned, it is not pursued any further.

This chapter will demonstrate the use of CAPRE technology as a mechanism for continuously improving processes. CAPRE technology allows multiple reengineering scenarios to be tested before they are implemented, eliminating those that will not be productive and permitting fine-tuning of those that will be. It is simply not possible to perform such analysis manually since it is too expensive and too time-consuming.

The work involved in migrating from Level 1 to Level 4 represents about 90 percent of a process reengineering effort. The development of models and simulation of a process represent about 10 percent of the work; however, the application of CAPRE technology to process reengineering provides 90 percent of the payback of a process reengineering effort. This payback comes in the form of costs avoided.

Costs can be avoided in the following ways.

- The use of CAPRE tools eliminates the need for most pilot programs by predicting the effects of changes to processes. Pilot programs, particularly those that do not provide process improvements or that require an investment in technology, can be costly.
- The use of CAPRE tools avoids costs by reducing the possibility of implementing nonproductive or counterproductive changes to processes.
- The use of CAPRE tools provides a mechanism of testing alternative changes to processes. This allows companies and organizations to look for the best change to a process, not only obvious changes to a process.

Business and government processes are too complex to be reengineered without taking all the steps described in the last few chapters. The effects of changes on complex processes can best be understood through modeling and simulation. Reengineering experiments can only be performed in a cost-effective manner by using modeling and simulation.

Origami Process Migration To Level 5

So far, the Origami Process has evolved through process maturity levels 1, 2, 3, and 4 by applying Process Reengineering Rules 1 through 4. The following activities have taken place so far:

- Process participants (students) have met on a continuing basis to discuss the process and suggest changes to it.
- Process tasks were documented so they could be repeated.
- A flowchart, or activity diagram, was developed to present an overall view of the process.
- Process parameters were defined and measured. Rules of behavior were also established, and IF...THEN...ELSE statements were defined to represent those rules.

These actions have resulted in a set of information that can now be applied to the development of a model of the Origami Process. In the discussion that follows, the dynamic modeling and simulation tool Extend will be used to demonstrate how that information is utilized in process modeling and simulation. Model development begins with the first tasks in the process, the paper-cutting tasks. There are two such tasks, white paper cutting and yellow paper cutting, represented in Figure 8.1.

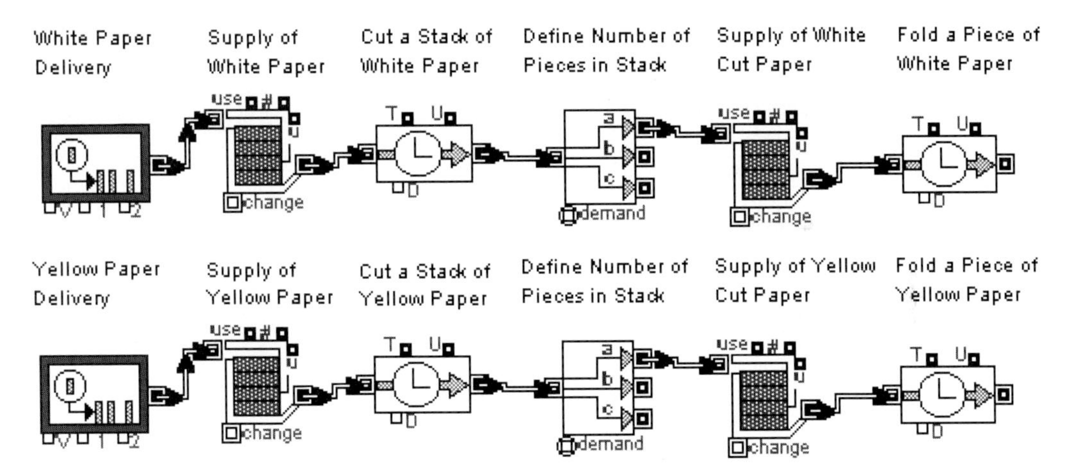

Figure 8.1 First Model of Paper Cutting Using Extend.

The iconic blocks used in the model require some explanation. The Generator block, shown in Figure 8.2, represents a timed event. In this case, it is used to model the periodic arrival of a delivery of a stack of either white or yellow paper. This iconic block controls the availability of raw materials and may be used in future models to test the effectiveness, for example, of just-in-time concepts.

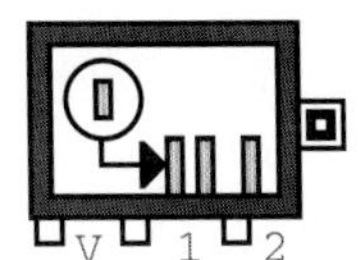

Figure 8.2 Generator Block.

The Resource block, shown in Figure 8.3, represents a Repository, in this case, the supplies of uncut yellow and white paper. These have been given an initial value, since each work period should start with raw materials available.

Figure 8.3 Resource Block.

The Activity, Delay block, shown in Figure 8.4, represents the actual task of cutting the paper. In this iconic block, an amount of time is established to represent the amount of time required to complete the task. In the model shown in Figure 8.1, the time has been set to 1, representing one minute.

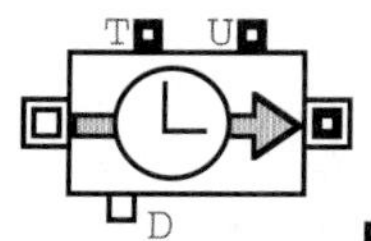

Figure 8.4 Activity, Delay Block.

The Unbatch block, shown in Figure 8.5, has many uses, one of which is to separate an item into its original components. Another is to make copies of an item. This iconic block is used in Figure 8.1 to separate a stack of paper into ten individual pieces. In other words, a stack of paper, when cut, yields ten pieces of cut paper.

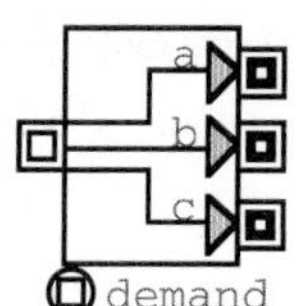

Figure 8.5 Unbatch.

These, at this point, are the only iconic blocks required to start modeling. At first glance, this would appear to be an accurate representation of the cutting task, but in reality it is not. The cutting tasks are shown moving directly into the folding tasks. This is not quite accurate, since this portrays a *serial* process and we already know (through discussions of flowcharting and Process Reengineering Rule 3) that the cutting and folding tasks (as well as the others) operate in parallel.

This parallelism can be demonstrated by adding some more iconic blocks, as shown in Figure 8.6.

Two blocks have been added to the model shown in Figure 8.6: a Labor Pool block to represent a person cutting white paper, and a Batch block, which is used to define all the items necessary to perform a task. The Unbatch block has also been modified to output an object representing the paper cutter.

As shown in Figure 8.6, the Batch block is used to specify that the cutting task requires both a stack of uncut paper *and* a person to cut the paper. Similarly, the model will specify that the folding task requires an input of cut paper and a person to fold the paper. If the person cutting the paper is the same person folding the paper, then the

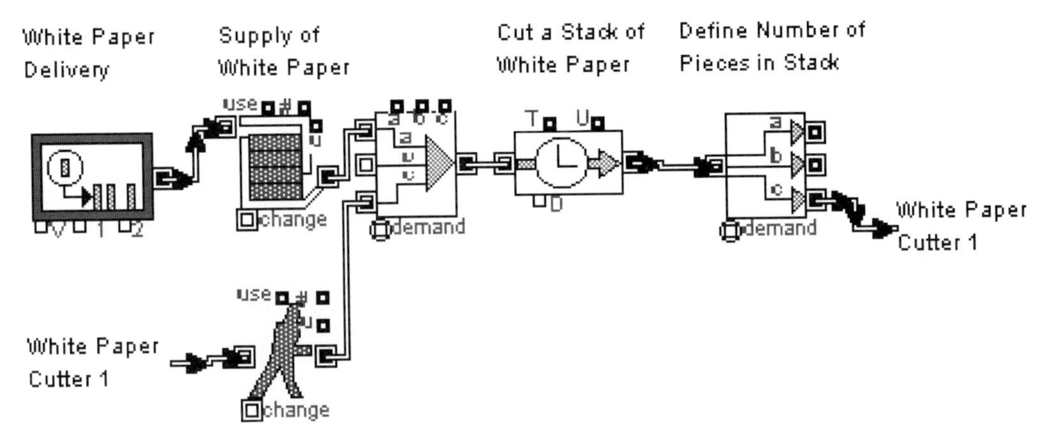

Figure 8.6 Second Model of Cutting Task.

tasks are being performed serially. If the cutter and the folder are two different individuals, then the two tasks operate independently, and therefore, in parallel. By specifying that one individual is associated with each task, parallelism is built into the model.

When developing a model, it is useful to test portions of the model periodically to demonstrate that they are functioning correctly. Figure 8.7 depicts a model in which the initial supply of white paper is set to 100 stacks, and in which the animation feature of Extend has been turned on. With this feature on, the model will display the count of items available after each step of the simulation. The numerical displays in Figure 8.7 show the results of simulating the process for a short period of time.

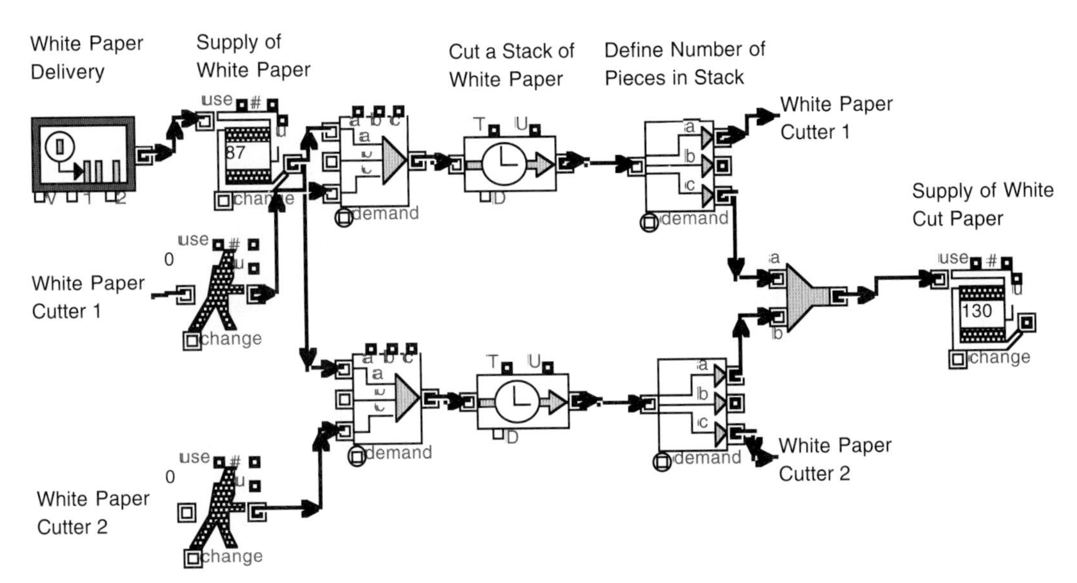

Figure 8.7 First Simulation of Cutting Task.

Of the initial 100 stacks of paper available, 13 have been used, resulting in an output of 130 sheets of paper. Since this portion of the overall model appears to be functioning correctly, it can be enclosed in a hierarchical block, and a model of the folding process can be developed. Figure 8.8 depicts the cutting task hidden in a hierarchical block and the added folding task.

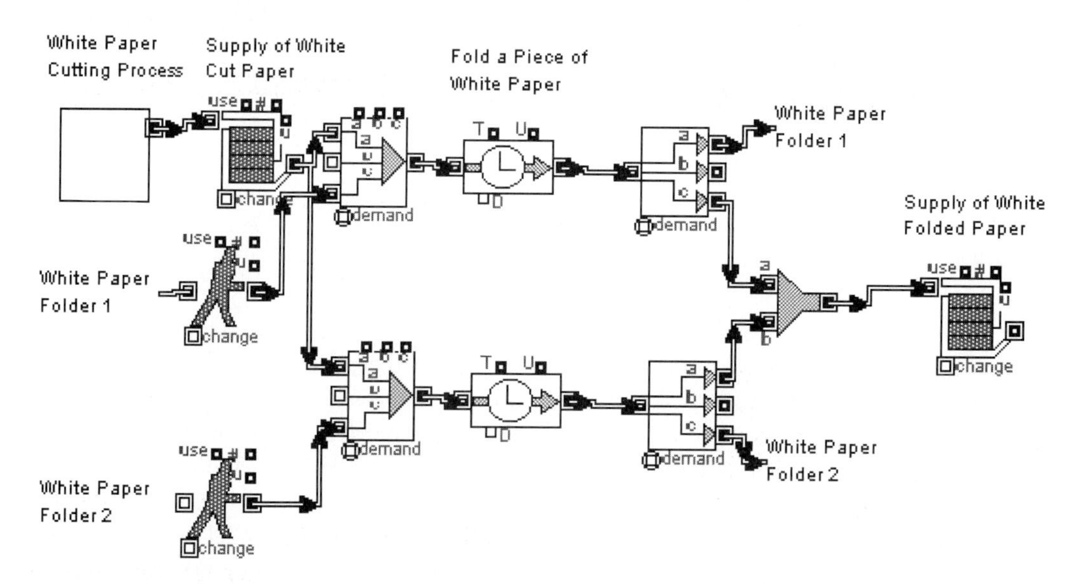

Figure 8.8 First Model Utilizing Hierarchical Blocks.

Executing the model shown in Figure 8.8 for a simulated ten minutes reveals that there are 80 folded pieces of white paper completed, but also 128 pieces of paper waiting to be folded. Figure 8.9 shows the results of this simulation.

The amount produced and the amount waiting to be folded indicates that there is a possible mismatch between the amount of paper being cut and the amount that the folders can actually use. When the simulation is run for a simulated 60 minutes, there are 480 pieces of paper folded, but 538 pieces of paper in the input stack waiting to be folded. Clearly more paper is being produced than can be used, so there is an opportunity for improvement. Several approaches can be taken:

- Add one or two more folders.
- Slow down the cutting process.

The development of the model is continued in Figure 8.10 so the effect of proposed changes on the overall process can be predicted. The overall view of the process is shown with the details hidden in hierarchical blocks.

Note that the hierarchical blocks have been modified to include graphics depicting the tasks they are simulating. This is a valuable feature of Extend that adds meaning to process models. Other than the specialized graphics, Figure 8.10 looks nearly

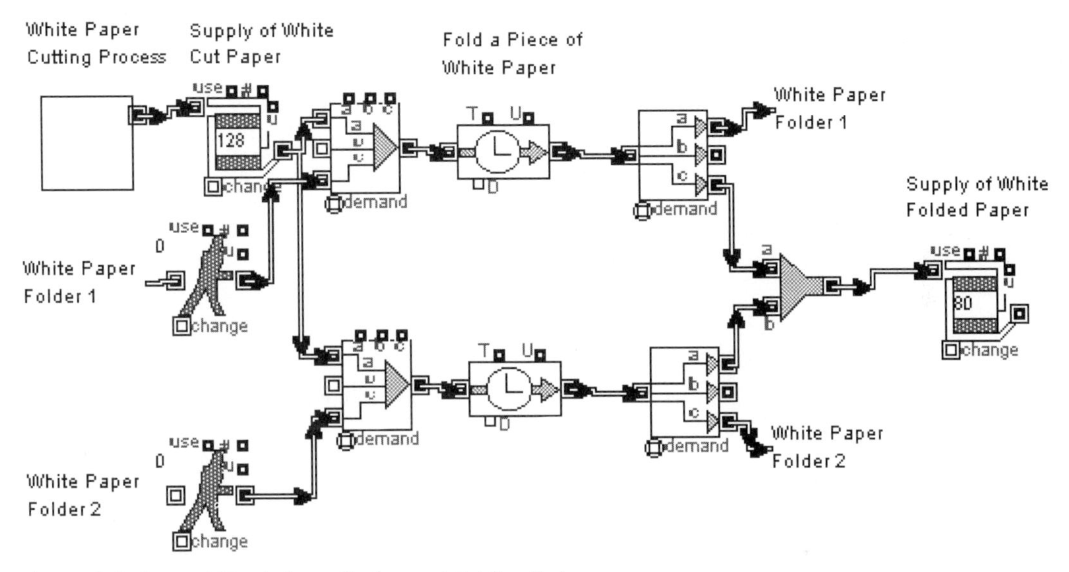

Figure 8.9 Second Simulation—Cutting and Folding Tasks.

identical to flowcharts developed for the process; however, underlying each hierarchical block is a great deal of information (process parameters), such as,

- Each task has two workers operating at approximately the same rate of speed and in parallel.
- The paper cutters cut one stack of 10 pieces of paper every minute (so 20 pieces of cut paper are produced every minute).
- The folders require 15 seconds to fold a piece of paper.

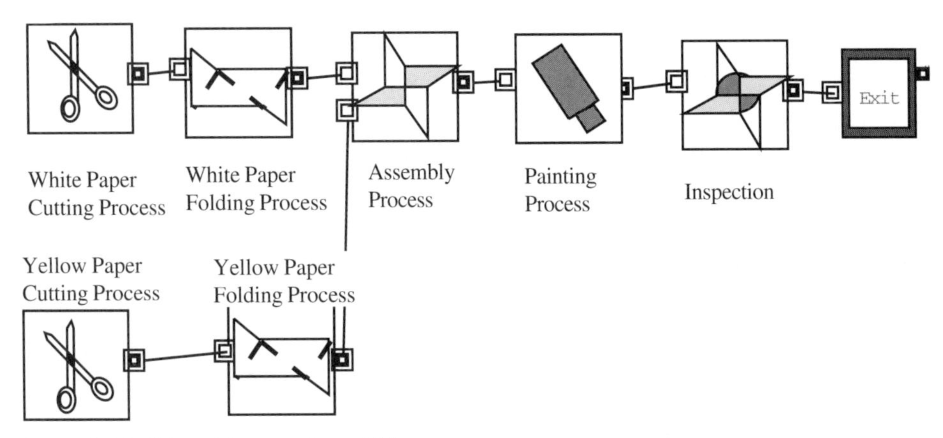

Figure 8.10 Origami Process Overall View.

- The assemblers require 30 seconds to assemble a product.
- The painters require 30 seconds to paint a product.
- The inspectors require 30 seconds to inspect a product.

A simulation of the process representing one hour reveals that

- Two hundred forty products are completed.
- Eight products are ready to be inspected.
- Eight assembled products are ready to be painted.
- Almost *500* sets of yellow folded paper and white folded paper are ready to be assembled.
- Almost *800* pieces each of yellow paper and white paper are ready to be folded.

It would seem that paper is being cut faster than it can be utilized, and that it is also being folded faster than it can be utilized.

Possible solutions to the problem can be discussed (Process Reengineering Rule 1). Some possible solutions include the following.

- Eliminating one white paper cutter and one yellow paper cutter.
- Assigning one cutter to the white paper folding task and one to the yellow folding task.
- Assigning the paper cutters to the painting tasks.
- Combining these three suggestions.

For the purposes of demonstrating the utility of modeling and simulation, the second suggestion was chosen. The model has been changed so only one person is cutting paper and three people are folding paper. The changed paper-folding tasks are shown in Figure 8.11.

When the model is changed to represent these changes, there are now only 7 pieces of each color paper to be folded, but about 800 pieces waiting to be assembled. The number of final products remains at 240. This shows that there has been a good match made between the paper-production tasks and the folding tasks, but that a bottleneck still exists at the assembly task.

The model could be changed to try more approaches to labor allocation, but there is enough information to make the following points:

- Even seemingly simple processes are quite complex and attempts at reengineering do not always yield expected results.
- Changes in productivity of tasks that occur early in a process do not necessarily affect productivity of tasks that occur later in the process, and vice versa.

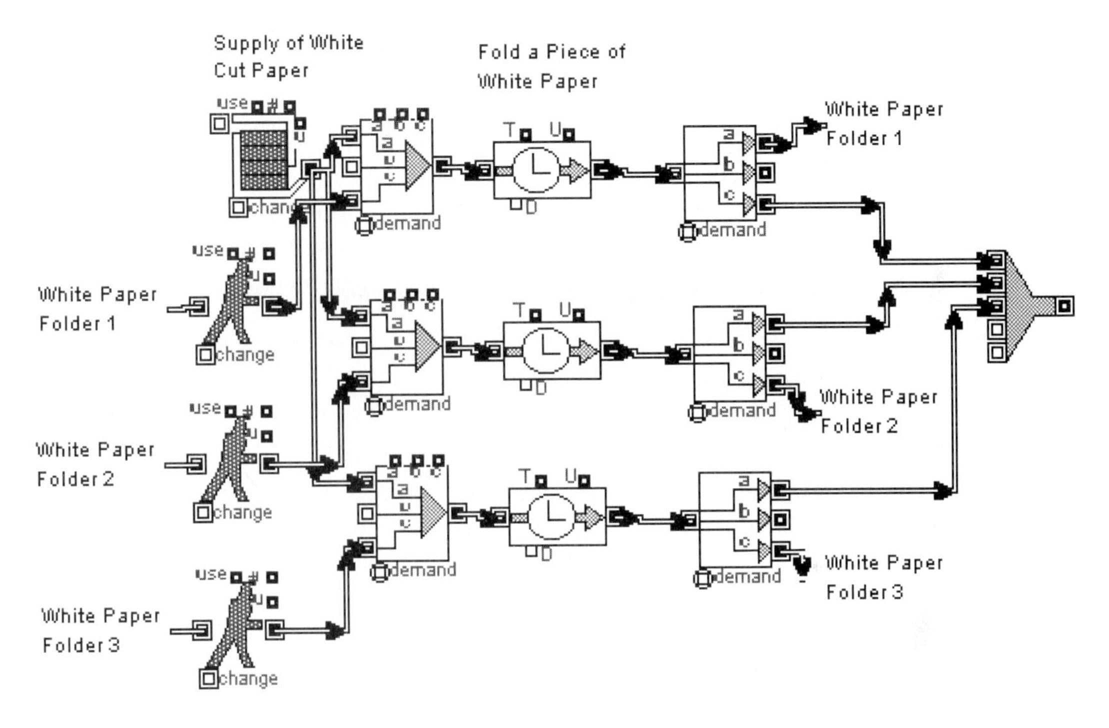

Figure 8.11 Process with New Division of Labor.

- Changes to process parameters are not reflected in flowcharts or other types of activity diagrams. This demonstrates the limitation of activity diagrams.
- CAPRE technology is very effective in predicting the results of changes to a process. CAPRE is a cost-effective alternative to pilot programs.

The initial, overall model of the Origami Process represents an *archetype* of the process, or a model upon which all other models are based. An archetype can be used to fine-tune a process by altering the behavioral rules associated with it. In this case, the basic model remained the same, but the amount of work done in each task was changed by moving elements of the model around. Archetyping is a powerful concept that facilitates what-if analysis, and when a process archetype has been developed, the investigation of process improvements can be accomplished very rapidly through the use of simulation.

It is important to remember that when changes to a process are made, (1) those changes must be reflected in the documentation of the affected tasks, and (2) process parameters must reflect rule changes. In addition, the process flowcharts developed while migrating from a Level 2 process to a Level 3 process can be replaced by the model developed using the CAPRE tool, particularly if the tool permits the use of hierarchical blocks.

The above discussion demonstrates the power of CAPRE technology to predict the effects of process changes. A Level 5 process is one that uses process modeling and simulation to influence change. Process Reengineering Rule 5 is *simulate the process.* Simulation not only allows testing of the effects of proposed reengineering changes but also provides a mechanism to migrate rapidly through process levels.

Purchase Order Process: Fifth Attempt at Reengineering

In the last chapter, we described how a process reengineering facilitator took action to rapidly migrate the purchase order process through maturity levels. Although the process participants wanted to institute change, the process reengineering facilitator wanted to test their assumptions through simulation before recommending changes. The facilitator wanted to simulate the current and proposed processes and utilized Extend to develop a model of the process.

Since a flow diagram of the process had been developed previously, the process reengineering facilitator established the rules that determined the timing of events in the process. Those rules of behavior are described as follows.

- Input purchase orders are allowed to accumulate during the day.
- At some point in the day, they are reviewed for accuracy and passed along via interoffice mail to the person who will review them against the budget.
- Budget reviews are handled in a similar fashion and are also passed to the next person by interoffice mail.
- After a purchase order passes budget review, it is sent back to the original processor, who then must obtain a signature from management.
- There is a rule of thumb by which the processor operates: If ten purchase orders have accumulated *or* three days have passed since the last occasion of getting signatures, then get signatures; otherwise, continue to wait until the conditions are met.
- Once management signatures have been obtained, the purchase orders are passed to the Purchasing Department where the amount of processing time varies over a fairly wide time period.

(*Note*: The models which follow do not include the tasks associated with rejections. This was done to simplify the discussion for the purposes of the book.)

The process reengineering facilitator began the modeling effort by first defining the time frame for the work. It was decided to represent the process in terms of days.

The first activity that was modeled was the arrival of purchase orders and the activity of the purchase order processor.

It would seem that this could be simply modeled, but it is actually quite complex. For example, the processing of purchase orders was done on a time-available basis. The processor had other tasks to perform, so they had to be accounted for in the model. This was accomplished by setting up two parallel processes and treating the processor as an object that had to be available for either task to execute. This is shown in Figure 8.12.

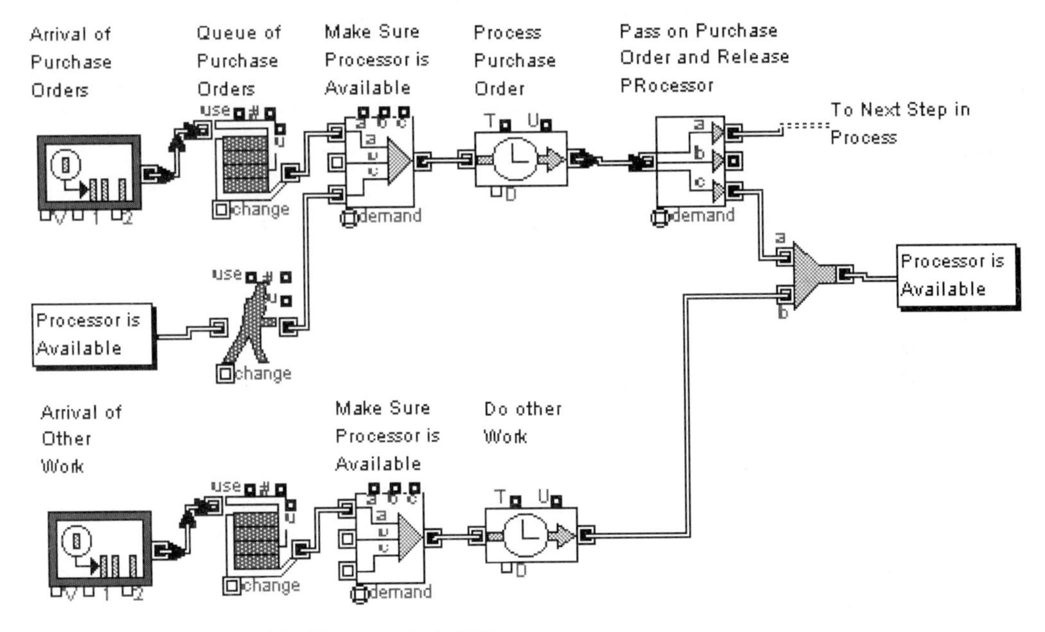

Figure 8.12 Extend Model of Processor's Activities.

There is a lot going on here! First, the arrival of purchase orders and the arrival of other work are determined using the Generator block, as shown in Figure 8.13.

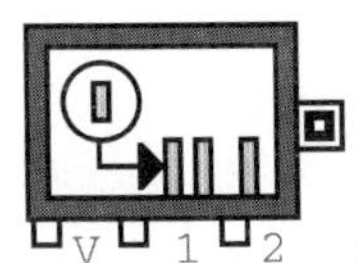

Figure 8.13 Extend Generator Block.

This block is used to generate *items* for the simulation. In the model depicted in Figure 8.12, these were specified to generate one purchase order in a period of time between one hour and one day and to generate other work in a time period of one-half day to one day. In other words, as few as one purchase order or as many as eight purchase orders could arrive in a day, and the requirements for one or two other tasks could also arrive in a day.

The outputs from the Generator block were held in a first-in-first-out (FIFO) queue, which is represented by the block in Figure 8.14.

Figure 8.14 FIFO Queue.

Next, a Batch block is used to make sure that an item is available to be processed *and* the processor is available to do the processing. The Batch block is shown in Figure 8.15.

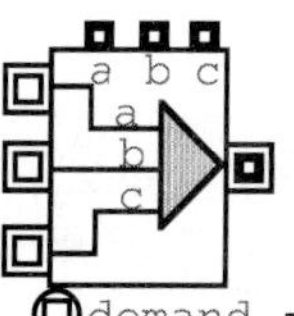

Figure 8.15 Batch Block.

The Batch block is defined for the model to execute only when both a purchase order *and* the processor are available. If there is no purchase order available, the block will not hold the processor but will allow it to be used to do other work.

Finally, the definition of the tasks is contained in the Activity, Delay block. This block is shown in Figure 8.16.

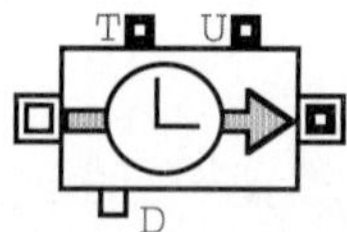

Figure 8.16 Activity, Delay Block.

This block is defined to execute for one (simulated) hour for purchase order processing and two hours for other work. When these activities have been completed, the processor is released to be available for either task, using the Unbatch block, shown in Figure 8.17.

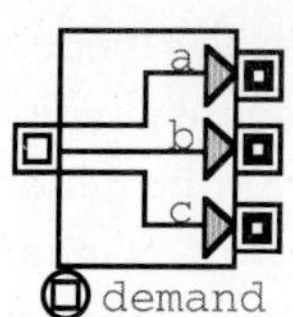

Figure 8.17 Unbatch Block.

This is a complex model for a seemingly simple task, but, as has been emphasized in this book, even simple tasks are complex. Moreover, this is not the type of behavior that can be captured in a flowchart or process map. In fact, the use of flowcharts alone can lead to unwarranted expectations of productivity.

Since the process reengineering facilitator wanted to determine how best to reduce cycle time, the facilitator decided to capture the amount of labor time that was actually used to process purchase orders. He did this by attaching an Accumulate block to the Activity, Delay block that defined the task of processing purchase orders. This is shown in Figure 8.18.

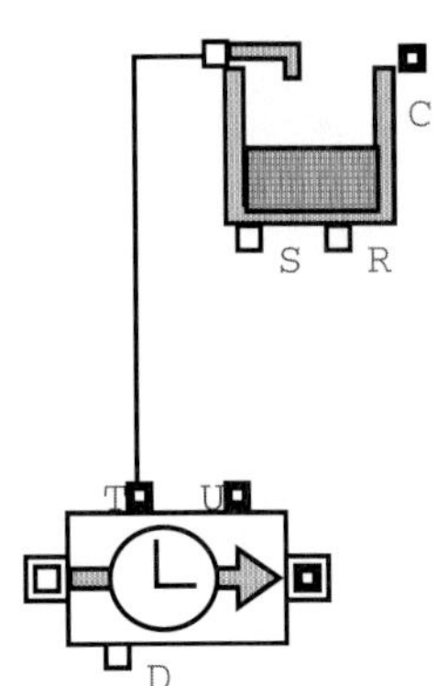

Figure 8.18 Capturing Labor Time.

Every time the processing task executed, it input the processing time into the accumulator, where it was summed. The processor then added another Activity, Delay block to the model to represent the transfer of the purchase order to the next person in the process via interoffice mail. The time for this transfer was also captured and accumulated and stored as *delay time.*

This development of the model was repeated for each step in the process, and both the labor and delay times were captured in ongoing calculations. To depict delay associated with getting management signatures, a customized block was developed. Figure 8.19 shows this custom block and its use in the model.

The behavior of the custom block is defined as

> IF (ten purchase orders need signatures) OR IF (three days have elapsed since the last time signatures were obtained), THEN (get signatures) ELSE (wait).

This rule was established by the processors who were trying to deal with management as little as possible. The custom block also has two readout connectors. One specifies the time at which the block is available to accept input, and the other specifies the time at which the block completed executing. The readout of the two connectors are input into an Equation block, which is used to calculate the wait associ-

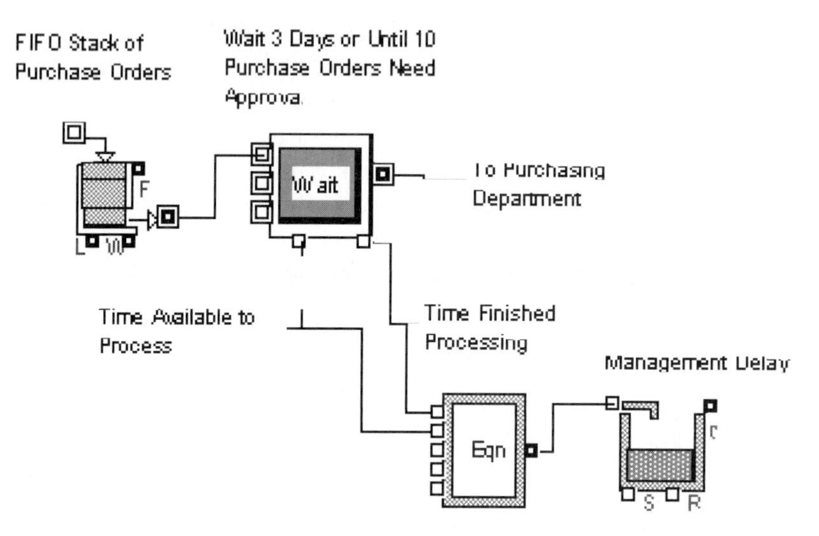

Figure 8.19 Use of Customized Block to Represent Management Delay.

ated with getting management signatures. This amount of time is added to the overall delay time in the model.

After the entire model was simulated, the process was simulated for a period of time that represented one calendar year. The results showed that

1 The average amount of time spent processing a purchase order was *18 days*.
2 The average amount of actual work time spent processing a purchase order was less than *8 hours*!

Then the process reengineering facilitator developed a model of the suggested reengineered process. That process, as suggested by the purchase order processors, was very simple: The processors would do all the reviews, including budget reviews, and would also be delegated approval authority. The development of this model followed the same basic steps as the development of the first model, except all the delays caused by interoffice mail transfers were eliminated, and the activity of getting management signatures was also eliminated. When a simulation of this model was run for the same period of time as the other model, the following information was revealed:

1 The average amount of time required to process a purchase order became *six calendar days*.
2 The average actual work time spent on a purchase order was reduced to less than *four hours*.

The new process was based on delegation of authority, and implementation of the improved process would require management concurrence with the suggestions of the participants. To provide more information on the effects of the change to the process, the process reengineering facilitator (1) translated the difference in processing times between the two processes to lost productivity in terms of dollars, and (2) calculated the difference in labor dollars due to work performed between the two processes.

This provided information on what the current process cost and how the new process could save money. Thus, a new *process parameter*, the cost of the process, was created and measured.

The process reengineering facilitator then added one more element to the models—a prediction of how often the system would be bypassed. He interviewed users who had bypassed the system to determine their reasons for doing so. He concluded that the incidents of bypass were related to the amount of time required to process a purchase order. The longer it took, the more likely it was that the system would be bypassed.

The Extend modeling and simulation tool provided a mechanism for tracking the time required to process each purchase order in the model. This was done using an object-oriented capability of Extend which allows *attributes*, or values that describe an object, to be assigned to each individual object in a model. In this case, an attribute called Time so Far was created for each purchase order. Figure 8.20 shows how this calculation was performed in the model.

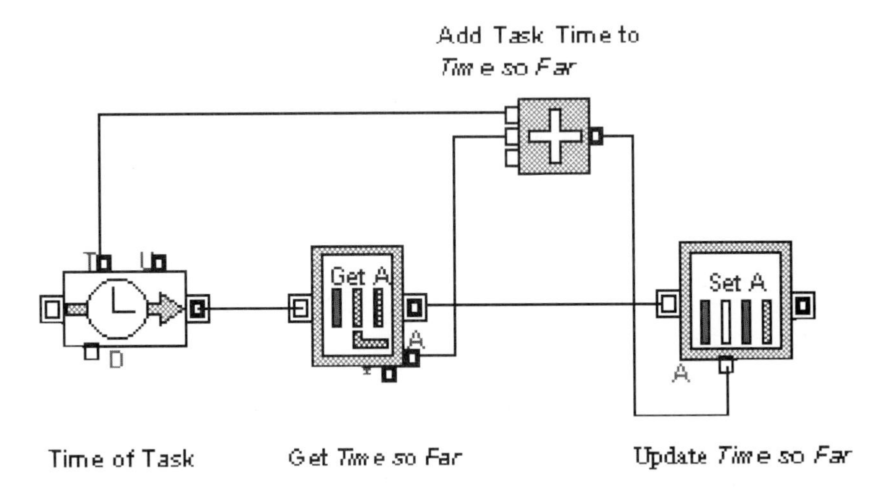

Figure 8.20 Updating the Time so Far Attribute for Each Purchase Order.

The execution time of a particular task in the process was read through the T connector on the Activity, Delay block, and the Time so Far attribute of the purchase order was read using a Get Attribute block. The two numbers were added, and the result replaced the old value of the attribute using a Set Attribute block.

The process reengineering facilitator then inserted a rule into the model that generated a system bypass whenever a processing time threshold was exceeded, as shown in Figure 8.21.

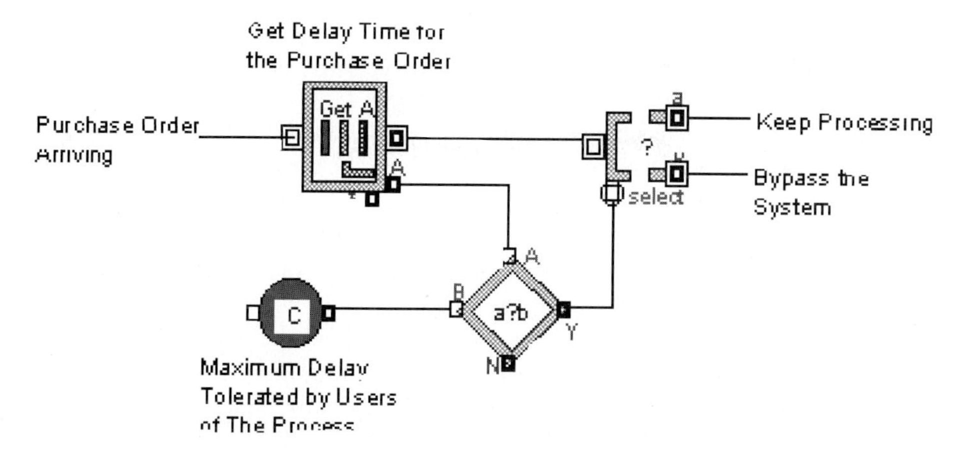

Figure 8.21 Decision to Bypass the System.

Here, the Time so Far attribute of the purchase order was read and compared to a constant value representing the maximum delay that would be tolerated by users of the process. The comparison was done using a Decision block, shown in Figure 8.22.

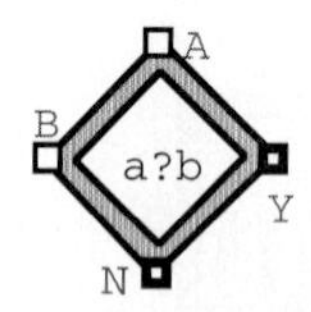

Figure 8.22 Decision Block.

This block compares the inputs (in an equation defined by the modeler) at A and B and outputs at 1 or 0 at the Y and N connectors based on the comparison.

The value of the Y connector was then input to a Select DE (Discrete Event) block, which determined the path that would be followed. Since the amount of delay associated with a purchase order was dependent on the amount of other work a processor had to do, the amount of time the purchase order was held up waiting for management approval, and so on, the model was able to accurately predict the number of times users would bypass the system.

When the two models were compared, it was not surprising to learn that the new process would rarely be bypassed, whereas users would continue to bypass the current process.

The process reengineering facilitator then drew the causal loop to explain the dynamics of the current process and to show how adopting the new process would change that behavior. Figure 8.23 shows this causal loop.

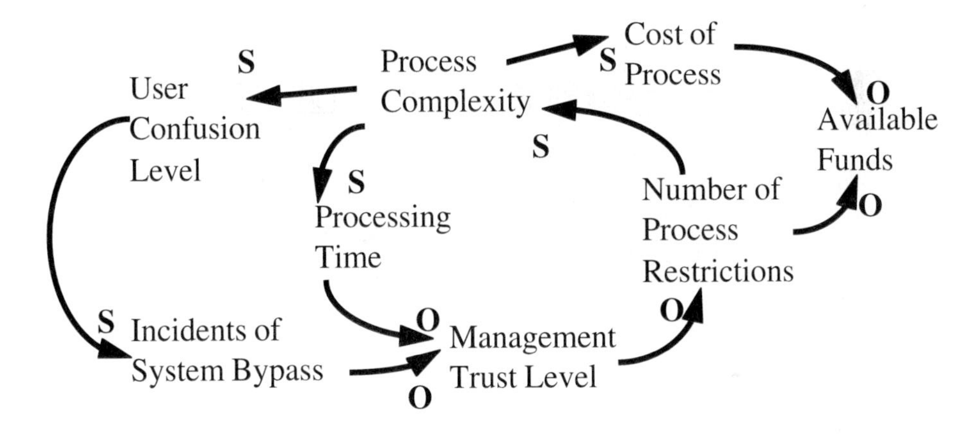

Figure 8.23 Purchase Order Process Causal Loop.

This loop is interpreted as

- Management confidence decreases;
- The number of process restrictions increases;
- Then the complexity of the process increases;
- Then the user confusion level increases and the cost of the process increases;
- Then the incidents of system bypass increase as management trust decreases;
- Available funds decrease;
- Then the loop repeats itself.

This is a true vicious loop; however, if management trust begins to increase, then the same loop becomes virtuous and the incidents of system bypass decrease.

The facilitator used the models he had developed and the causal loop in a presentation to management, the point of which was that the process should be changed and that authority should be delegated. Since the information was presented in a strictly factual manner and did not deal with emotional issues such as territorial boundaries, the suggestions were accepted and a new process was put into place. The model, now an archetype, was further used to determine the savings that could result from using an electronic form, and ultimately the paper purchase order was redesigned and simplified.

Summary

Migration from a Level 4 process to a Level 5 process requires modeling and simulation of the existing and proposed processes. The process models are based on the information gathered by applying Process Reengineering Rules 1 through 3, and the simulation is based on the behavioral rules established in the Level 4 process and the measurement of process parameters. Using dynamic modeling and simulation technology, a graphical, *enactable model* of a process can be created. Eventually, an archetypical process model can be created to fine-tune a process and test new assumptions.

Process modeling and simulation provide a mechanism to rapidly migrate through process levels. All of the Process Reengineering Rules stated as part of an evolutionary process must still be applied; however, by utilizing CAPRE technology, the need to actually implement a process in an evolutionary manner is eliminated. Process migration using CAPRE technology would be performed as follows:

1. Apply Process Reengineering Rule 1 by developing an informal description of the process.
2. Apply Process Reengineering Rule 2 by documenting the process tasks.
3. Apply Process Reengineering Rule 3 by providing an overall view of the process using diagrams. The diagramming method of a CAPRE tool can be used to provide such a view.
4. Apply Process Reengineering Rule 4 by developing a behavioral model of the process. Determine the rules by which the process operates and measure process parameters of interest.
5. Apply Process Reengineering Rule 5 by simulating the process.

Refinements of the process can be achieved by changing the order of the steps described above, namely,

1. Discuss process options (reuse Rule 1).
2. Change the model (reuse Rules 3 and 4).
3. Simulate the changes (reuse Rule 5).
4. Document the changes (reuse Rule 2).

The iterative process just described is an example of Deming's point 5, (constantly improve the system of production and service). It is only through process modeling and simulation that this iterative reengineering can be achieved. The cost of running iterative process reengineering pilot programs is prohibitive, and since those programs are always run under controlled conditions, the results are not always accurate. The use of CAPRE technology is the best method available for an iterative approach to process reengineering.

9

Modeling and Simulation Terminology and Techniques

Overview

As the popularity of modeling and simulation increases, the number of products being offered for BPR simulation is increasing as well.

Without a firm understanding of the different aspects of modeling and simulation, an uninformed buyer may wind up with an attractive tool that, unfortunately, provides little insight into the business process being studied, or worse, provides misleading information about those processes.

This chapter will present modeling and simulation concepts in an easy-to-understand manner so readers will be able to ask meaningful questions to BPR modeling and simulation tool vendors.

The chapter avoids using highly technical terms and explains simulation from a business perspective. This chapter is *not* an in-depth discussion of simulation, but a discussion of simulation as it applies to BPR.

Visual Paradigm

A *process model* is a graphical representation of a process—a process map, for example. A *simulation* is executable software that mimics the behavior of the process being investigated. A tool that combines the capability of presenting a graphical representation of a process (i.e., a modeling capability) with a simulation capability is known as a tool that uses a *visual paradigm.* To be precise, a visual paradigm provides the capability of building *enactable models*, or graphical representations of processes, that also can be simulated. It has become the practice to use the terms "model" and "simulation" interchangeably.

It is important that BPR modeling and simulation tools utilize a visual paradigm for the following reasons.

- We are accustomed to seeing a process in the form of a flowchart or map. Language-oriented simulation tools do not provide such a view and, therefore, have fallen out of popularity in the business community.
- Visual paradigms typically provide *predefined functions*, thereby making the construction of a model easier than with a language-oriented tool. These predefined functions are referred to as *blocks*, that is, blocks of code. Blocks that are represented graphically are known as *iconic blocks.*
- Models built with visual paradigms are easier to understand and maintain than models built with language-oriented tools. For example, compare the Extend+BPR model[1] shown in Figure 9.1 with the language-oriented model shown in Figure 9.2 (created using a language called SLAM[2]).

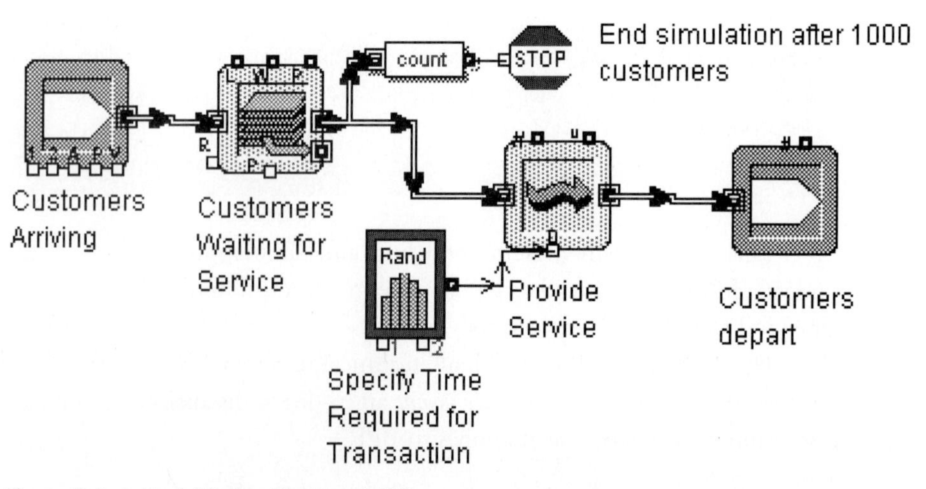

Figure 9.1 A Model Built with Extend+BPR.

[1]Diamond etal, "Comparing Extend" Model, Imagine That!, San Jose, CA.

[2]Law and Kelton, *Simulation Modeling & Analysis, 2nd ed. (McGraw-Hill, 1991). Also, thanks to Imagine That, Inc., for providing this example.*

```
GEN,A. LAW, M M 1 QUEUE, 7/13/1989,1,,,,,,72;
LIM,1,1,100;
;
NETWORK;
;
   RESOURCE/SERVER(1),1;              Define the resource server
;
   CREATE,EXPON(1.0,1),1,1;           Create arriving customers
   AWAIT(1),SERVER;                   Wait for/seize server
   COLCT,INT(1),DELAY IN QUEUE,,2;    Collect delay in queue
   ACTIVITY,EXPON(0.5,2),,DONE;       Delay for service
   ACTIVITY,,,CNTR;                   Send "dummy" entity to counter
DONE  FREE,SERVER;                    Release server
   TERM;                                 Customers Depart
;
CNTR  TERM,1000;                         End simulation after 1000
delays
   END;
;
INIT;
FIN;
```

Figure 9.2 The Same Model Built with SLAM.

Without understanding Extend at all, one can at least see the flow through the process and understand, to some degree, what is going on in the model. In the SLAM model, one must completely understand the language to determine what is being simulated.

Icons

An *icon* is a graphical representation of something that provides meaning or understanding. For example, the stop sign and hand gesture shown in Figure 9.3 have the same meaning as do the stoplight and flowcharting decision symbol shown in Figure 9.4.

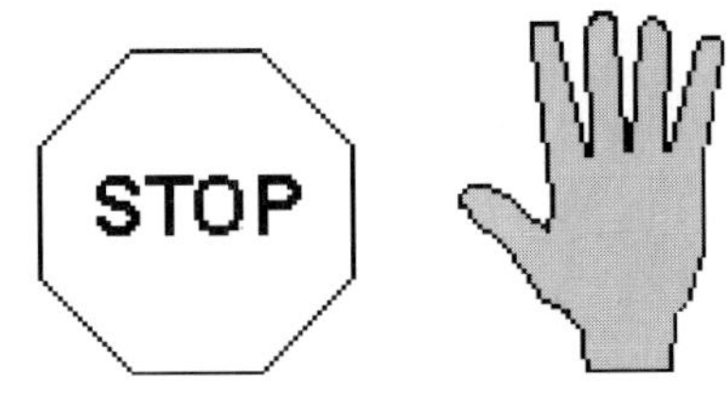

Figure 9.3 Iconic Representation of a Stop Command.

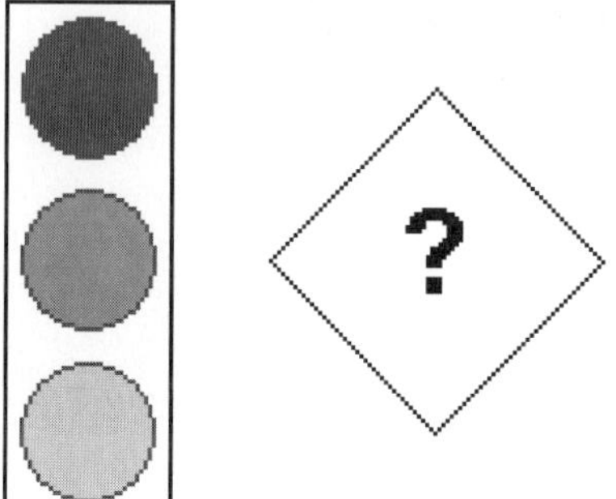

Figure 9.4 Iconic Representation of a Decision.

These symbols, in and of themselves, have meaning to the viewer, regardless of the language the viewer is accustomed to using.

Iconic Blocks

An *iconic block* is a graphical, or iconic, representation of executable code. If, for example, we wanted to represent a decision with three possible paths ("yes," "no," and "maybe"), we would add process flow connectors (Flow In and Flow Out) to either of the symbols shown in Figure 9.4, add some decision, and the result would be the iconic block shown in Figure 9.5.

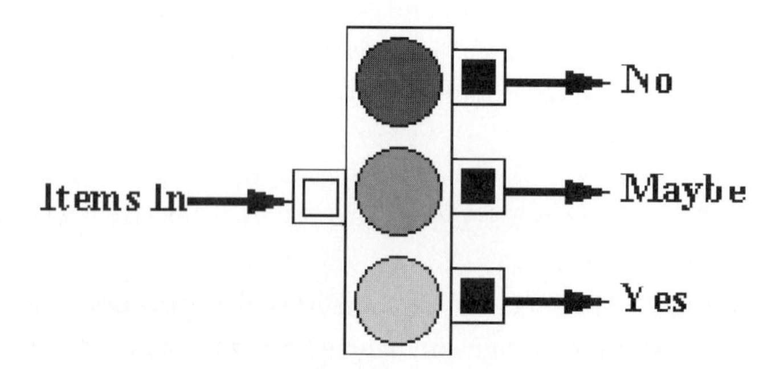

Figure 9.5 An Iconic Block.

This symbolic representation of executable simulation code provides immediate intuitive meaning—there is a decision to be made at some point in a process and, based on the results of that decision, there are three possible paths that can be taken as the process proceeds. The value of visual paradigms is clear from this simple example.

Types of Simulation: Continuous Simulation

Continuous simulation is used to model systems that are not controlled by events but rather happen in smooth increments of time. For example, consider the environmental model shown in Figure 9.6.

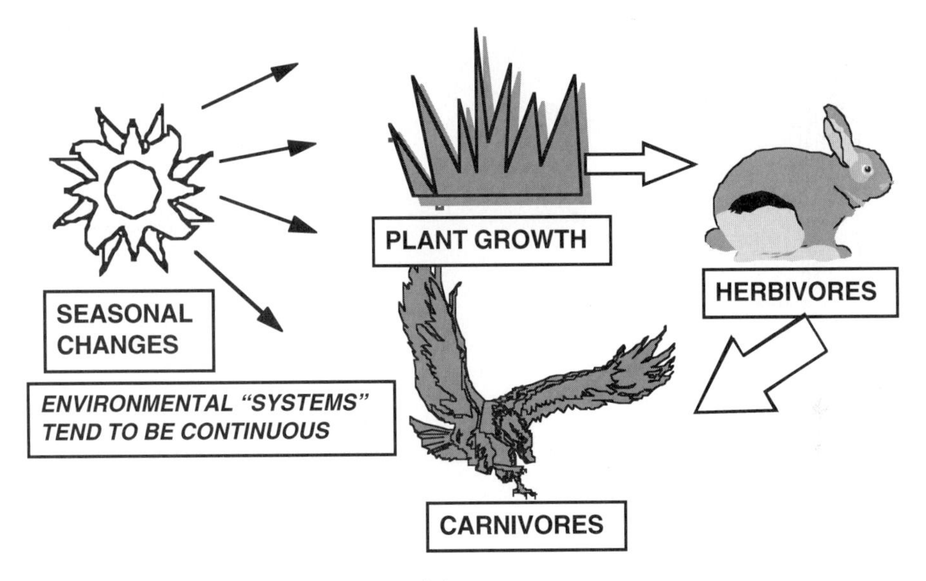

Figure 9.6 Example of a Continuous Model.

This example shows the effect of seasonal changes on plant-growth, plant-eating animal populations, and meat-eating animal populations. We cannot say with certainty that plant growth stops on a certain date, thus leading to a decrease in plant-eating animals. Instead, it is more accurate to say that plant growth slows as available sunlight decreases and speeds up as available sunlight increases.

Continuous models are valuable in BPR activities when they are used to determine the relationships between business "drivers," that is, those factors that affect the viability of the business. Consider the example shown in Figure 9.7.

In this model, we are examining the relationship between various Cost of Business factors, Revenues, and Profit. We can assume that an increase in advertising spending will lead to an increase in revenues, but also that an increase in advertising spending will lead to an increase in Cost of Business. We have defined *relationships* between business factors (drivers). It is not correct to say that an increase in advertising on a particular day will lead to an increase in sales on the same day, nor is it accu-

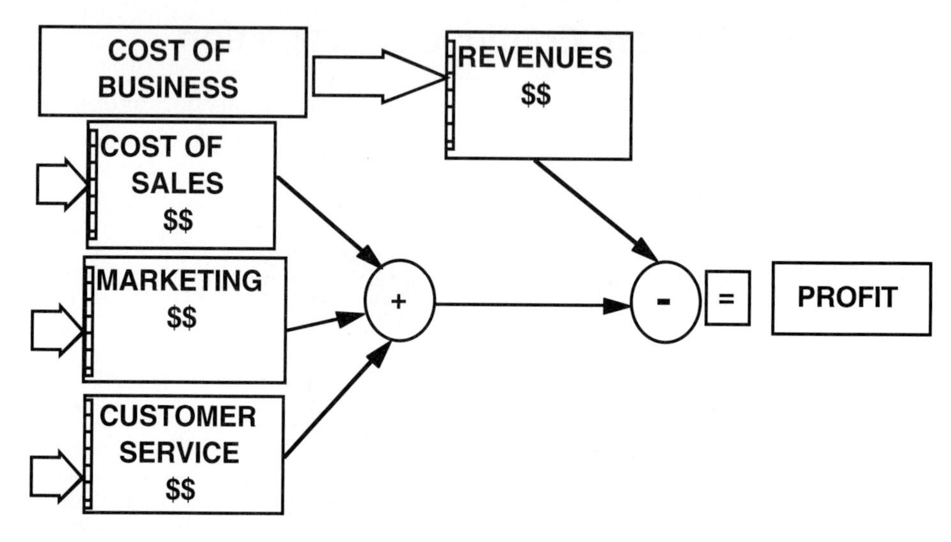

Figure 9.7 Continuous Business Model.

rate to say that the increase in advertising expenditures will happen on one day only. It is more accurate to determine a formula that relates the increase in advertising expenditures to the Cost of Business and Revenues over time. This formula would represent a continuous relationship. Continuous modeling and simulation are based on a flow of information through a model, that is, a flow of values through the model.

Continuous modeling and simulation can be very valuable for *enterprise modeling*, or the modeling of the economics of an entire business. This type of modeling is very useful for high-level decision making.

Types of Simulation: Discrete Event Simulation

Discrete event simulation is time oriented and event oriented. A discrete process is one in which process steps occur when a need arises, such as the completion of a task, the expiration of some time unit, a decision, and so on. Discrete models and simulations have *items* or *objects* flowing through them. For example, a manufacturing process creates finished goods, a loan processing process produces completed loan applications, and so on. Consider the example shown in Figure 9.8.

This example shows that at noon, people in office buildings tend to leave the buildings and go to restaurants, and that at one o'clock they tend to go back into those office buildings. The entering and leaving of the office buildings and the entering and leaving of restaurants are triggered by an event, that is, the "event" of time

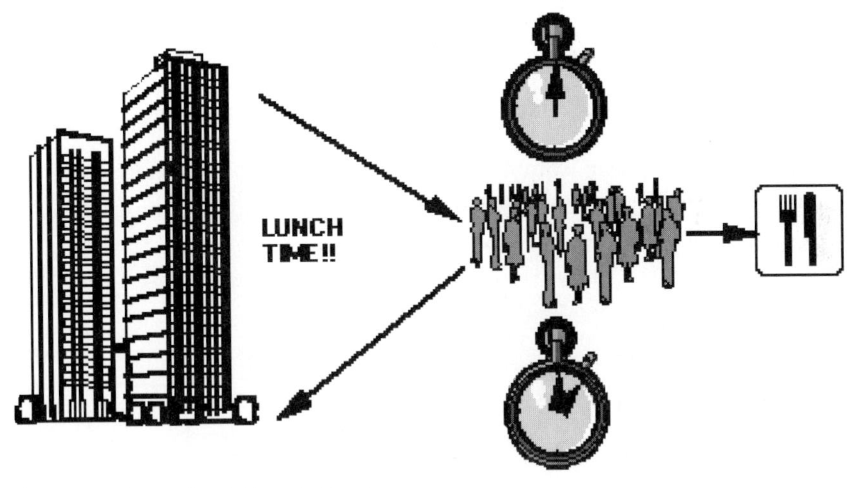

Figure 9.8 Example of Discrete Events.

becoming "lunch time." Restaurants staff according to such events, cleaning crews are assigned in office buildings based on such events, and so on.

Discrete event models are used to determine *process measures*. For example, consider the model shown in Figure 9.9.

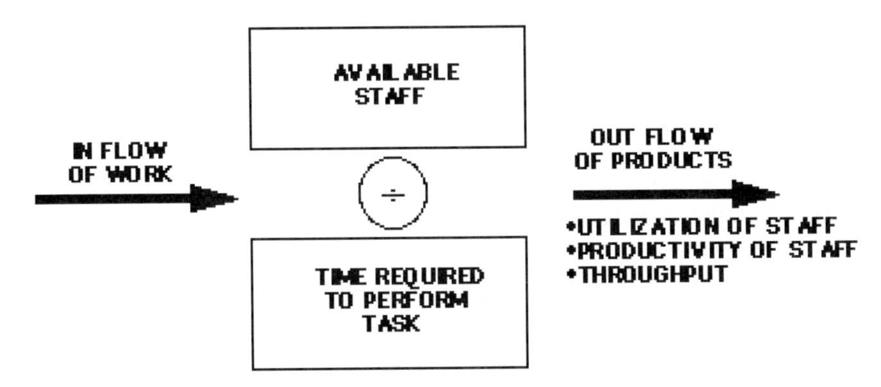

Figure 9.9 Discrete Event Simulation.

In this model, the inflow of work and time required to perform a task are time-oriented parameters, as is the outflow of finished products. Utilization of Staff, Productivity of Staff, and Process Throughput are measures of the process. These measures *cannot* be determined using continuous models; therefore, to determine the *effectiveness* of a process in terms of cycle time, cost, productivity, and so forth, it is necessary to use discrete event modeling.

Types of Simulation: Hybrid Simulation

Hybrid simulation combines both continuous and discrete event simulation capabilities. This is very important for the accurate analysis of business processes. For example, consider the model shown in Figure 9.10.

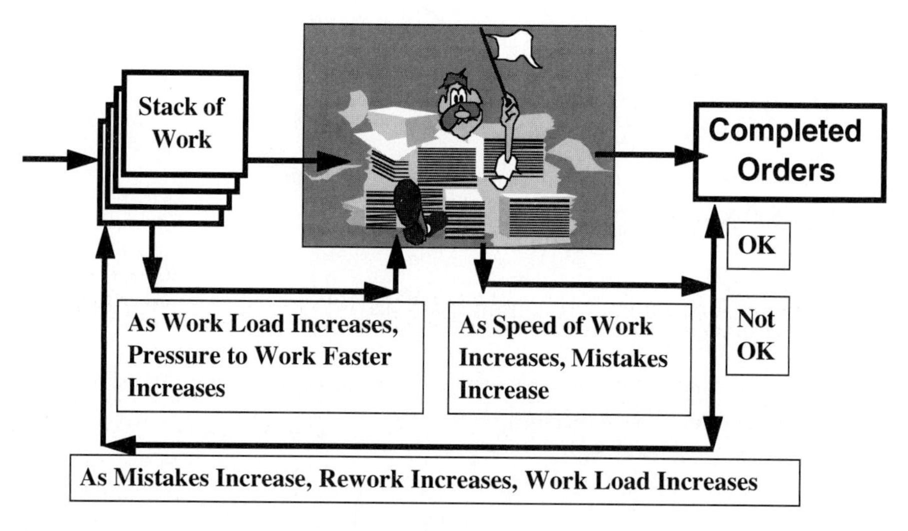

Figure 9.10 An Example of Hybrid Modeling.

In this model, the harried worker has a flow of work coming into his in-box. His task requires a certain amount of time, but he is subject to stress. Therefore, as his work load increases, his stress increases, and he tries to perform his task faster. When he tries to work faster, he is prone to errors, and errors, unfortunately, are routed back into his in-box. Rework increases the pile in his in-box, his stress increases, he tries to work faster, there are more mistakes, and so on. This is a classic *vicious circle*, that is, a causal loop in which one event causes an increase in another event, which in turn causes an increase in the first event, and so on.

In this model, the rate at which work arrives is discrete, and the amount of time required to perform the task is discrete. The relationship between the amount of work to be done and the stress the worker feels is continuous, as is the relationship between the speed at which the work is performed and the probability of an error. Therefore, this model uses both discrete modeling techniques and continuous modeling techniques. It is a hybrid model, as are most business models.

Figure 9.11 shows how continuous and discrete event modeling can be combined to create a hybrid enterprise model. Assume you have created an enterprise model that allows you to perform multiple "what-if" scenarios. You have determined the cause-and-effect relationships between cost of sales and revenues, and now you want to determine the actual cost of each functional process in the enterprise. To do so, you have built a discrete event model of the advertising function, analyzing the effect of adding staff, investing in equipment, outsourcing, and so on. The data from the discrete event model can now be used in the continuous model to increase the reality of the model.

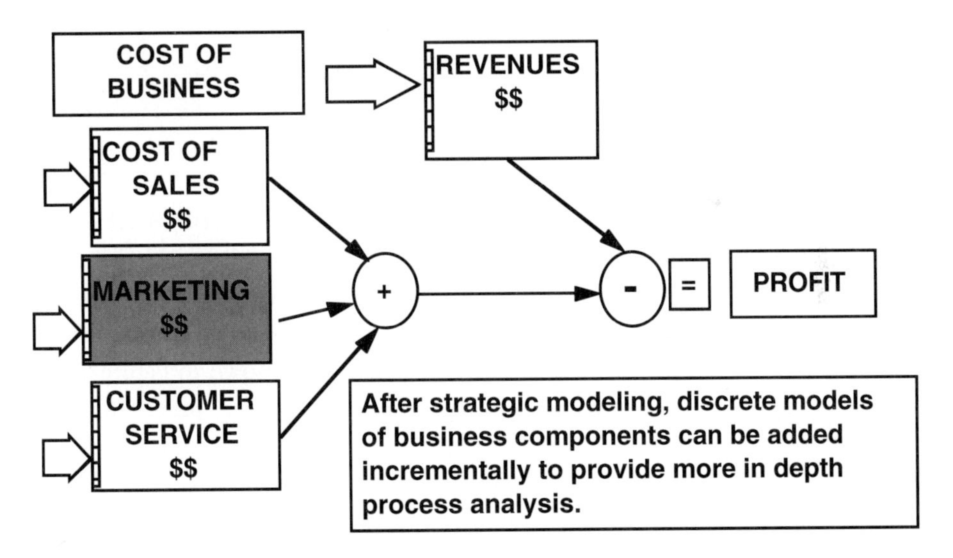

Figure 9.11

Object Orientation

Object orientation is a term that has a number of different definitions and varies from person to person. For business process analysis, the definition is simple: Objects in a simulation or model are assigned *attributes*, or descriptors, that are used in the model to determine cycle time, cost, and so forth. These descriptors can represent the type of object flowing through the model, the size of the objects, and so on. More importantly, each item of the same type will have the same characteristics.

Figure 9.12 demonstrates the use of object orientation.

In this model, orders arrive either in the mail or by phone. There is no differentiation of orders until they are processed, at which time the type of order is determined. The *type* of order received is an attribute, and this attribute may determine the

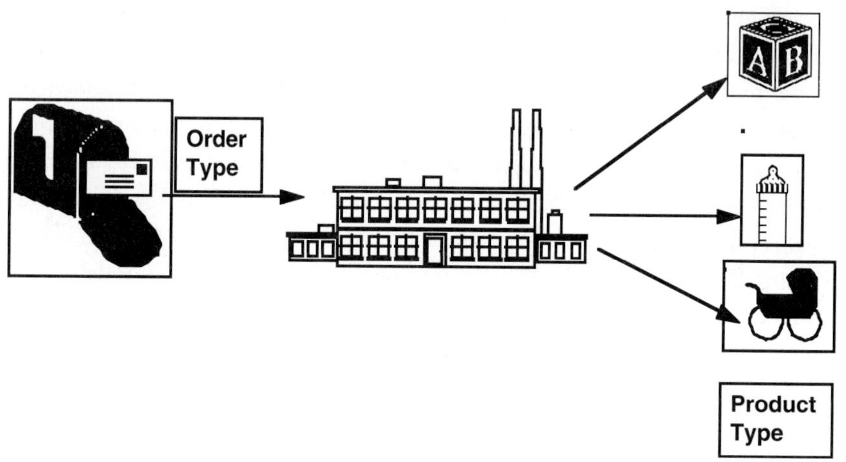

Figure 9.12 Example of Object Orientation.

time required to build a type of product, the raw material needed, labor hours required, and so on. This type of detailed analysis is not available in nonobject-oriented tools.

Object orientation lends itself to a number of process analysis techniques, such as *Activity Based Costing* (ABC). Consider the model shown in Figure 9.13.

In this example model, the cost of making (servicing, processing, etc.) an item is carried as an attribute by that item as it moves through the process. After each step in the process, the cost of the step is added to the accumulative cost of the item, so that at the end of the model, we know exactly how much every item cost to make.

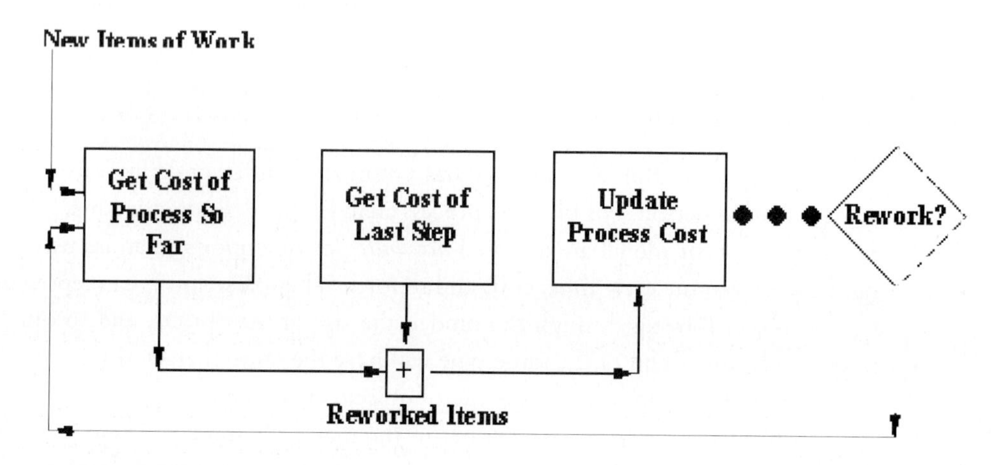

Figure 9.13 A Process with Rework.

More importantly, if there is rework associated with an item, the cost of the process for that item is not lost but accumulated, no matter how many times the item passes through the process. This is an important consideration, since rework raises the overall average cost of all the items in the process. Again, this type of calculation is not possible without object orientation.

Requirements-Based Analysis

Requirements-based analysis is a concept that is growing in popularity. It suggests that you cannot determine how a task or process will behave until you know the requirements for that task or process. Consider the model shown in Figure 9.14.

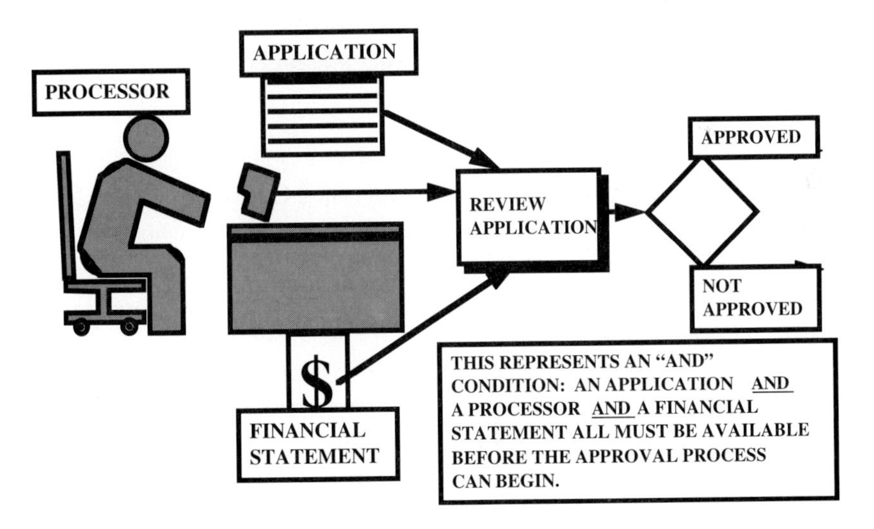

Figure 9.14

Many simulation tools will simply have a model segment called "loan processing" and not worry about what is really required to perform that task. The model in Figure 9.14 shows that a loan processor, an application, and a financial statement associated with the application are all required to perform the task. The total cycle time for the loan application process is dependent upon the timely submittal of an application, the timely processing of a financial statement and credit check, and the availability of a loan processor. If mistakes are made or information is missing, the time required to complete a loan review will be extended.

Requirements-based analysis is necessary to create detailed models of processes, as opposed to simple approximations of processes. Tools that do not effectively merge process streams or flows will not be able to provide this type of detailed analysis.

Hierarchical Decomposition

Hierarchical decomposition is a mechanism for building models in layers, so details that are revealed as layers are "peeled" away. The terms "drill down" and "dive down" are terms that refer to the incremental display of hierarchical layers in a model. The hierarchical layers themselves are called *hierarchical blocks.* Hierarchical decomposition is very important in a model, since it helps reduce the visual complexity of a model. Consider the model shown in Figure 9.15.

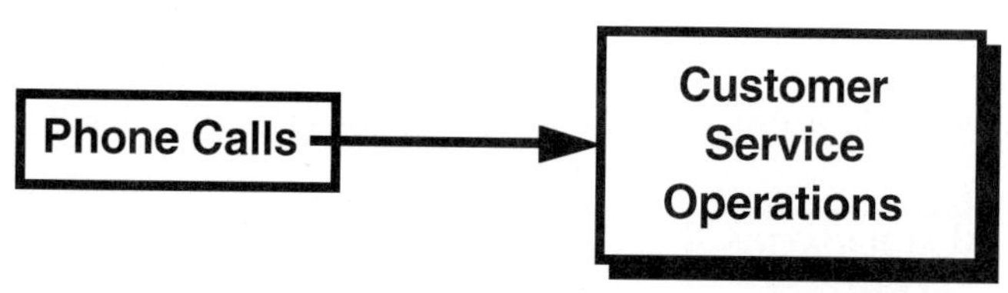

Figure 9.15 First Layer of Hierarchy.

This model shows that there is a process called Customer Service Operations. At this point, we do not have details about that process; however, when one more layer of the hierarchical block is revealed (typically by double-clicking on the model component itself), we see that the Customer Service Operations process consists of three subprocesses, as shown in Figure 9.16.

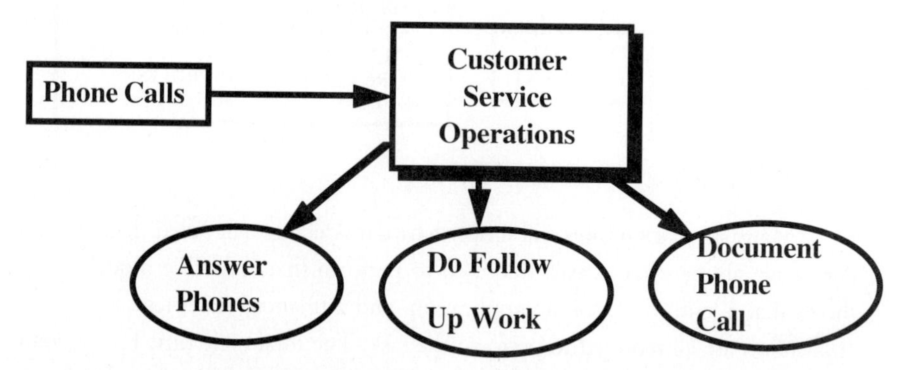

Figure 9.16 Second Layer of Hierarchy.

These three subprocesses can be further analyzed in detail by revealing another layer of hierarchy, in this case, a layer of the Do Follow Up Work subprocess. The model would then appear as shown in Figure 9.17.

This revelation of details can continue indefinitely. The value of hierarchical decomposition is that it allows models to be built from the top down, it allows details

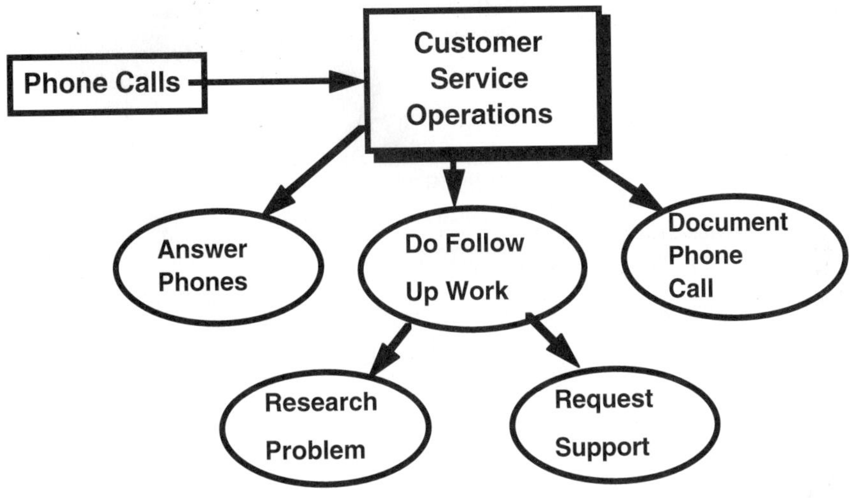

Figure 9.17 Third Layer of Hierarchy.

to be added to a model as required, and it also helps determine the relationships between processes. For example, the Research Problem subprocess shown in Figure 9.17 may require assistance from another functional group.

Customization

There are two types of customization that tool vendors refer to: customization of graphics and customization of the tool itself. Very few tool vendors allow customization of their tools, and that discussion is beyond the scope of this book; however, some tool vendors do allow customization of graphics. Graphical customization is most valuable when applied to hierarchical blocks. The model shown in Figure 9.18 is an Extend model of a public utility's Accounts Receivable Department. The model consists of hierarchical blocks that have been given graphical representations of *data flow diagrams* so viewers of the model can immediately feel comfortable with it.

Customization facilitates acceptance and understanding of modeling and simulation in BPR, and since resistance to technology is often a limiting factor in the application of technology to anything, this is an important feature of any simulation tool.

Block Libraries

A *library* of blocks contains executable blocks of code that can be changed and updated over time, just as books in a library are updated and changed over time. In

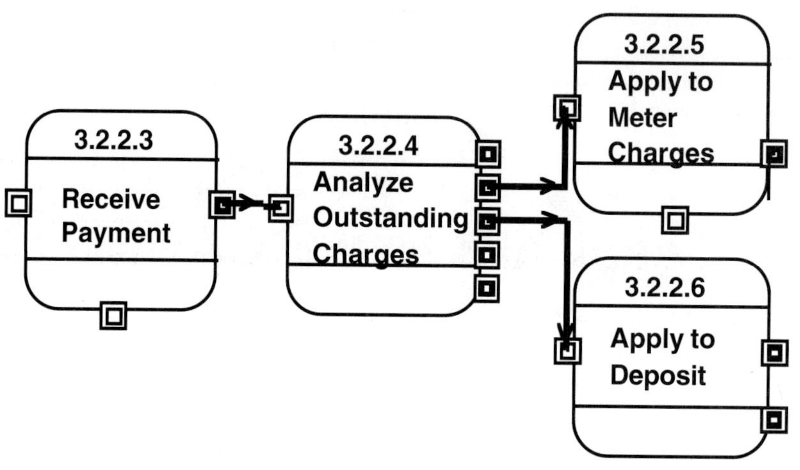

Figure 9.18 Customized Graphics.

addition, blocks can be added to libraries to enrich the modeling and simulation tool being used. Libraries add flexibility to a tool, since a user is not limited to a restricted set of functions.

Hierarchical Block Libraries

Hierarchical blocks represent process functions and, as such, can be used to represent repeatable functions. The storage of reusable hierarchical blocks in libraries provides another mechanism of enriching a modeling and simulation tool and also facilitates model development.

Open Architecture

A tool is considered to have an *open architecture* if it allows users to make modifications to basic building blocks. Since no tool can provide functional blocks to meet all possible simulation possibilities, the ability to modify blocks is another mechanism for enriching the tool in use. This capability is also known as *extensibility*.

10

Requirements for Effective CAPRE Tools

Progressing through process maturity levels and applying the Rules of Process Reengineering yield both a *functional* view of a process and a *behavioral* view of a process. The functional view tells us *what* is happening and *what* is being produced. The behavioral view tells us *how*, *why*, and *when* the process flows. *Process parameters*, which govern the behavior of the process and allow us to determine the effectiveness of the process, have also been established.

To be effective, a CAPRE tool must be able to provide both the functional and behavioral views of a process and incorporate process parameters into those views.

To provide these views, a CAPRE tool must be graphical, since it will build upon the activity diagrams that were developed while migrating to Level 3. The most effective way of providing both views within a single graphical representation is through the use of *iconic blocks.*

Minimum Iconic Block Set for CAPRE Tools

The following sections discuss the minimum set of iconic blocks (hereafter simply referred to as blocks) that are required to engineer a process effectively.

Operations. An operation is an activity that takes input, performs some function, and generates output. There are two types of operations that a CAPRE tool must simulate: simple operations and batch operations. Simple operations behave according to the following rules.

- A simple operation takes *n* input items and creates one (and only one) output item. For example, an operation may be defined as taking as input 12 spokes, one rim, and one tire to create one wheel. If more than one output is created by an operation, more than one operation is taking place.
- A simple operation requires some time to complete. The amount of time can be fixed or variable and can be broken into waiting (or collection) time, operating time, and output time. Since human processes rarely proceed in an orderly, consistently timed manner, an effective CAPRE tool will have the ability to alter the amount of time required for an operation while the model is executing.
- When a simple operation requires more than one type of input, an AND condition exists. In other words, if an operation requires inputs *A*, *B* and *C*, then input *A* AND input *B* AND input *C* must be present for the operation to take place.
- Some action is taken to physically or logically change the characteristics of the inputs. For example, spokes are *assembled* into wheels, a document is *printed*, and so on. An operation, therefore, represents either a transformation of an object into a new object, or a *state change* of an object.
- Batch operations follow the same rules, with the following exception: The output of a batch operation is *n* copies of some item. For example, 12 bottles and 12 bottle caps are input to an operation to create 12 capped bottles.

Transactions. A transaction is movement from one operation to another operation. Transactions are typically displayed as arrows in activity charts. A transaction is a special type of operation, in that it accepts *n* items of different types and, after a period of time, outputs the same items with no changes to those items. The rules of behavior for transactions are as follows:

- A transaction takes some time to complete.
- A transaction requires one or more inputs.
- A transaction does not modify the inputs in any way.

Stores (stocks, reservoirs). A store is a collection of items. The store may (1) contain items of the same or different types (it is easier to model behavior if all the items are of the same type, but real life is not always so organized)and, (2) have a positive or negative value. Negative values of stores are required since the store may represent some intangible object, such as a credit balance, temperature, and so on.

There is no time required to fill or empty a store; that time is captured in either an operation or a transaction. A store should also have the following features:

- A store should have a defined maximum capacity.
- If that capacity is met or exceeded, the CAPRE tool should output a message specifying that condition.

Decisions. A decision is a check of some *condition* in the process. The condition may relate to the characteristics of some object in the process or may relate to the state of the process itself. Good CAPRE tools will implement condition checking in the form of IF...THEN...ELSE statements.

Event. An event is something that causes a condition to change. Events can be *implicit* or *explicit.* Examples of implicit events are

- The start or end of an operation.
- The start or end of a transaction.
- A decision.
- The expiration of some set period of time.

Implicit events do not have to be captured in a graphical representation, since they are incorporated into other graphical representations.

Explicit events are events that happen outside of the normal flow of a process. For example, if we want to check the effect of a critical employee being unavailable for some reason (i.e., sickness, vacation, and so on), an effective CAPRE tool will allow a user to *force* that condition to exist through the use of an Event Generator block. An Event Generator block will generate an explicit event and operate in the following manner: An explicit event will cause a condition to exist until some other event occurs.

An *interrupt* is a particular type of explicit event that stops one or more processes so different processes can execute. For example, one may wish to simulate the effect of telephone calls on a process. A modeling tool should have the capability to generate interrupts that initiate a phone call handling process and shut down another process. When the phone call process ends, the other process should resume.

Mathematical and Logical Operations. An effective CAPRE tool must have a rich array of mathematical operations available, such as add, subtract, multiply, log, square root, and so on. The logical operations of AND and OR must be present as well as a logical NOT operation. A random number generation feature is extremely important for simulation purposes.

Rather than present a list of mathematical and logical operations, it is sufficient to say that the more operations available, the better. There are two ways of presenting these operations to a modeler: They can be incorporated into the blocks described previously, or they can exist as blocks on their own. While you may have a personal preference, the following points should be considered.

- By requiring mathematical and logical operations to occur in blocks outside of other blocks, a viewer of a model is alerted to the fact that logic is being applied to the model.
- Requiring separate blocks for mathematical and logical operations can crowd a model; on the other hand, once one is accustomed to seeing the blocks, the model may actually be easier to understand.

Queues (stacks). A queue is a type of store, except that the items stored in a queue are organized in a first-in-first-out (FIFO) order, last-in-last-out (LIFO) order, or priority order. An iconic block for a queue should be different from the iconic block for a store, since queues have some implied behavior, whereas stacks do not. The behavior of queues is defined as

- A queue receives input items.
- These items are placed in an order defined as LIFO, FIFO, or priority.
- The items are held until they are removed by an operation or a transaction.

There is no explicit delay time associated with a queue, nor does a queue alter the items in it in any way. However, items moving through a simulation may be delayed, or stuck in a queue, if the next step in the process is not ready to receive them. Since this is the case, a queue should have the following features:

- A queue should have a defined maximum length.
- If that length is met or exceeded, the CAPRE tool should output a message specifying that condition.

Other CAPRE Tool Requirements

Printing of Models. A CAPRE tool should be able to print to any printer with graphic capabilities and also to plotters. When printing to a page printer, the CAPRE tool should ensure that iconic blocks are not broken over page boundaries.

Displays and Interaction. An effective CAPRE tool will have the following capabilities:

- The capability of displaying process parameters as a model executes.
- The ability to interact with the viewer. Interaction is minimally defined as displaying messages and requesting input.
- The capability of animating the simulation so that process flow can be observed.

Input/Output. An effective CAPRE tool will be able to accept input parameters from a file and output (during model execution) to a file.

Data Encapsulation. A tool must have the capability of encapsulating or tracking data in tabular and graphical form.

Scenario Analysis. Scenario analysis is a term describing the testing of multiple "what-if" conditions of a process. For example, a company might want to determine the effects of adding staff, of reducing staff and adding technology, or some combination of adding staff and adding limited technology.

Time Orientation of Input. Many processes have an inflow of data that changes with time. For example, customer service centers receive more phone calls during the middle of the day than at night. An effective CAPRE tool must be able to allow a user to determine process parameters by time; otherwise important process details may be lost.

Other Desirable Features

Ease of Use. Iconic blocks are, in essence, programmed to perform functions according to specifications. In previous generations of modeling and simulation tools, this programming was often done in complex specialized languages, and the skills of a software engineer were usually required to create enactable (i.e., executable) models.

The emergence of object-oriented software development techniques has eliminated the use of specialized languages in many computer-based tools, such as computer-aided software engineering tools. These tools now utilize what is commonly known as *dialog boxes*, in which information which defines the behavior of an iconic block is placed. An effective CAPRE tool will utilize dialog boxes to specify the behavior of a model, and the language used in the dialog, when necessary, will be a simple IF...THEN...ELSE language.

Customization of Blocks. No tool will have all the iconic blocks required to meet every possible modeling situation. Therefore, a tool with an open architecture that allows modifications of blocks, or the development of custom blocks, will be superior to a tool with a closed architecture. The tool should permit not only customization of operating code for blocks but also customization of dialog boxes and the iconic (graphical) representation of a block.

On-line Documentation and Help. Any effective CAPRE tool will be accompanied by well-written documentation; however, most computer-based tools now have on-line documentation and Help, and CAPRE tools should not be an exception. Documentation and Help information should be obtained from either menu selections or from the dialog boxes that are used to set specifications for a block.

Summary

CAPRE tools can have other built-in capabilities, but the preceding sections represent the set of requirements for business process reengineering efforts. Since this book is about these efforts, features which may make a tool desirable in other types of modeling efforts will not be discussed.

11

Dynamic Modeling and Simulation Tools

One factor limiting the introduction of computer-aided process reengineering has been the cost of traditional modeling and simulation systems. Typically, these systems reside either on a workstation or a mainframe and can cost more than $10,000 per copy. Since the cost of a workstation is currently about $15,000, the total cost for a simulation system can exceed $25,000. Not surprisingly, management personnel have been very reluctant to make such a large investment in technology they are unsure of.

In addition to their cost, the complexity of traditional modeling and simulation tools frightens many managers. They assume (rightly or wrongly) that their employees would be reluctant to invest the time required to learn the tools and, therefore, any investment that was made would soon be wasted. Fortunately, there are some emerging PC-based and Macintosh-based tools that can be used to model and simulate human processes and cost less than $1000 per copy. The only tool that currently meets the requirements put forth in previous chapters is Extend, a dynamic modeling and simulation tool from Imagine That, Inc.

Other tools being promoted as business process reengineering tools fail to meet the requirements previously described and will not be discussed in this chapter.

The Extend Modeling and Simulation Tool

The Extend tool

- Utilizes iconic (graphical) blocks that are connected to form a model.
- Utilizes an underlying IF...THEN...ELSE language to determine the behavior of the model.

- Uses dialog windows to specify behavior and to enter process variables.
- Provides a capability to animate iconic blocks so process flow can actually be viewed as a simulation executes.
- Runs on Macintosh and IBM PC computers and clones.
- Maximizes the use of color to enhance models.
- Can be used to perform continuous (time flow) modeling.
- Provides a discrete event modeling capability.
- Provides predefined connectors on blocks, and connection rules are defined by the connection type.
- Is object oriented and allows attributes to be assigned to items. Attributes are characteristics of items, such as color, height, quality, and so on.

Extend has a comprehensive set of iconic modeling blocks that can be used for continuous and discrete event modeling and simulation. As the full set of Extend iconic blocks approaches 100, they cannot all be presented in this evaluation.

The philosophy of Extend is that objects flow through a model based on either a change in time, the completion of an event, or a change of state—either in an object or the model itself. States of objects are described by attributes, or values that describe them in some manner, such as color, priority, cost, and so on. This is an object-oriented concept that is useful when developing complex models whose behavior changes as the objects in the model change.

The following sections describe the capabilities of Extend as related to the requirements of CAPRE tools presented in the previous chapter.

Operations. Operations in Extend can be modeled several ways. A *simple operation* is programmed using either an Activity block or a combination of the Batch block and the Activity block. (There are several types of Activity blocks, but for the sake of simplicity, they will be referred to collectively by one name.)

The Activity block takes one input and holds it for a period of time. The block looks like the one in Figure 11.1.

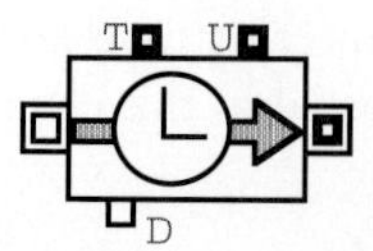

Figure 11.1 Extend Activity Block.

When more than one input is required to make one output, a modeler must use the Batch block in conjunction with the Activity block. The Batch block looks like the one in Figure 11.2.

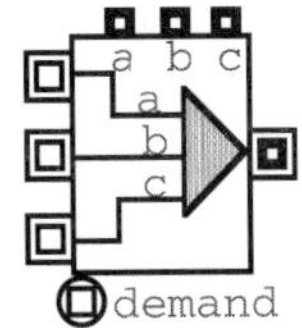

Figure 11.2 Extend Batch Block.

A dialog box for a Batch block is shown in Figure 11.3.

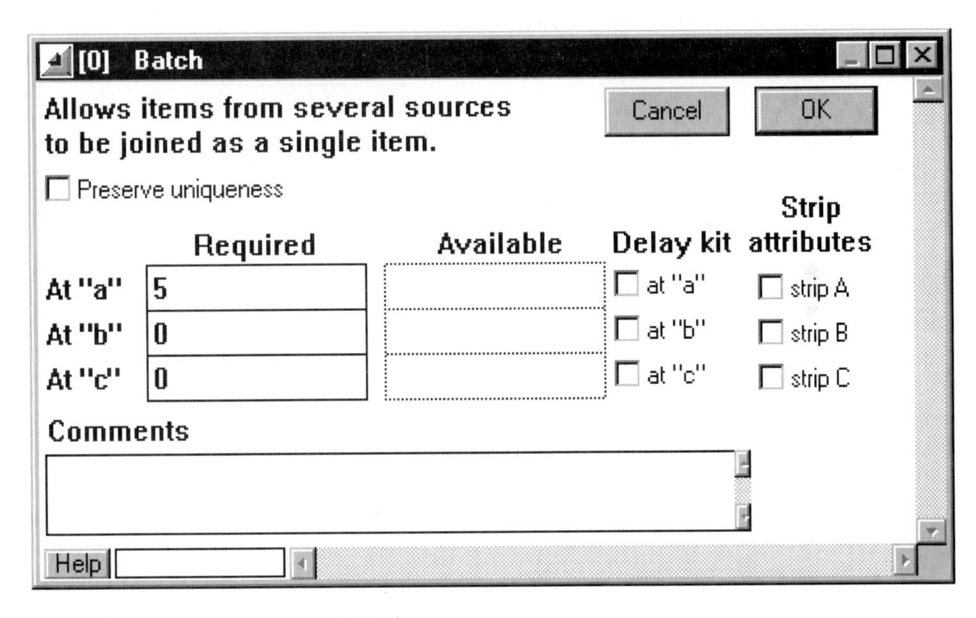

Figure 11.3 Dialog Box for Batch Block.

Note the "a" connector has been set to specify five items. Therefore, an operation requiring five items would be modeled using a combination of the two blocks. Figure 11.4 is an example of how this would be done in Extend.

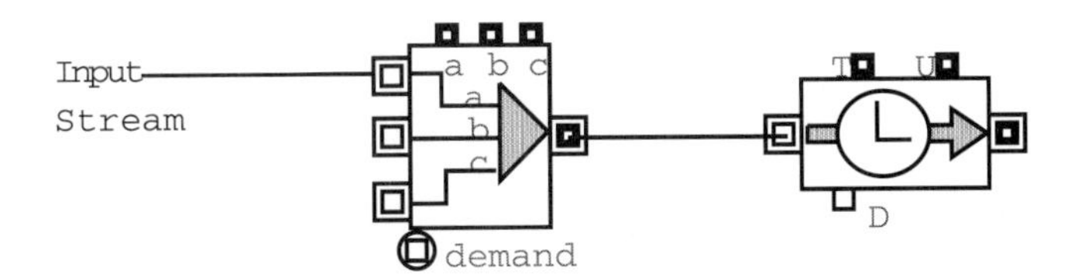

Figure 11.4 Operation with Combined Blocks.

The Batch block can also take up to three input streams, combine them into one item, and pass them to the Activity block, which holds that item for a specified period of time. The Batch block defaults to an AND condition, so all three inputs must be available to create an output item.

The combination of the Batch block and Activity, Delay block represents a *batch operation* and is shown in Figure 11.5.

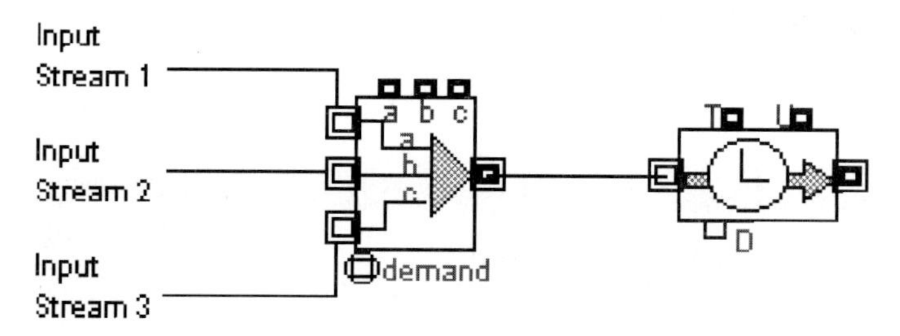

Figure 11.5 Batch Operation in Extend.

An Unbatch block can now be added to the model to decompose the output item into the original input streams. This is depicted in Figure 11.6.

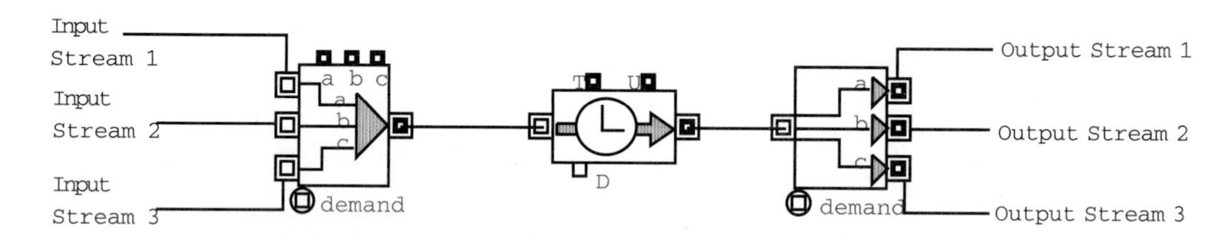

Figure 11.6 Batch Operation with Unbatch Following.

The amount of time associated with both a simple operation and a batch operation is set in the dialog of the Activity, Delay block and can be any amount of time. For example, an operation can be set to execute for 1.75 simulation time units.

In addition, the amount of time required for a process step is dynamically modifiable in Extend and, using object-oriented concepts, can be associated directly with the type of object being processed. For example, consider Figure 11.7.

The Extend Activity, Delay block has a connector that allows a modeler to dynamically alter the amount of time an item or items are delayed while migrating through the model. This is valuable in modeling efforts, since conditions may determine the rate of flow of objects and the model must be able to adjust to those conditions.

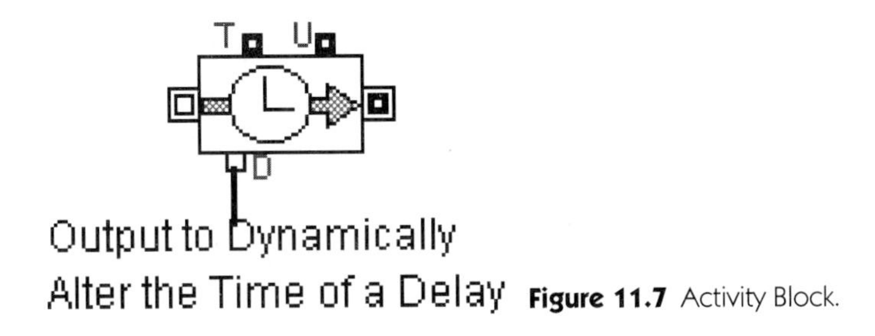

Figure 11.7 Activity Block.

Besides separating an item into its original components, the Unbatch block will preserve the attributes associated with those components. For example, Figures 11.8 and 11.9 show how this occurs.

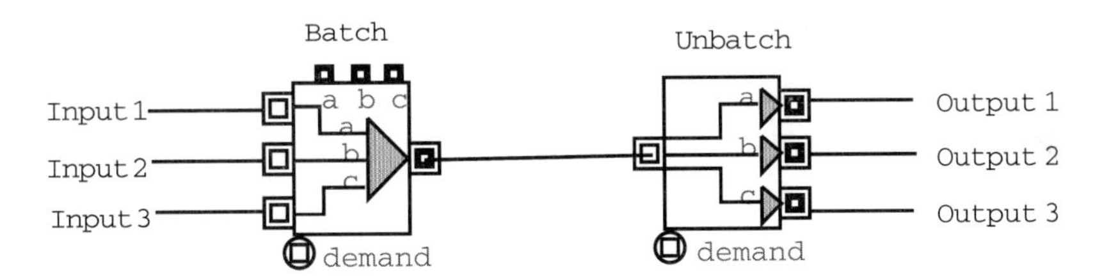

Figure 11.8 Batching and Unbatching.

The dialog for the Batch block would appear as shown in Figure 11.9.

[0] Batch

Allows items from several sources to be joined as a single item.

Cancel OK

☑ Preserve uniqueness

	Required	Available	Delay kit	Strip attributes
At "a"	1		☐ at "a"	☐ strip A
At "b"	2		☐ at "b"	☐ strip B
At "c"	3		☐ at "c"	☐ strip C

Comments

Help

Figure 11.9 Batch Dialog Box with Preserve Uniqueness.

Note that the Preserve uniqueness box has been checked. This means that any attributes associated with the input items, such as cost, weight, color, and so on, will be preserved and related back to the items after they are unbatched. This is an object-oriented capability that is useful when developing complex models.

Transactions. A transaction, or movement of objects from one point in a model to another, is a special type of operation. It is broken out in this discussion because the time to complete transactions is typically overlooked in traditional process representation techniques, such as flowcharts.

Extend provides several blocks for representing a transaction. One such block is the Activity, Multiple block, shown in Figure 11.10.

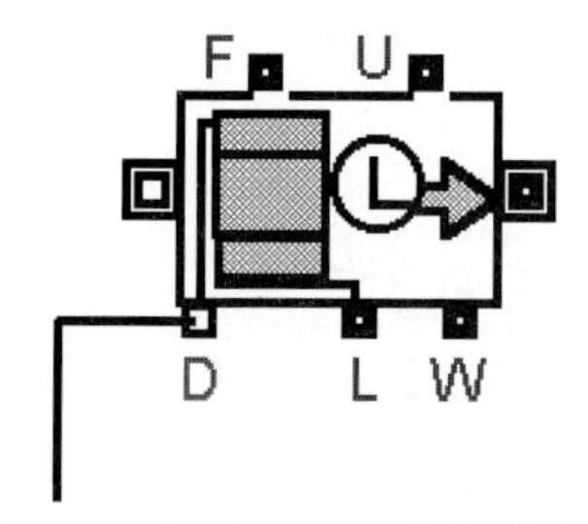

Figure 11.10 An Extend Activity, Multiple Block.

The Activity, Multiple block will take up to *n* items (as defined in the dialog) and hold them for a period of time, releasing them at the end of that time. The amount of time items are held can be dynamically changed using the D connector on the block.

Another block that can be used to represent a transaction is the Activity, Delay (Attributes). This block reads an attribute of an item entering and uses the value of that attribute as the delay for that item. Therefore, the amount of time an item is delayed is dependent on the item itself. Figure 11.11 shows the dialog box for this block.

In this case, an item has an attribute called Speed which determines the amount of time spent in the activity. This attribute, or any other attribute associated with the item, can also be modified within the activity itself for further use elsewhere in the model. All attributes can be defined and modified using mathematical and other operations in Extend.

This is an example of how object-oriented techniques can be applied to process analysis and improvement. Assume that a model has been developed that accounts for the abilities of employees in a trucking firm to complete their deliveries. The speed in which they perform that task (i.e., complete a transaction) is dependent upon the

[1] Activity, Delay (Attributes)

Activity delay that interacts with attributes.

OK

Delay (time units) = 1

Cancel

Delay is the: first attribute / attribute named: speed

Modify the: first attribute / attribute named: newspeed

Modify attribute by: subtracting the delay / adding the delay / changing by -10 %. / adding 0.5

Utilize blocking

"T" connector is true/false

Arrivals

Departures

Utilization

Comments

Help

Figure 11.11 Activity with Delay Specified by Attributes.

number of years experience they have and their knowledge of the area they are working in. These factors can be used in a calculation to create a Speed attribute so the amount of time for a trucker to complete his deliveries would be defined by the Speed attribute associated with the object in the model that represents the trucker.

Stores. There are a number of blocks defined in Extend that represent storage, some of which follow. These storage areas can contain items of different type or items with attributes that define type. For example, the Labor block shown in Figure 11.12, could be used to store employees whose skills are differentiated by attributes.

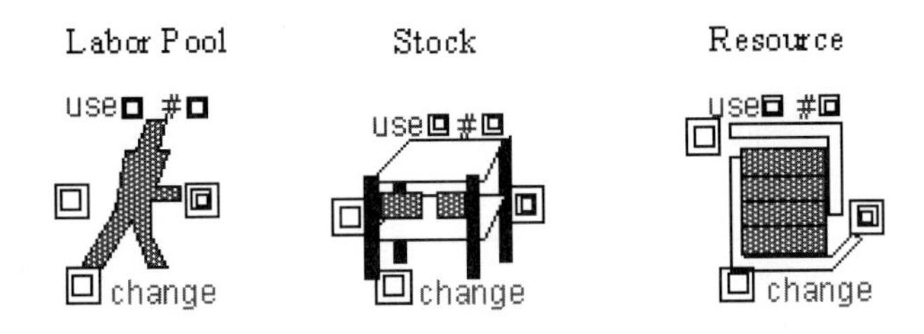

Figure 11.12 Examples of Storage Items in Extend.

Each of these blocks operates in essentially the same manner. Extend has provided separate blocks for different types of stores to give the visual representation of the model enhanced meaning.

Decisions. A decision in Extend is modeled by using a Decision block, which compares two variables, and a Select DE Output block, which chooses a path based on the value it receives. In this case, a True (1) or False (0) value from the decision is used to select a path. This is shown in Figure 11.13.

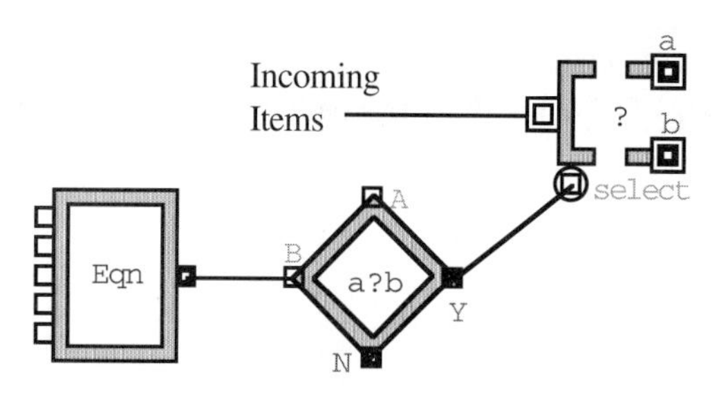

Figure 11.13 Decision in Extend.

In this example, some condition is represented by the Equation block, checked by the Decision block, and based on that check, a path is selected by the Select DE Output block. The Decision block can be combined with many other blocks, so there are multiple ways to depict a decision in Extend.

Events. Extend supports implicit and explicit events. In addition, Extend has the capability of generating events that will shut down other operations *while they are executing*. Interrupts can shut down operations, change the value of attributes associated with items, restart operations, and so on. Figure 11.14 is an example that demonstrates this capability.

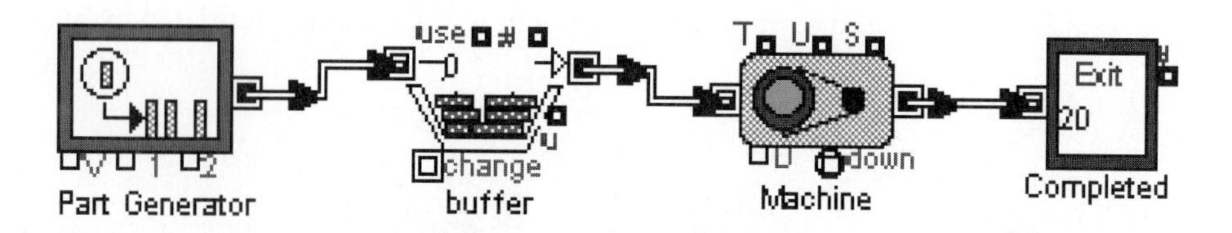

Figure 11.14 Extend Example Using Machine and Bin Blocks.

In this example, a part is generated every 5 simulation ticks, and used by a machine for four time units. When the simulation is run for 100 ticks, one would expect 20 items to flow through and none to be stored, which is exactly what happens. The blocks shown are from the Manufacturing and Discrete Event Libraries offered with the Extend product.

Figure 11.15 shows the machine operation being interrupted interactively by the user, causing an item to be held in the machine until the machine is restarted.

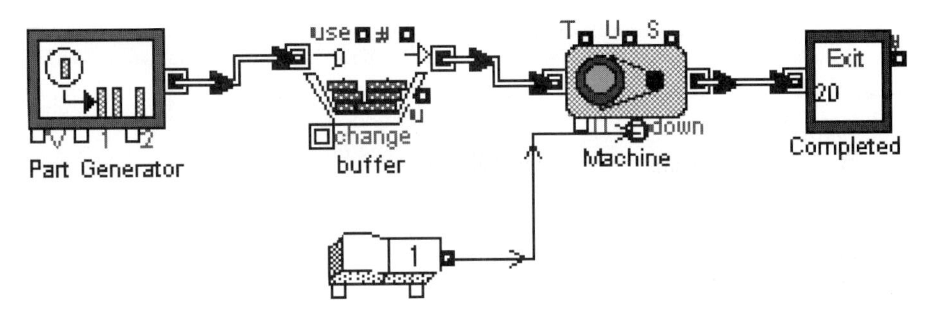

Figure 11.15 Shutting Down Machine Interactively.

Because the Machine block has changed appearance, it is showing that there is an item being held in it for processing. This means that the operation has begun but was suspended. Figure 11.15 demonstrates only one of many ways processes can be interrupted in Extend and how blocks can interrupt other blocks at any time.

The use of interrupts is an interesting feature of Extend and can be used to model processes that are interrupt driven. For example, a software development process could be interrupted by reviews, meetings, sickness of employees, and so on. The use of interrupts is related to the exactness with which one wants to represent a process.

Mathematical and Logical Operations. Extend has a large library of mathematical and logical operations available. The logical and mathematical blocks that control the behavior of a model can be implemented in one of two ways. The first is visual, using graphical iconic blocks, shown in Figure 11.16.

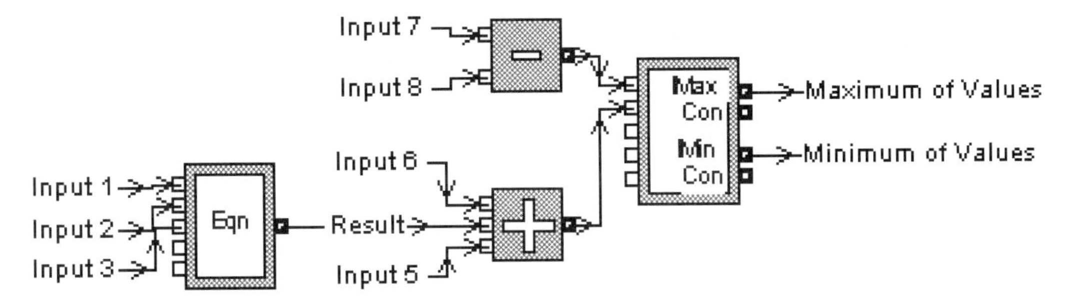

Figure 11.16 Extend Mathematical / Logical Operators.

The Subtraction, Addition, and Max/Min blocks are used to perform mathematical and logical operations. The Eqn, or Equation block, is used to perform more complex logical and mathematical operations. Those operations are defined in the dialog box for the Equation block, shown in Figure 11.17.

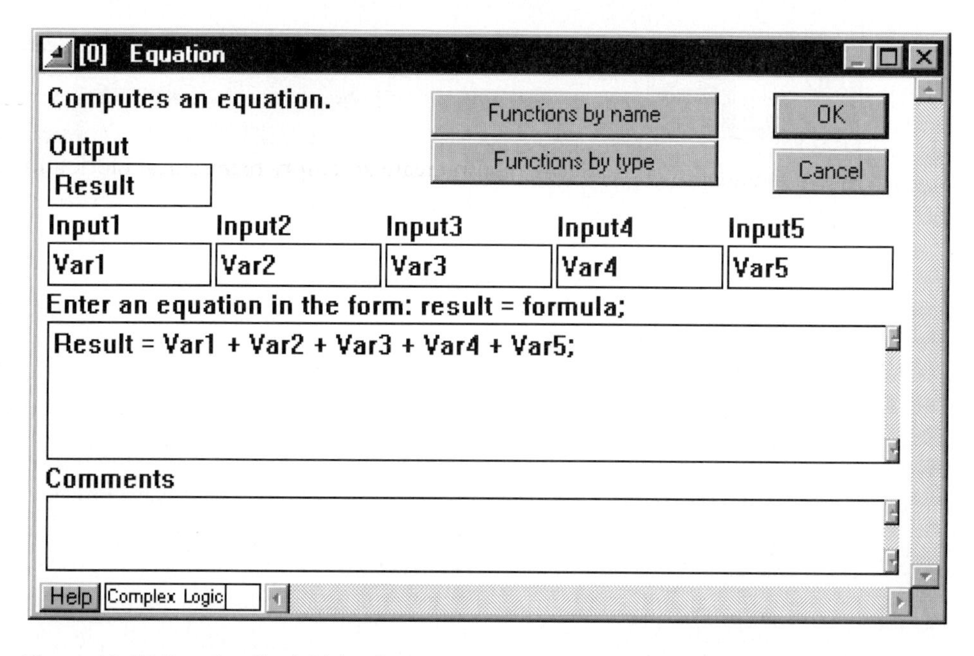

Figure 11.17 Equation Block Dialog Box.

Any type of mathematical or logical operation can be represented in this block. Note that built-in functions are available both by type and in alphabetical order.

Therefore, an Extend model can contain logical and mathematical operations that are either hidden from the viewer or presented visually as part of the model itself. The choice of displaying or hiding logic is dependent on the modeler and audience, and different modelers will prefer different approaches.

Queues. Extend supports *first-in-first-out queues*, *last-in-first-out* queues, and *priority* queues. Extend provides current information about the number of items in a queue and the length of a queue, and additional features of calculating the average amount of time items spend in the queue and the average length of the queue.

Other CAPRE Tool Requirements

Time Representation. Extend allows arbitrary time intervals (or fractions of time) to be assigned to operations. This is a feature of discrete event modeling and, since most business processes are event driven, the specification of time in this manner is important to business process analysis. In addition, intermediate storage areas between process steps do not have to be defined in Extend models. This is an artificial concept common with continuous modeling tools applied to business process modeling.

Hierarchical Decomposition. Extend supports both *top-down* and *bottom-up* hierarchies. A modeler can select a portion of a model and make it a hierarchical block (top down). A modeler can also create an empty hierarchical block and build a subprocess within it (bottom up). There are an unlimited number of layers of hierarchy possible, so that hierarchical blocks can be composed of other hierarchical blocks.

In addition, hierarchical blocks can be saved for use in other models, and hierarchical blocks can be graphically customized.

Printing Models. Extend supports a number of printing options and also supports output to plotters. Extend has a Show Page Breaks feature which will display where page breaks are in relation to the model so the model can be adjusted before being printed.

Displays and Interaction. Extend supports displays of model parameters during execution. Extend also has a graphical animation display capability, and a read-out capability that permits the numerical display of variables at particular points of interest in the model. In addition, some blocks display their contents numerically while the model executes. Extend also provides the capability to display model parameters through graphs and tables.

Extend supports interaction with the user. There are several mechanisms for doing so, but the most interesting is the capability of bringing input and output boxes contained in a dialog box to the surface of the model.

Input / Output. Extend permits both input and output directly to/from multiple files during execution.

Data Encapsulation. Extend supports data encapsulation in many different types of tables and graphs. The most applicable to computer aided process reengineering is the Discrete Event Plotter Block, which plots values over time in both graphical and tabular form.

Object Orientation. This term has a slightly different meaning depending on who is defining it. The concept of object orientation that is important to modeling is that an object (item) can be assigned attributes that define its behavior in the model. For example, an object can have an attribute of color, priority, time to completion, and so on. Extend provides the ability to assign numerous attributes to an object, including identifiers and priorities. This provides the capability of *uniquely* defining each individual object in the model. By changing the value of attributes, a modeler can change the behavior of a model. The logic of the model does not have to be changed; therefore, Extend provides the ability to build *archetypes* of models.

Scenario Analysis. Extend provides the capability to modify process parameters by specified amounts, random amounts, or from file input for an unlimited number of executions of a simulation. Therefore, a modeler can test the effects of changing many parameters in many combinations. This allows for an in-depth analysis of the relations between process parameters.

Desirable Features

Ease of Use. Extend uses simple dialog boxes to specify the behavior of blocks and, therefore, of the model. For example, Figure 11.18 is an example of a dialog box for the Activity, Delay block.

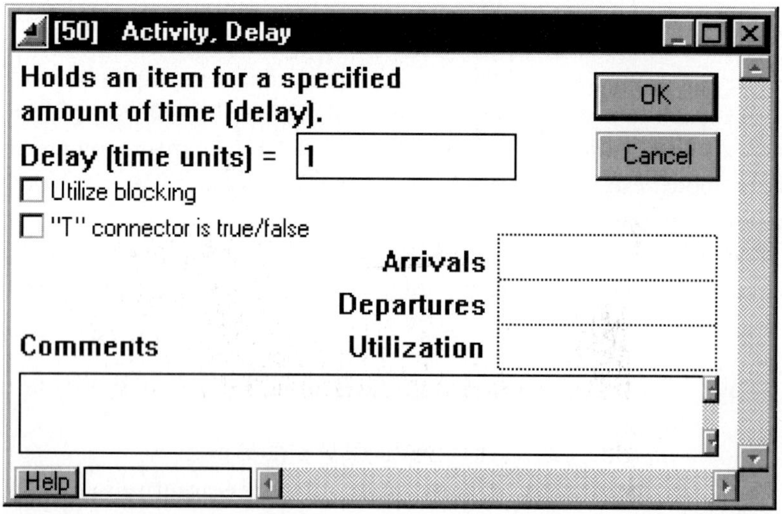

Figure 11.18 Example of Dialog Box.

The Activity, Delay block holds an item flowing through the model for a specified period of time. The amount of time for the delay is specified by inputting a value into the Delay (time units) box. Other options, such as Hold item until pulled, are used when creating more sophisticated models. The Arrivals and Departures boxes are display boxes used to show how many items have entered and left the iconic block.

More sophisticated models require more sophisticated blocks. For example, Figure 11.19 shows an Extend Equation block.

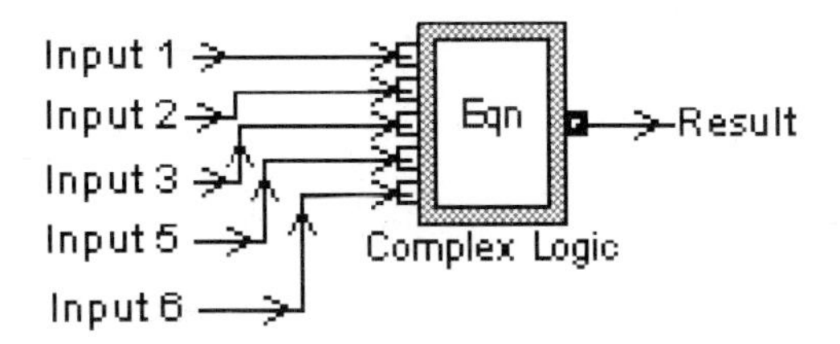

Figure 11.19 Extend Equation Block.

It has five connectors for input variables and one output connector. The different connector type symbolizes its use whether it is an input or an output. Figure 11.20 shows how an Equation block might be utilized.

[0] Equation

Computes an equation.

Functions by name

Functions by type

OK

Cancel

Output

Result

Input1	Input2	Input3	Input4	Input5
Var1	Var2	Var3	Var4	Var5

Enter an equation in the form: result = formula;

```
IF( Var1 == Var2) Result = Var3;
ELSE Result = Var4 + Var5;
```

Comments

Help

Figure 11.20 Extend Equation Block Dialog Box.

The IF...THEN...ELSE language has been simplified to an IF...ELSE... language, and programming is straightforward. The functions that can be used in the dialog box can be selected from a list that is interactively available to the user, either alphabetically or by type. The inputs can be referred to as Var1, Var2, and so on, or can be given names to enhance the understanding of the model.

Extend has another feature that allows users to interactively determine the values of parameters contained in dialog boxes. This is the Grab and Drag feature of bringing dialog inputs and outputs to the surface of a model. For example, Figure 11.21 shows the dialog box of an Activity, Multiple block.

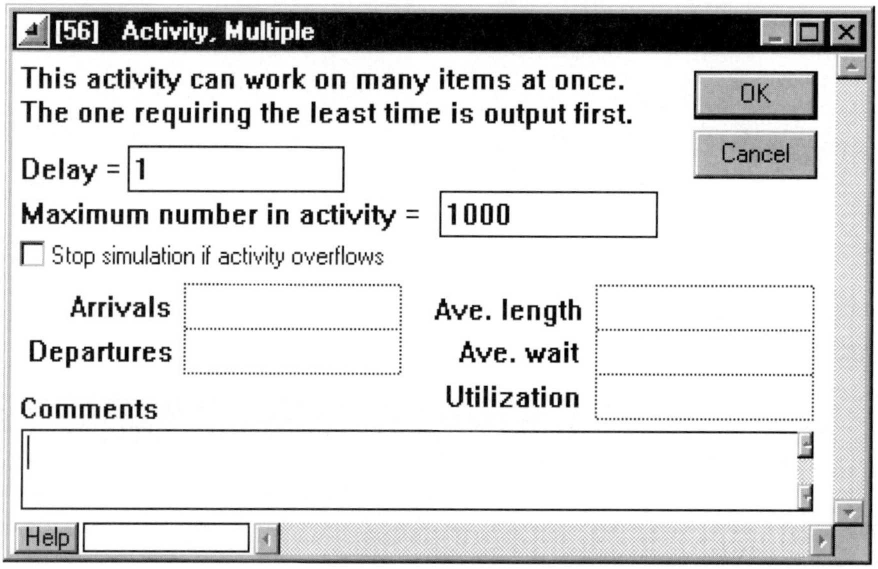

Figure 11.21 Dialog of Activity, Multiple Block.

Any of the input or output boxes within the dialog box can be "grabbed" using a special pointer from the menu bar and "dragged" to the surface of the model, so the displays can be seen while the model executes and the parameters of the block can be changed dynamically, also while the model executes. This allows a user to fine-tune a model in real time to test the response to changes in conditions.

For example, Figure 11.22 is a simple example of a model using an Activity, Delay block. In the dialog box of that block, described earlier, there is an input parameter specifying the length of the delay, and an output parameter specifying the number of arrivals and departures. In Figure 11.22, the input and output boxes of the dialog have been brought to the surface, so the user can view the results of the simulation in real time and, if so desired, change the delay in real time.

Customization of Blocks. Extend has an open architecture so users can build custom blocks or modify existing blocks. Users can also build libraries of custom blocks and libraries of hierarchical blocks. This is a useful concept when building models that require repetition of tasks.

On-line Documentation and Help. Extend dialog boxes all have a Help option that provides information about the block being used. In addition, users may provide documentation in hierarchical blocks and attach Quicktime movies to models.

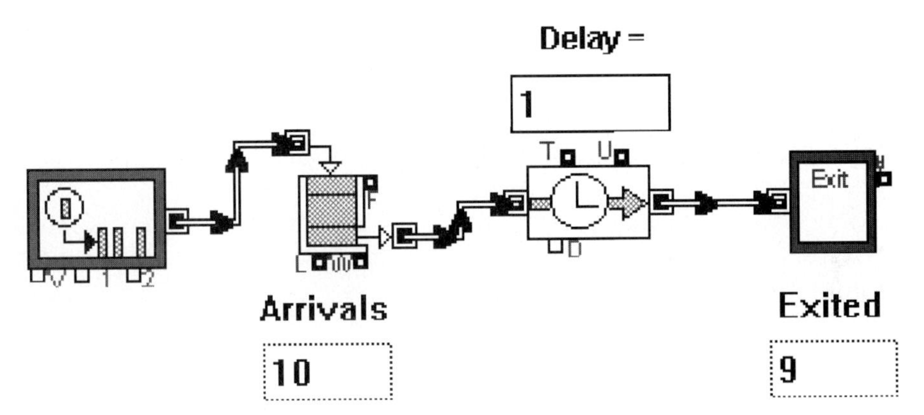

Figure 11.22 Interaction in Extend.

Extend also has a Help block that can be used in a model to provide information to a viewer or user of the model in real time. The icon looks like the one in Figure 11.23.

Figure 11.23 Help Block.

When a user selects this block, he or she is presented with information provided by the developer of the model, as shown in Figure 11.24.

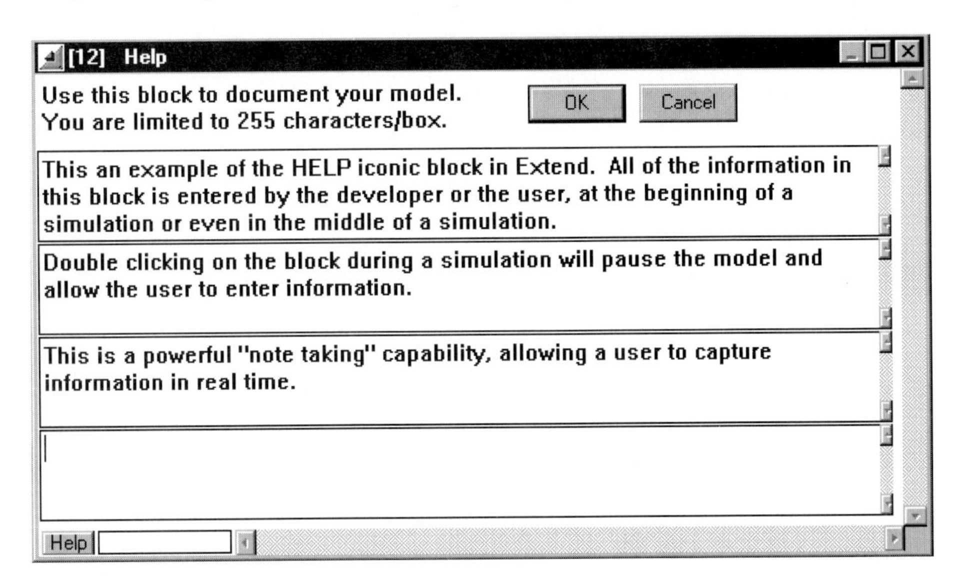

Figure 11.24 Dialog of Help Block.

Conclusions

Low-priced CAPRE technology is the wave of the future. Extend represents, in the opinion of the author, the best tool currently available for Business Process Reengineering. Perhaps the most important feature of Extend is that a user need not be an expert in simulation to develop a simulation of his or her business processes.

Extend is a complex tool that consists of a comprehensive set of iconic blocks. Iconic blocks are often grouped into libraries that are targeted to particular markets such as manufacturing, electronics, and others. Since Extend can be customized for any market, the Business Process Reengineering (BPR) Library was developed to facilitate the analysis and simulation of business processes. The BPR Library utilizes a set of iconic blocks that will be recognized by individuals familiar with process mapping and/or flowcharting. The BPR Library will be detailed in the next chapter, and examples of its use in process reengineering modeling efforts will be presented in later chapters.

12

The CAPRE Toolset Extend+BPR

Extend is a powerful simulation tool; however, it is, in its "native" form, somewhat difficult for individuals not familiar with simulation to understand. Therefore, Extend had to be adapted to a process reengineering audience, which consists mainly of management and professional personnel. That adaptation was the BPR Library.

Extend plus the BPR Library, hereafter referred to as Extend+BPR, is an extremely powerful and flexible tool for the analysis of business processes, yet one that provides a visual, intuitive insight to the dynamics of business processes.

Commercial Real Estate Example

The following example is a case in point. This book has emphasized the fact that business processes tend to evolve over time. They are typically neither planned nor designed, tending to grow as new functional requirements appear. When changes are made to those processes (i.e., they are reengineered), the results can be disastrous.

One such process was the process used by a major diversified corporation which, as part of its capital business, managed a commercial real estate leasing operation. Commercial real estate agents routinely dealt with contracts that fell into three categories:

1. Contracts with no changes to terms and conditions.
2. Contracts with minor changes to terms and conditions.
3. Contracts with major changes to terms and conditions.

Procedures were written that described who would handle each type of contract. The official procedure dictated that any change to terms and conditions had to be handled by the Legal Department of the business; however, as the business grew over time, an *unofficial* set of rules evolved. To expedite the processing of contracts, agents, under certain conditions, negotiated both minor and major changes to terms and conditions. The agents used a set of mental rules to determine the types of changes they would handle, with all others being sent to the Legal Department.

Unfortunately, the inevitable happened, and an agent made an error that could have proven costly to the business. Although the recipient of the lease agreed to accept additional changes to the terms and conditions after the lease had been signed, the company decided that the unofficial process could no longer be tolerated. The management of the commercial real estate business decided that all changes, no matter how minor, had to be referred to the Legal Department. The real estate agents and the Legal Department knew that this would be a disaster but had no way of communicating this to management, except by offering their opinions.

One agent asked that the change in policy be postponed until she had a chance to model the existing process and the proposed process. Using Extend, she developed an as-is model of the process, providing information about the time required to process changes under varying conditions. The agent used available statistics to determine how often customers changed terms and conditions, the categorization of the changes, how often agents handled the changes and how often Legal handled them, and how long it usually took to handle each type of change. Figure 12.1 is an overview model of this process.

The agent then modeled the proposed process, in which all changes to terms and conditions were referred to the Legal Department. Figure 12.2 shows a model of the proposed process.

This model revealed that, on average, the amount of time required to handle minor changes decreased, while the amount of time required to handle major changes remained approximately the same. The model also showed that the backlog of work in the Legal Department had grown tremendously. To reinforce the fact that the proposed changes would lead to delays in processing, the agent added some more information, using a *plotting function* from Extend to show the trends that the Legal Department would be facing. The first plot, Figure 12.3, shows the average time that would be required to process a contract from the time it was received to the time it made it out of the Legal Department, for both major (plot labeled "1") and minor (plot labeled "2") changes to terms and conditions.

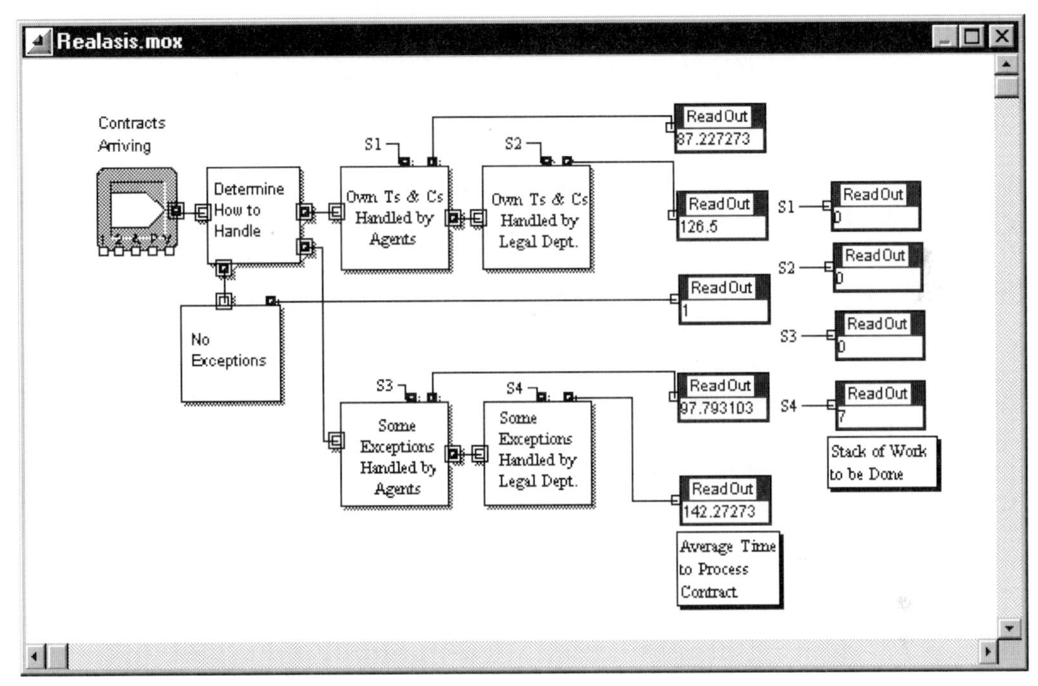

Figure 12.1 Overview of Real Estate Contract Processing.

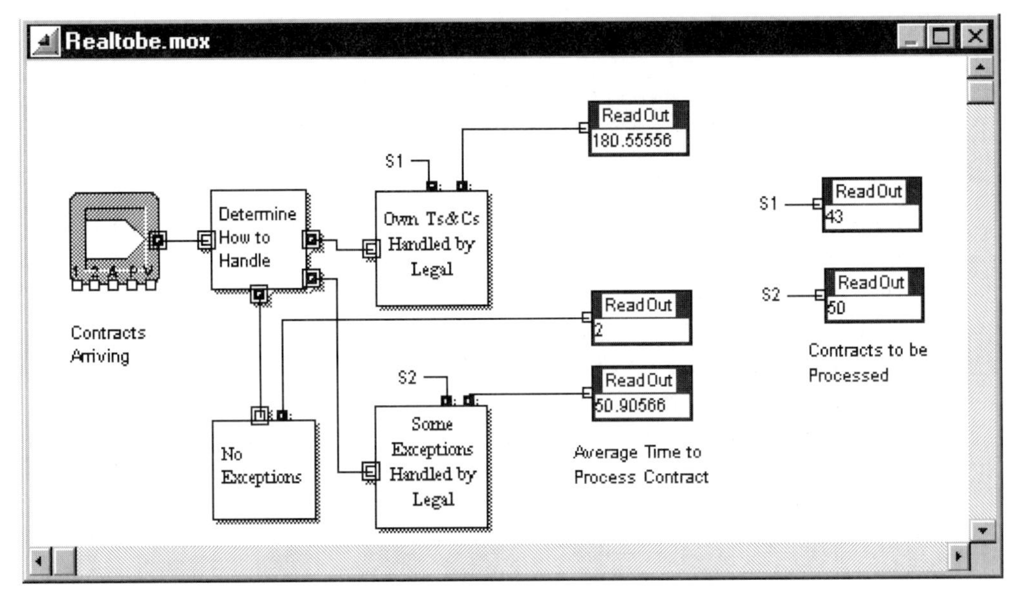

Figure 12.2 Proposed Process.

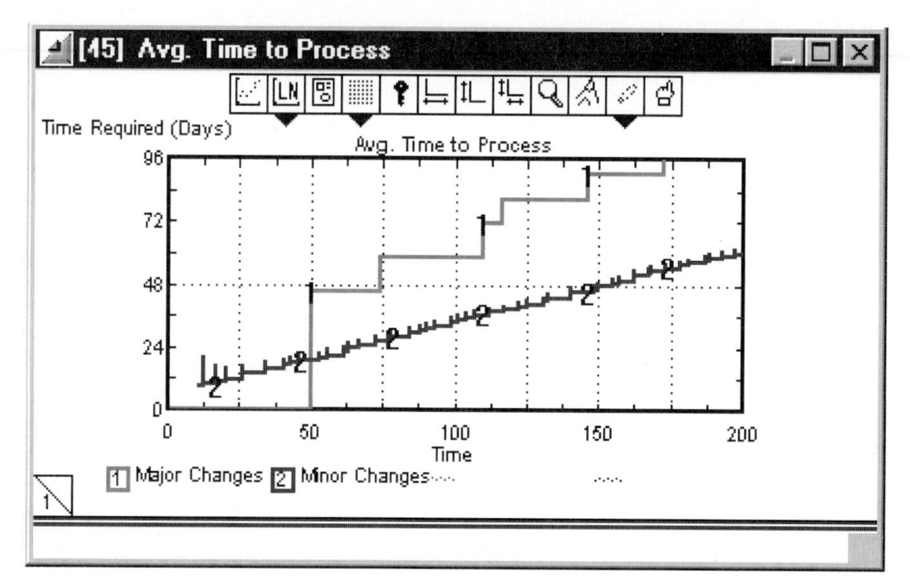

Figure 12.3 Plot of Average Time to Process Contracts in Proposed Process.

The second plot, Figure 12.4, showed the number of contracts that would be waiting to be processed in the Legal Department.

Each plot reveals an increasing slope, meaning that the number of contracts waiting for processing and the time required to process them would increase continuously.

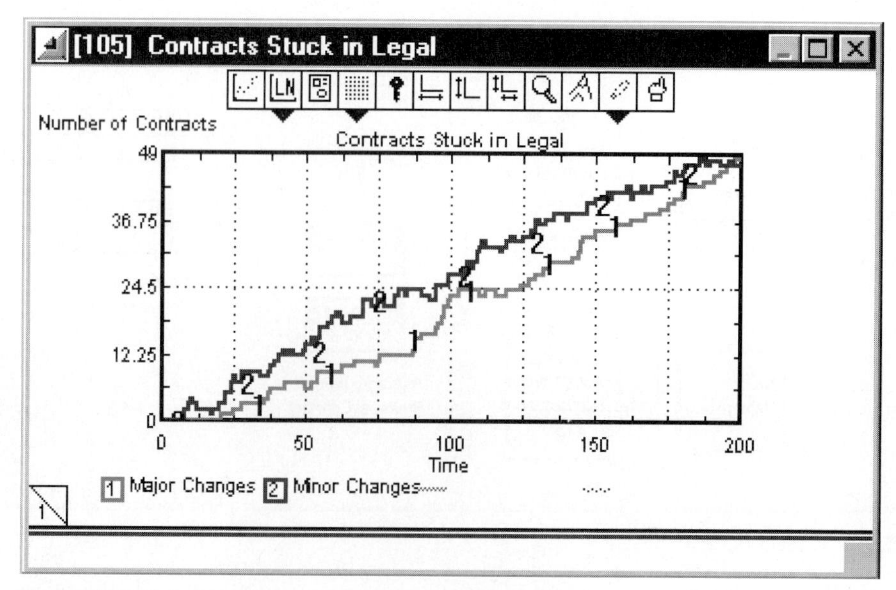

Figure 12.4 Plot of Number of Contracts to be Processed.

Particularly disturbing were the trends associated with contracts that had minor changes to the terms and conditions. Since they occurred more frequently, they would stack up faster, and the time required to process them would increase accordingly. It was very unlikely that potential customers who submitted what they thought were minor changes would wait more than 60 days to have a signed contract. The agent suggested that the company faced the possibility of losing business if the new procedures were adopted. To institute the new process, the agent claimed, the company would have to add staff to the Legal Department.

The models shown in Figure 12.1 and Figure 12.2 were developed using hierarchical blocks and appear rather simple. Therefore, before management accepted the information it had been presented, it wanted to know more about the modeling tool that had been used. After all, they reasoned, it was possible that the agent had "rigged" the model so it provided the information she wanted. The first detailed portion of the model the agent demonstrated was the hierarchical block called *Determine How to Handle.* This part of the model assigned contracts to the other parts of the model based on a percentage basis derived from statistical analysis. Figure 12.5 shows this model.

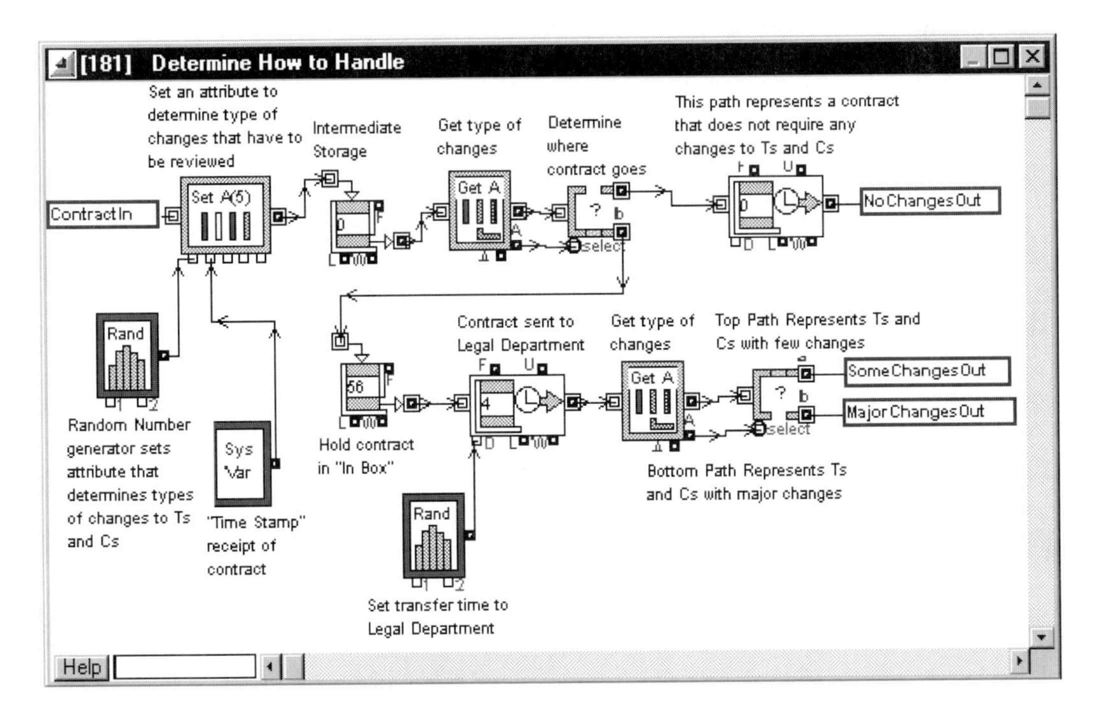

Figure 12.5 Model of Real Estate Contract Routing Logic.

When the agent showed this to management, she was greeted with blank stares. While this model is perfectly acceptable and meaningful to individuals familiar with simulation, it had no meaning to a business manager with no modeling or simulation experience. Since management could not understand what was happening, they were ready to reject both the model and the results of the model. The agent had to redevelop the model in a way that management could relate to, or face the possibility that it would be ignored.

BPR Library Description

Fortunately, Extend provides developers with a customization capability. The libraries are open, and both the source code and iconic representations are modifiable. To meet the needs of management and professional personnel, a library of *computer aided process reengineering* blocks was designed to match, as much as possible, the symbols used in flowcharts. Those iconic blocks and their functions are explained in this chapter and from here on will be referred to as the CAPRE Toolset Extend+BPR, Extend+BPR, or the BPR Library. In addition to making the iconic blocks resemble flowcharting symbols, the apparent complexity of Extend was reduced in the BPR Library by combining the functions of multiple blocks into single blocks.

For example, the Toolset contains an Operation block shown in Figure 12.6.

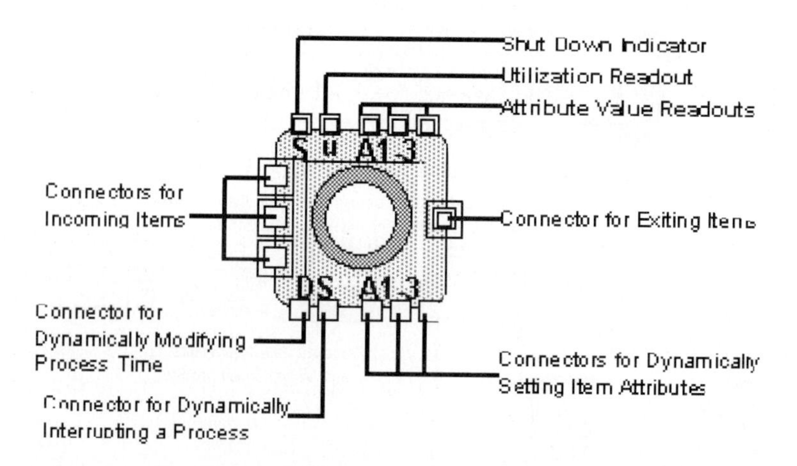

Figure 12.6 BPR Library Operation Iconic Block.

This block combines the function of the Extend Batch block, Activity block, Set Attribute block, and Get Attribute block into a single block. It is called a *simple* operation since the number of input items, once set, remain fixed. The parameters that define the behavior of the operation are set in the dialog box, shown in Figure 12.7.

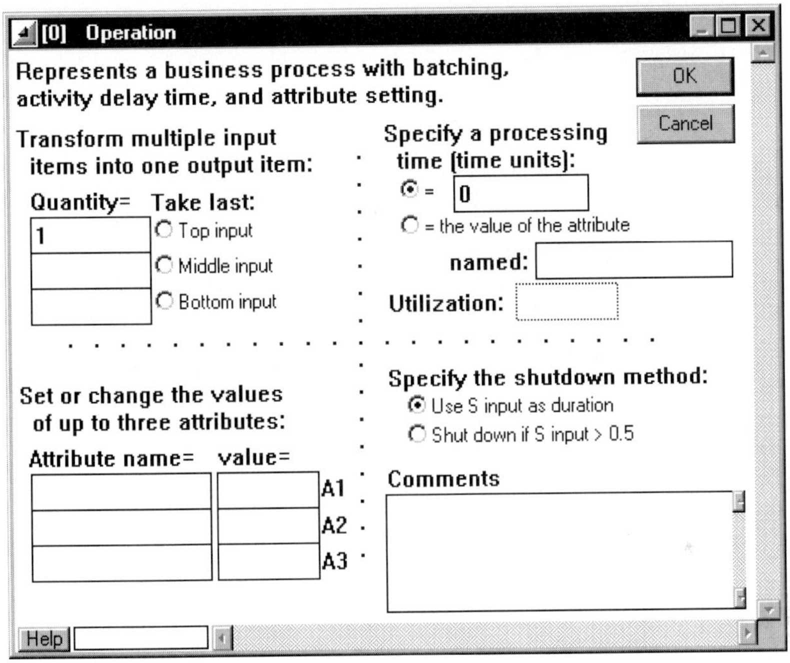

Figure 12.7 Operation Block Dialog Box.

The following information can be specified or found in the dialog box for the Operation block.

- The processing time, or the amount of time the Operation executes during a simulation. The processing time can be defined in the dialog box as a fixed number, it can be defined by an attribute associated with an item entering the block, or it can be dynamically set by entering a value through the D connector.
- The length of time for which an operation can be dynamically shut down. This can be defined as either an indefinite or specified amount of time, depending upon the dialog option selected and the value being input through the S connector.
- The value of attributes. These can be fixed, or dynamically modified by inputting values through the A1–A3 connectors on the bottom of the block.
- The input requirements for the Operation.
- The utilization of the block. If this block represented the activities of a single worker, then the utilization would represent the utilization of the worker.

Providing these capabilities in a single block,

- Reduces the complexity of models, making them easier to understand by business managers.
- Reduces the possibility of error by reducing the number of blocks required to represent some part of a model.
- Increases the appeal of modeling and simulation to those individuals familiar with flowcharts and process maps.

The BPR Library contains another type of operation block called the Operation,Variable block. It is similar to the Operation block except that the number of input items required can be dynamically modified. Figure 12.8 explains the Operation,Variable block.

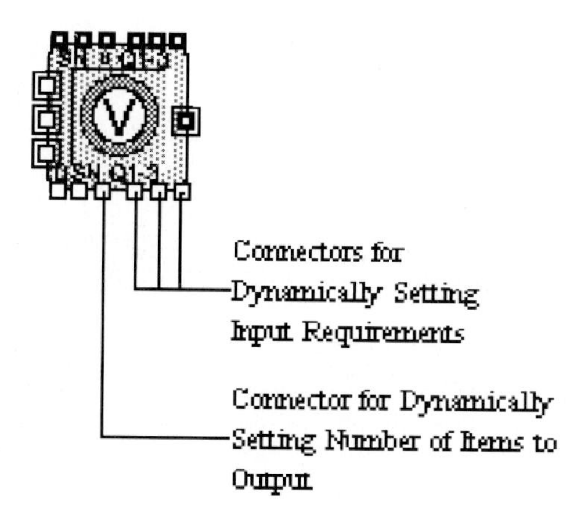

Figure 12.8 BPR Library Operation,Variable Block.

Figure 12.9 describes the dialog box for the Operation,Variable block.

In this block, the Q1–Q3 connectors are used to modify the inputs required. These values are carried forward as attributes and can be interrogated later in the model. This is useful for modeling process tasks in which the number of input items changes under certain conditions, such as stacking grocery shelves, moving household goods, and so on.

Two other blocks show how the BPR Library simplifies modeling and simulation for business purposes. These are the Stack and Repository blocks, shown in Figure 12.10.

These blocks are used to provide temporary storage for items flowing through the process. The Stack block looks like a stack of documents, which is what a stack in business processes would most likely be. This block can be used to define first-in-first-out, last-in-first-out, and priority queues, thereby combining the functions of three

[11] Operation, Variable

An operation where the number of output items, the batch quantities, and the processing times are variable.

OK

Cancel

Transform multiple input items into 1 output item(s).

Quantity input= Take last:

1 [Q1] Top input

[Q2] Middle input

[Q3] Bottom input

Specify a processing time (time units):

= 0

= the value of the attribute named:

Utilization:

Sets the value of the following attributes to the quantity above:

Attribute name= Value:

Q1

Q2

Q3

Specify the shutdown method:

Use S input as duration

Shut down if S input > 0.5

Comments

Help

Figure 12.9 Dialog Box for Operation,Variable Block.

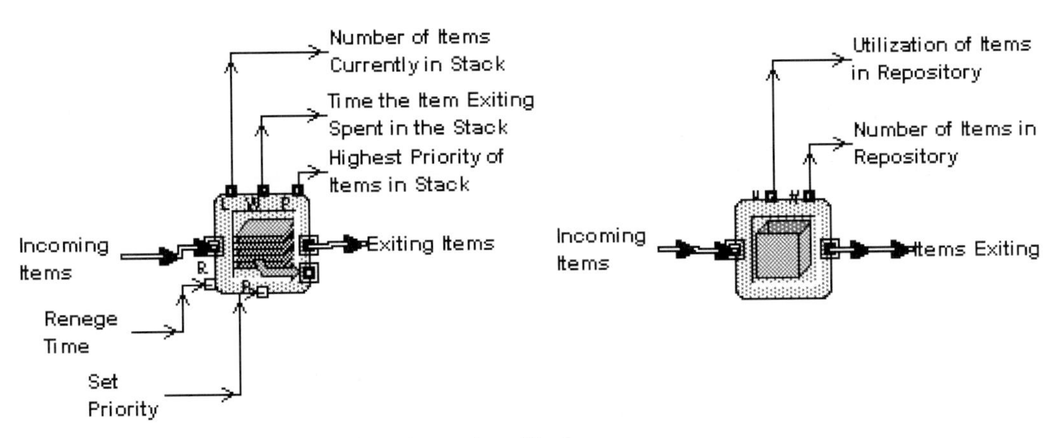

Figure 12.10 BPR Library Stack Block and Repository Block.

Extend blocks into one. The Repository block was designed to look like a box and replaces several Extend blocks, which serve as temporary storage areas.

Two other blocks, the Decision(2) and Decision(5) blocks, have combined the functions of several Extend blocks into one and, more importantly, allow the user to utilize decision symbols in the same way they are used in flowcharts. These blocks are shown in Figure 12.11.

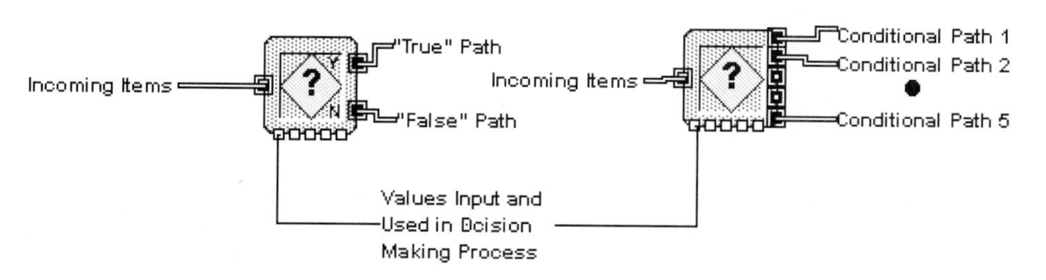

Figure 12.11 Decision(2) and Decision(5) Blocks.

In a flowchart, a modeler determines the status of some process or object when it enters the symbol, and this is how these blocks function. This is more intuitive to a user than the decision-making method described in the previous chapter. In addition, the logic used to determine a decision or path to follow is simple IF…THEN…ELSE logic, as shown in Figure 12.12.

[0] Decision (5)

Decides which path to take based on conditions.
The decision takes 0 time.

OK

Cancel

Name the value inputs to use in the equation

V1	V2	V3	V4	V5

Enter an equation in the form: if(condition) Path = PathName;

Functions by name

Functions by type

```
if(V1 < V2)
  Path = Path1;
else if (V1 == V2)
  Path = Path2;
else
  Path = Path4;
```

Name the paths

Path1
Path2
Path3
Path4
Path5

Comments:

Help

Figure 12.12 Decision(5) Dialog Box.

IF…THEN…ELSE statements are very familiar to software development personnel and represent the manner in which decisions are actually made in business situations. One will often discuss alternatives and say: "IF this condition exists, THEN we will take alternative A, ELSE (or otherwise) we will take alternative B." Once someone who is learning modeling and simulation understands this concept, the creation of Decision(2) and Decision(5) dialogs becomes quite easy.

Entry into and exit from models built with the BPR Library are defined using the Import and Export Blocks. Figure 12.13 depicts these blocks.

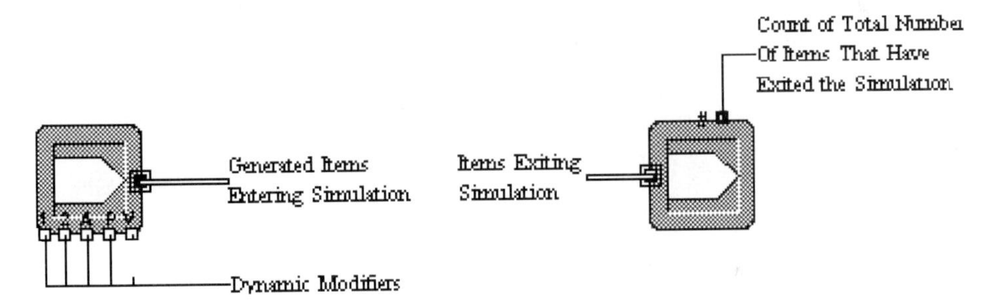

Figure 12.13 BPR Library Import and Export Blocks.

The icons used for the Import and Export blocks are standard flowcharting symbols. The Import block is not only used to specify the entry of an item into a model but also defines the rate of arrival of items from outside the model.

The dialog box of the Import block is shown in Figure 12.14.

[0] Import

Generates items according to a distribution.

OK

Cancel

- Constant
- Erlang
- Exponential
- Integer, uniform
- Normal
- Real, uniform
- Triangular: is most likely.
- Weibull

(1) Constant = 1

(2) (unused) 1

No item at time zero

Set attribute, priority, or number of items (optional):

Attr. name =

Attr. value =

Priority =

# of items(V) = 1

Use block seed offset 1

Comments

Help

Figure 12.14 Dialog Box of Import Block.

Items can be generated to flow through the model using a number of distribution curves, using a constant flow, or using a random flow. The parameters associated with the flow can be dynamically modified, as can the number of items that will be produced every time the Import block generates items. In addition, a default attribute and priority can be applied to each item generated; for example, a cost can be applied to each item of raw material in a manufacturing model, or a priority assigned to items of paperwork.

Three other blocks, not typically found in flowcharts, are included in the BPR Library and are shown in Figure 12.15. They are the Measurement block, Transaction block, and Operation, Reverse block.

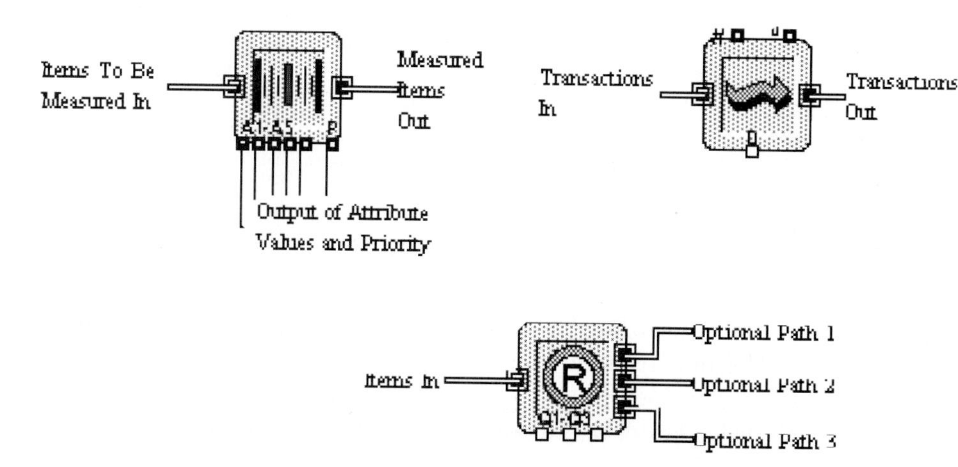

Figure 12.15 BPR Library Measurement, Transaction and Operation, Reverse Blocks.

The functions of these blocks are explained as follows.

- The Measurement block is used to obtain the value of attributes associated with items entering and exiting the block.
- The Transaction block is used to represent the flow of items from one process step to another, otherwise known as a transaction.
- The Operation, Reverse block is used to decompose an object into its original components, make copies of an item, or to initiate several parallel paths in a process.

One other important block in the BPR Library is the Labor Pool block. This is a special type of repository and is used in the same manner as a repository. The iconic representation is that of a person, so the visual representation of a model is more easily understood by the viewer. The Labor Pool block is shown in Figure 12.16.

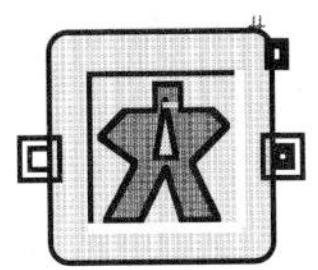

Figure 12.16 BPR Library Labor Pool Block.

The BPR Library blocks are intuitive in their graphical representation and flexible in their use. For example, the Operation blocks look like the letter "O"; a Stack looks like a stack of papers; a Repository looks like a box, and so on. The BPR Library, combined with Extend's Generic Library (a library of continuous simulation blocks), will make modeling and simulation available to business people at all levels, regardless of their experience with simulation. The only requirements for getting started with the BPR Library are an understanding of the Process Engineering Rules and a capability to define the behavior of processes.

Commercial Real Estate Example Continued

The agent who had developed the commercial real estate contract process simulation obtained the BPR Library and redeveloped the model. On the surface, the models looked the same. The details in the hierarchical blocks, however, were completely different. Figure 12.17 shows the hierarchical block Determine How to Handle described in detail earlier.

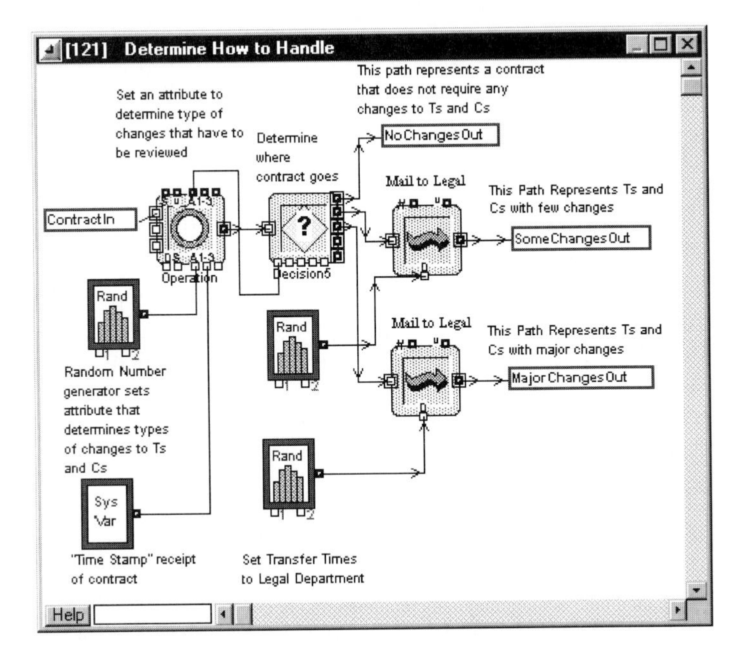

Figure 12.17 Determine How to Handle Modeled with the BPR Library.

While this model is not simple, it does utilize the flowcharting symbols of the BPR Library, and it is much easier to understand than the original model which used standard Extend blocks. The agent took the additional step of hiding "simulation stuff" from the viewing page, using named connectors to input data into the model. (Figure 12.18)

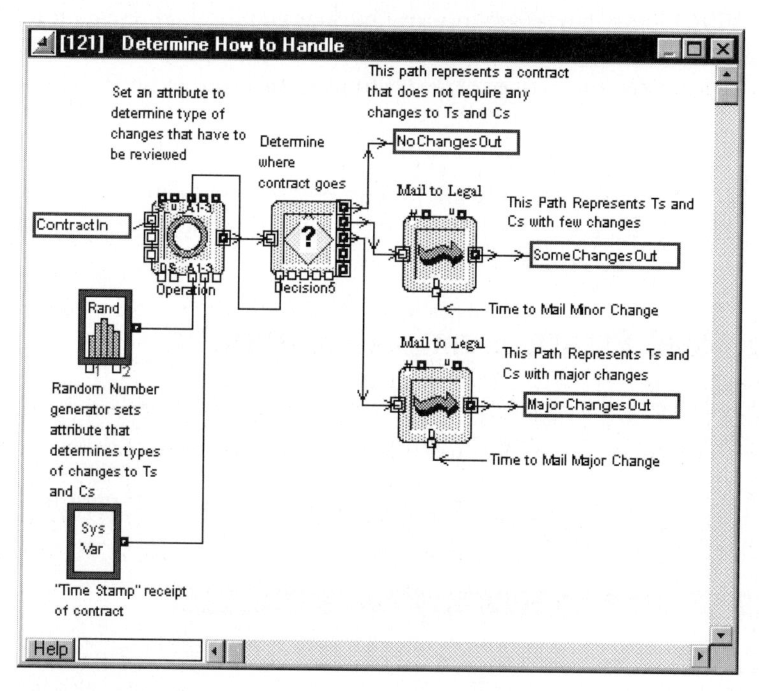

Figure 12.18 Real Estate Model Using Named Connectors.

This was a model that management could understand, if not in detail, at least conceptually. After the model of the contract process had been developed using the BPR Library, management felt more comfortable and accepted the results of the model. Instead of changing the process, they elected to document the unofficial process and allow agents to process changes under certain conditions. Management and the Legal Department also provided documented guidance to the agents to minimize the chances of errors occurring in the future.

This story is a success story that demonstrates several points.

1 Technological innovation, no matter how valid, will meet with resistance if it cannot be understood.

2 In business process reengineering, the KISS (Keep it Short and Sweet) approach is the best approach.

3 Computer aided process reengineering, in addition to being used as a predictive tool, can also be used as a means to increase communication between workers and management.

The BPR Library allows representation of complex models so they can be understood by nontechnological personnel. Although the models themselves may be complex, that complexity is hidden from the viewer. The BPR Library is a powerful mechanism for breaking down barriers between management and workers and for facilitating automated process analysis and improvement.

Object-Oriented Modeling Concepts

Object orientation is a term that emerged from the expert system and artificial intelligence boom of the mid-1980s. There is no single accepted definition of object orientation, but there are some accepted concepts that form the foundation of object-oriented modeling. The concept that is most important to modeling is that objects have *attributes* which provide information about that object.

An object-oriented model views operations, transactions, and the items that are the subjects of operations and transactions as objects. Decisions interrogate the attributes of an object, and events set the attributes of an object. Stores (repositories, stacks) are objects that are collections of other objects. The modeling philosophy adopted in Extend+BPR is that models are governed by objects and the rules of behavior of those objects. The following example may help clarify the concepts of object orientation and attributes.

Assume we have developed a model of a process in which a book is reviewed by an editor and, when approved, sent to a publisher. We have decided that the reviewing task will take one day if the book is 200 pages or less, two days if it is between 200 and 400 pages, and so on. We have just established a rule that sets the operation time of reviewing the book based on the length of the book. The length of the book is an *attribute of the book*, and the operation time is an *attribute of the operation.*

This can be modeled in the following manner by

- Using the Operation block to establish the length of the book.
- Using the Equation block from Extend's Generic Library to interrogate the length of the book.
- Using the length attribute of the book to set the length of time the review will require.

Figure 12.19 shows how the model would appear.

Figure 12.20 describes the dialog box of the Equation block for this model.

As in the Decision(2) and Decision(5) blocks, the logic is simple IF...THEN...ELSE logic, making the modeling of decisions straightforward.

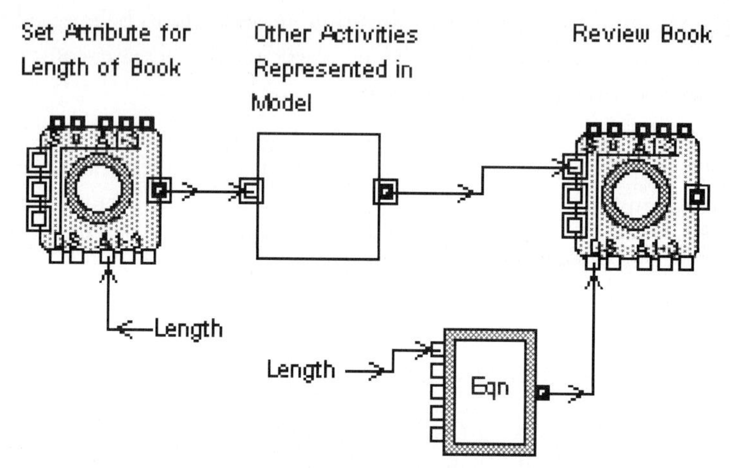

Figure 12.19 Processing Time Set by Relation to Book Length.

[12] Equation

Computes an equation.

Functions by name

Functions by type

OK

Cancel

Output

ReviewTime

Input1	Input2	Input3	Input4	Input5
Length	Var2	Var3	Var4	Var5

Enter an equation in the form: result = formula;

```
IF (Length < 200) ReviewTime = 1;
Else ReviewTime = 2;
```

Comments

Help

Figure 12.20 Dialog Box of Equation Block.

Now assume that if the book is in hard-copy form, it will be sent to the publisher by U.S. mail, which will take between two and five days. If it is in electronic form, it will take between two and five minutes. Here again, the attribute which specifies transaction time is related to the attribute that defines the form in which the book exists. In diagram form this process would look like Figure 12.21.

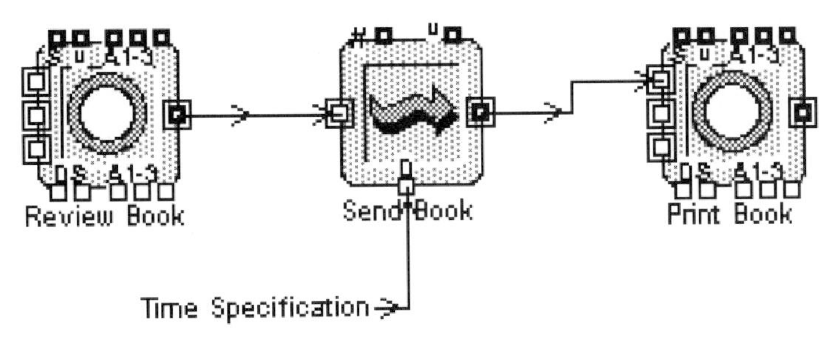

Figure 12.21 Model of Communication Between Reviewer and Printer.

To change the behavior of the process, we do not have to change the model. We only have to change the attributes of the objects that make up the model. For example, the transaction would be defined as having a process time of two days if the book is mailed, or a number of minutes if the book is transmitted electronically. As in the example depicted in Figure 12.19, this can be accomplished by setting an attribute for the book called Storage Type and using an Equation block to determine the length of the transfer based on the medium in which the book is stored.

What has just been described is an example of archetyping.

An *archetype* is a model upon which all other implementations of the model are based and can be used to investigate variations of similar processes. Archetyping is the development of reusable models that are changed only by changing simulation parameters.

Archetyping is extremely beneficial when modeling the effects of technology transfer, downsizing, and so on. For example, to simulate the effect of introducing technology that will be used in a task performed by a person, only the attribute defining the amount of time required to perform the task would be changed, not the model itself. Similarly, if a staff of three is downsized to two individuals, the availability of the remaining personnel can be modeled simply by changing the number of people in the process. Then the effects on schedules, cycle time, and so on can be predicted.

As another example of archetyping, suppose a model of a bicycle assembly process is being developed. One worker has the task of fitting brake mechanisms to a wheel and takes five minutes to attach the brakes for one type of bicycle and ten minutes for another type. In a traditional modeling tool, one would have to ask which type of bike it is and take one of two paths through the model. In an object-oriented tool, we can set an attribute called Brake Attachment Time and set that attribute to different values based on the type of bicycle. A model of the brake attachment operation would appear as shown in Figure 12.22.

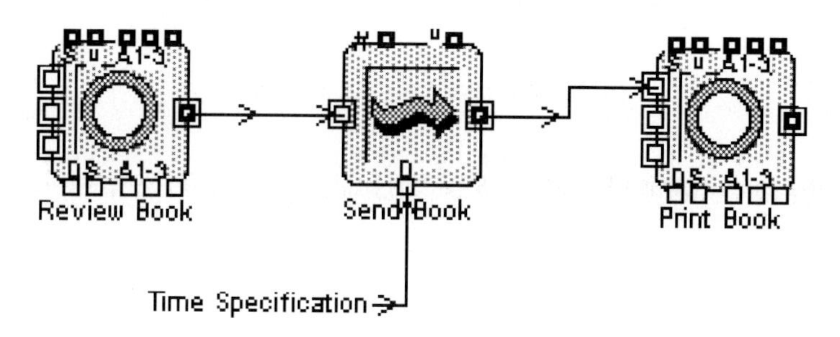

Figure 12.22 Setting Operation Time with Attributes.

Using a Measurement block from the BPR Library, we can get the value of an attribute, in this case the attribute called Brake Attachment Time. We can then use that attribute to define the amount of time required to complete the operation called Attach Brakes. Once again, the model itself has not changed visually, but the simulation will change based on the attributes associated with the objects in the model.

In reengineering activities, attributes are a useful mechanism for tracking costs. Since the cost of an operation can be defined as an attribute of the operation, each time the operation executes, the cost can be accumulated and ultimately summed with the costs of other operations to provide overall cost.

Understanding object-oriented concepts is essential for developing powerful process models. Attributes represent process parameters, and the values given to those attributes provide the process measures we are interested in. The use of object-oriented techniques allows a model developer to create models more quickly and accurately, since the task of calculating the values of process parameters explicitly is removed and performed by the modeling tool.

13

Developing Simulations: Step-By-Step Examples

Building a Simulation with Extend+BPR

This chapter is a detailed example of building a model using Extend+BPR and other Extend blocks. The process being modeled is a credit application review process, and scenario models will be used to explore the effects of downsizing on that process.

> The first consideration when building a model is the goal to be accomplished.

In this case, management has decided to reduce staff to save costs and has made this decision based on limited information. Management has *assumed* that a staff reduction will not have a negative effect on the process, so the *goal* of the modeler is to test that assumption. In other words, this modeling exercise will be used to determine if productivity (the number of credit applications processed each day) will be affected by a reduction in staff.

In this discussion, the iconic blocks that represent Extend+BPR and the dialog boxes associated with each iconic block will be shown and explained in detail. A dialog box appears when a user double-clicks on an iconic block, and it is used by the user to provide information that determines the behavior of the block in the simulation.

Description of the Process

The process is straightforward: Credit applications arrive in an office and are processed by two reviewers. A third person handles those applications that either lack sufficient information or require more investigation. Ultimately, applications are either accepted or rejected. The model in this example will represent a typical eight-hour workday.

The first task in developing the model is to determine what items will flow through it. In this case, the items are credit applications. The next task is to define the *process parameter* that determines how often the credit applications will arrive. Extend+BPR has a block, called the Import block, which generates items to be used in the model. The Import Block looks like Figure 13.1.

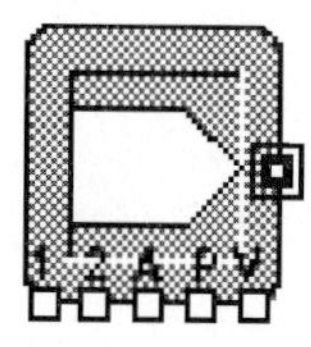

Figure 13.1 Extend+BPR Import Block.

The purpose of this block is to show that items are coming into the process from another process that is outside our scope of interest. In this example, credit applications can arrive by mail, fax, courier, and so on—the method in which an application arrives is not important for this model. What matters is that they arrive at some rate. That rate is defined through the dialog box of the Import block, which looks like Figure 13.2.

[1] Import
Generates items according to a distribution.
Constant
Erlang
Exponential
Integer, uniform
Normal
Real, uniform
Triangular:
Weibull
(1) Constant =
60
(2) (unused)
1
OK
Cancel
is most likely.
No item at time zero
Set attribute, priority, or number of items (optional):
Attr. name =
Attr. value =
Priority =
# of items(V) = 1
Use block seed offset
0
Comments
Help

Figure 13.2 Import Block Dialog Box.

The dialog box shows that there are many options available for generating the rate of arrival of items into the model. For this model, the rate of arrival of items into the model is set to a constant of 60, representing 60 minutes. In other words, the model has a built-in time reference of minutes. The Import block rate of arrival could have been set to a constant of 1, representing one hour, but that would have necessitated expressing time in fractional units later in the model, and it is easier to work in whole units of time.

Now that the arrival rate of credit reports has been specified, the number arriving each hour must be determined. The simplest approach is to assume that credit applications arrive at a steady rate, say 4 per hour. To simulate this rate, the Value of Item box in the dialog is set to 4; however, in reality, flows of information are rarely uniform, so the rate of arrival will be variable.

When simulating a variable rate, it is normal to make use of a *random number generator*. A random number generator is a computer function that creates a value that falls between preestablished limits whenever invoked. For example, if one were simulating the roll of a pair of dice, a random number generator that generated a value between 1 and 12 would be used. In this example, between two and four credit applications arrive every 60 minutes. To program this into the model, the Extend Input Random Number block will be used. This block looks like Figure 13.3.

Figure 13.3 Extend Input Random Number Block.

The Input Random Number block has a dialog box that looks like Figure 13.4.

An integer value between two and four will be output every time the block generates a number. The output of the random number generator is then connected to the V connector on the Import block. The V connector defines the number of items that will be input into the simulation by the Import block each time it executes, in this case, every 60 time units. Figure 13.5 shows how the number of items entered into the model can be dynamically altered.

Connecting the random number generator has the effect of overriding any value that may have been set in the dialog box for the number of items generated in each simulation time unit. The first step of building the model, the definition of the interarrival rate of the items to be processed, has just been completed.

The next modeling task is to specify how these items will be processed. First, they will be put into an in-box to await processing. This is specified using the Repository block, as shown in Figure 13.6.

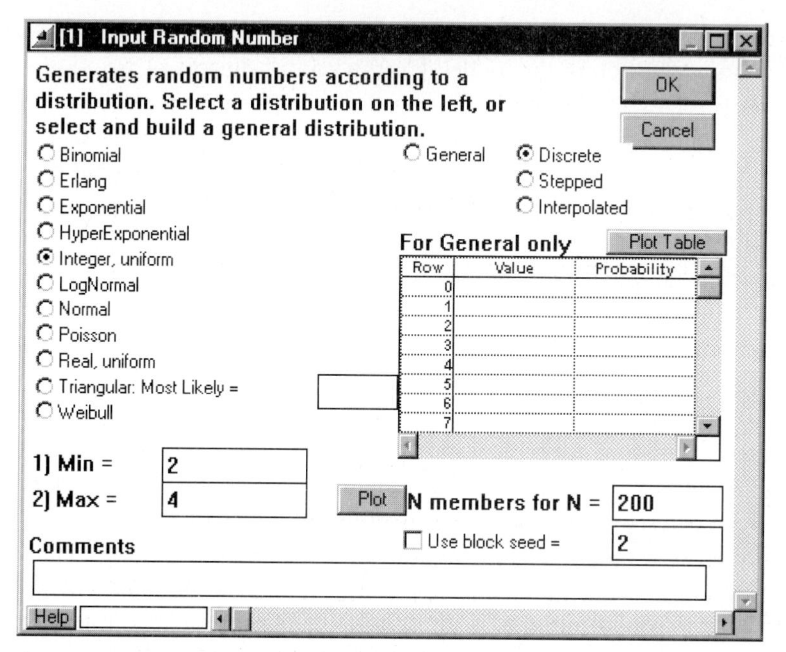

Figure 13.4 Input Random Number Dialog Box.

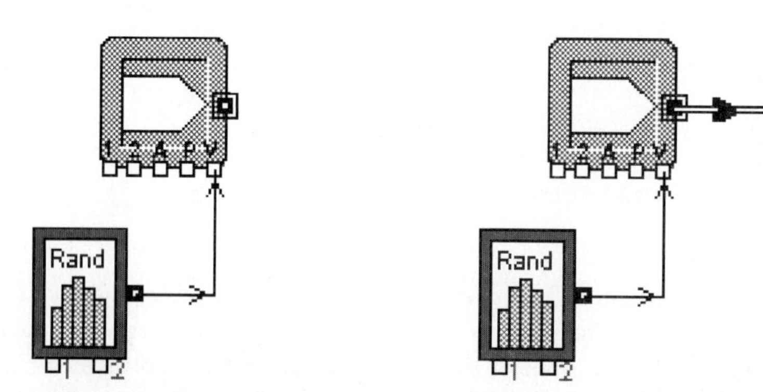

Figure 13.5 Modifying the Output of the Import Block Using Random Numbers.

Figure 13.6 Items Being Stored in an In-Box.

The dialog box of the Repository block is now set to an initial value, so the simulated workday starts with some credit applications waiting to be processed. Figure 13.7 shows the dialog box for the Repository block.

Note that there is a great deal of information that can be supplied via the dialog box. This other information is used when developing conceptually advanced models. The initial value of the Repository is a process parameter; that is, it describes how many items are in the "pipeline" when the model begins.

[4] Repository

Provides and stores items.

Initial number = 4

OK

Cancel

Set attribute, priority (optional):

Attr. name =

Priority =

Attr. value =

Strip attributes from incoming items

Items currently available: 0

Utilization rate:

Comments

Help

Figure 13.7 Repository Block Dialog Box.

The first activity modeled is an initial review of an application. The applications are reviewed one at a time and, typically, a review takes between 14 and 16 minutes, for an average of 15 minutes. This information is provided to the model using the Operation block (Figure 13.8).

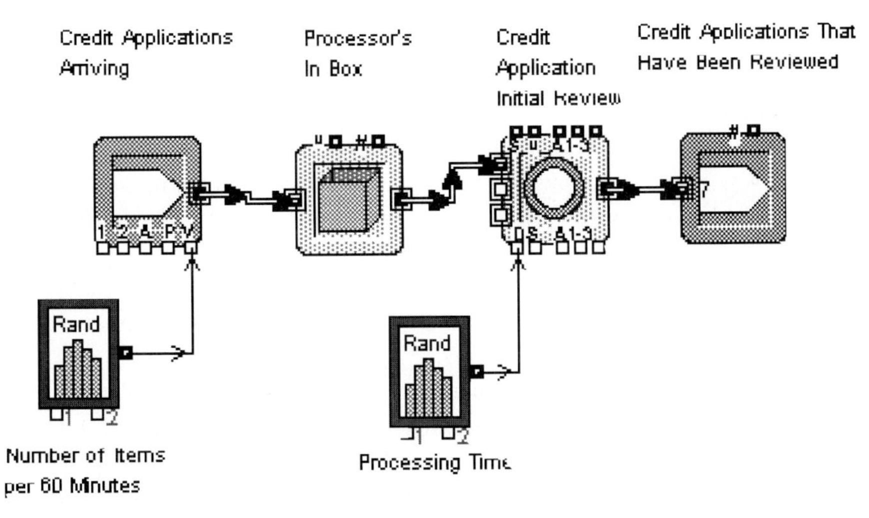

Figure 13.8 Processing Time Set by Input Random Number.

Again, the Input Random Number block from Extend is used to dynamically set the processing time for the activity block. This can be understood more easily by looking at the dialog box for the Operation block (Figure 13.9).

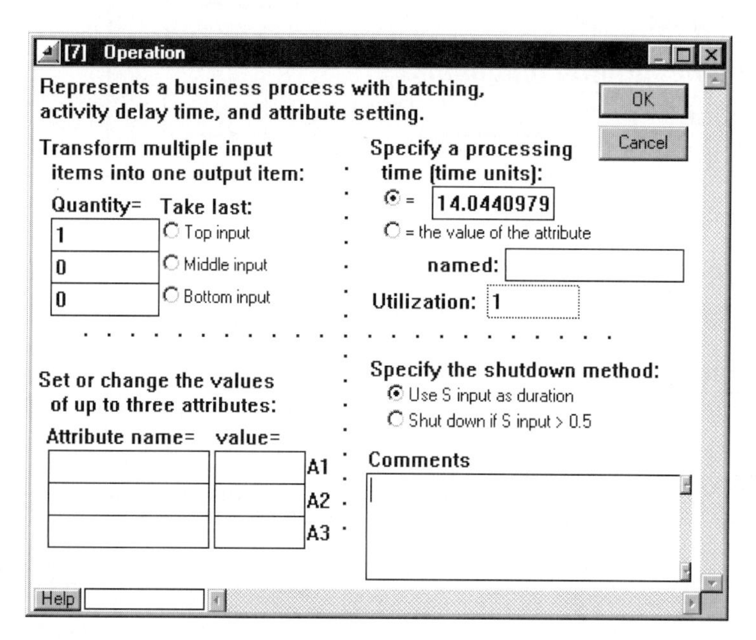

Figure 13.9 Operation Block Dialog Box.

As with the Import block, there is a lot of detail that can be supplied. In this example, the value of I2 is set to 1, instructing the simulation to process one credit application at a time.

The default value for the processing time of this block is set to 1; however, this number will be overridden by the value being input to the D connector. Therefore, this operation will execute for a period of time between 14 and 16 simulated minutes.

Now it is a good idea to check that what has been set up will actually execute, since doing so may save work later on. The model is set up to run for 120 time units, or a simulated two hours. An Export block is also added to the model, providing a termination point for the model. The Export block signifies that the items that have been processed are being forwarded to another process that is outside the scope of the process being studied.

Figure 13.10 shows the results of running the model for 120 simulated minutes.

These are the results that should be expected. Given the timing of the review operation, we can expect that about four credit applications per hour will be processed. After two hours, the Export block shows that eight applications have been processed. Also, the In Box has three credit applications in it, and the "bubble" at the

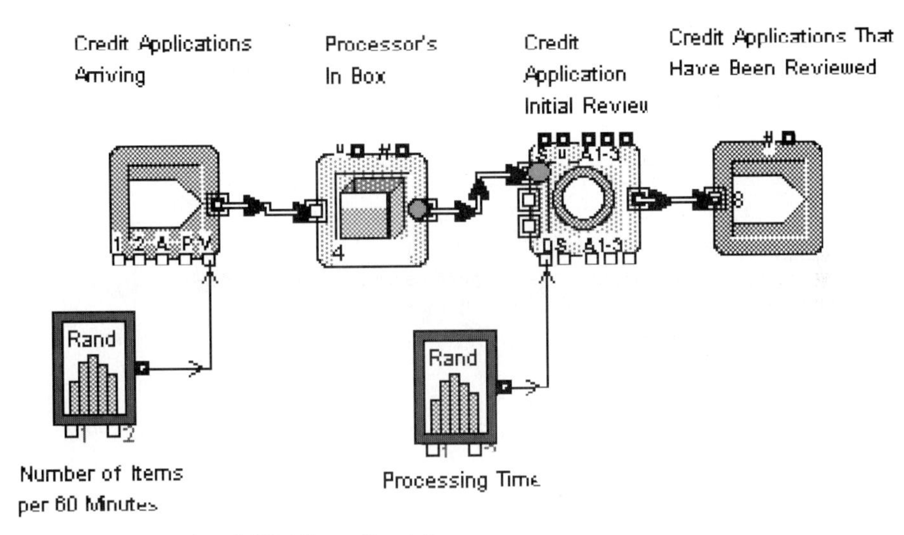

Figure 13.10 Results of 120-Minute Simulation.

Operation connector I2 shows that one more application is being processed. It appears that the model is functioning as planned.

What happens after a credit application is reviewed? There are two possibilities: It is approved, or it is sent to a third person for additional review. The test case is arranged so that 70 percent of the credit applications pass review and the remaining 30 percent fail review. The pass/fail test is made using the Decision(2) block from Extend+BPR. Figure 13.11 describes the Decision(2) block.

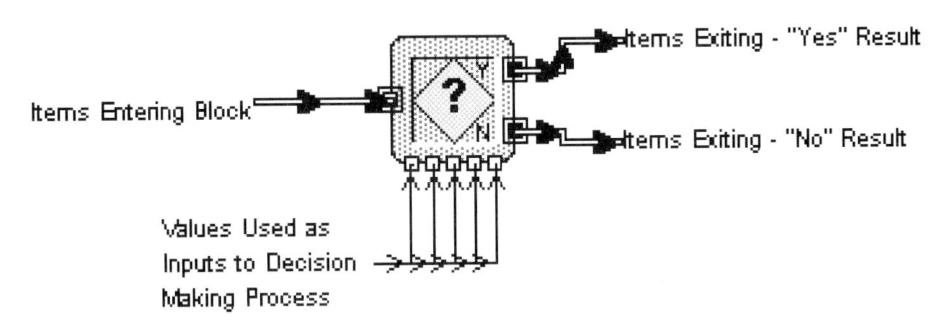

Figure 13.11 Decision(2) Block.

There are two mechanisms available to check for a pass or fail condition in this example:

1 A Pass/Fail attribute with a value between 1 and 11 can be associated with the credit application itself. This attribute can later be examined in Extend+BPR Decision(2) block.

2. A random number generator can be used to input a value between 1 and 11 into the Decision(2) block whenever an application flows through it.

The first approach is shown in Figure 13.12.

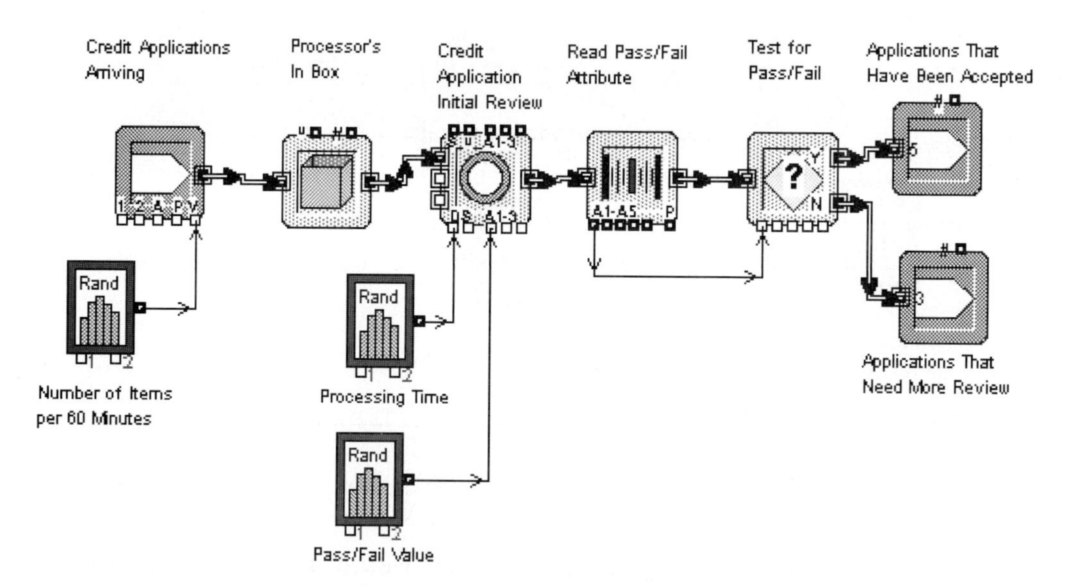

Figure 13.12 Model with PassFail Attribute.

A random number between 1 and 11 is input through the A1 connector, which can be used to dynamically set attributes of items passing through. In this case, the value of an attribute called Pass/Fail is set to the value of connector A1. When the operation is completed, the Pass/Fail attribute is read in a Measurement block, and the value of that attribute is passed to a Decision(2) block. This is a very powerful method of analyzing multiple attributes at the same time and is particularly useful when a decision is being based upon multiple conditions in a simulation.

Figure 13.13 describes the Decision(2) block dialog.

This example does not use the attribute method to determine passing or failing; however, if acceptance or rejection of a credit application were based on some formula that required information about several attributes of the application, such as amount, credit worthiness of the applicant, interest rates, and so on, the Pass/Fail attribute could be calculated over time. Then the use of the attribute in the Operation block would be very handy.

Instead, a random number between 1 and 11 is input directly into the Decision(2) block, and the pass/fail decision is based on the value of that number. The model would then appear as shown in Figure 13.14.

[25] Decision (2)

Decides which path to take based on conditions.
The decision takes 0 time.

OK

Cancel

Name the value inputs to use in the equation

PassFail	V2	V3	V4	V5

Enter an equation in the form: if (condition) Path = YesPath;
else Path = NoPath;

```
if (PassFail < 8)
  Path = YesPath;
else
  Path = NoPath;
```

Functions by name

Functions by type

Comments:

Help

Figure 13.13 Decision(2) Dialog Box Referencing Pass/Fail Attribute Value.

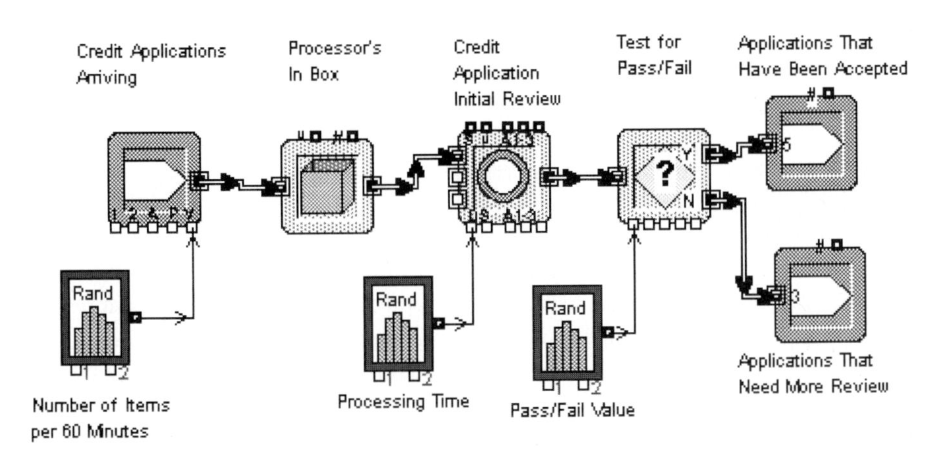

Figure 13.14 Decision(2) Block Using Random Number for Probability Check.

The equation in the Decision(2) block would be unchanged, regardless of the method used.

Note that a total of eight credit applications have been processed in this model; six have passed the initial review, and two have been designated as needing more review. In addition, one application is in progress, and three applications are in the Processor's In Box.

Credit applications that require more investigation are handled by another individual, who usually takes between 25 and 35 minutes to process them. Another intermediate Repository and an Operation block are used to represent this, again with the processing time of the operation established by a random number (Figure 13.15).

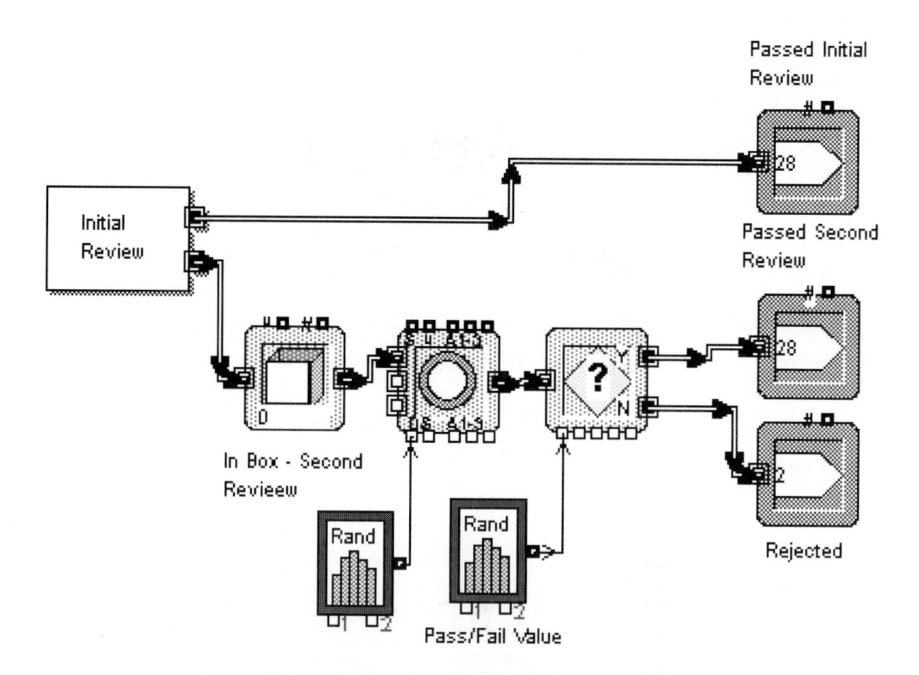

Figure 13.15 Model with Additional Review Operation.

The model is now becoming complex. The appearance of the model can be simplified by using the hierarchical feature of Extend to group process functions. By enclosing most of the details of the model in hierarchical blocks, the model appears as shown in Figure 13.16, which also shows the results of executing the simulation for a period of 480 minutes.

The next step in building the model is to add the second application processor. This is done simply by selecting the blocks of the model already built and duplicating them (Figure 13.17).

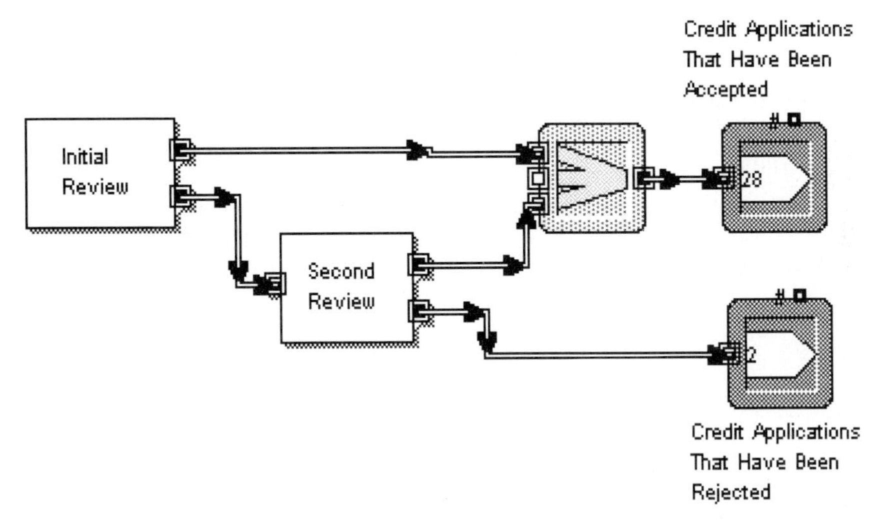

Figure 13.16 Model in Hierarchical Blocks.

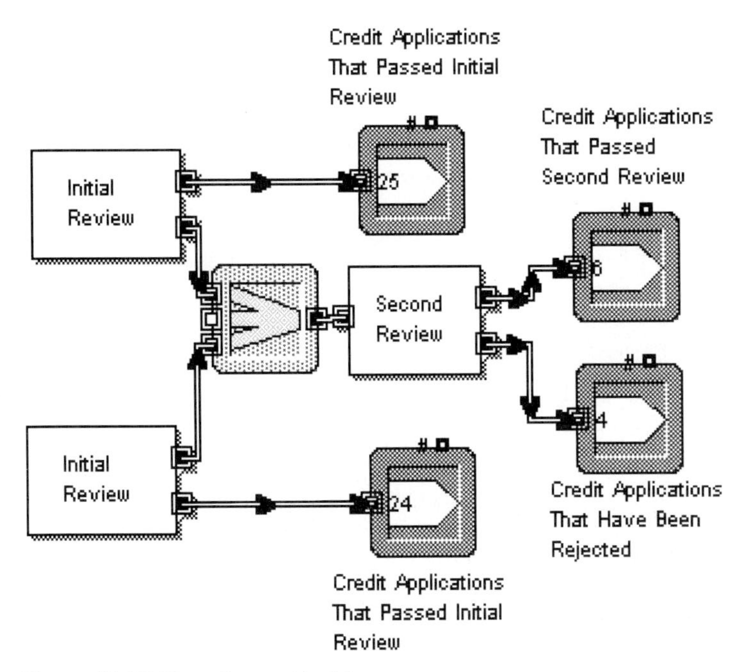

Figure 13.17 Three-Person Model.

Note that a few adjustments have been made. The output of the two Pass/Fail Decision(2) blocks has been combined to feed into the Need More Review process. This was necessary since the model assumes that only one person will perform the additional reviews. The numbers in the Export blocks represent the total number of credit applications that have passed through the various paths in the model.

Management would see these same numbers and probably reason that

1 Each processor handles about three applications per hour. Since an initial review takes an average of 15 minutes, a processor has about 15 minutes of idle time per hour. That comes to two hours a day.
2 There are only 10 or 11 additional reviews per day which, at one-half hour each, take about five hours to perform.
3 With a little extra effort, the two processors could also handle the additional reviews with no loss of productivity.

Therefore, management decides to downsize by laying off one person. When that happens, each processor's job will be modeled as shown in Figure 13.18.

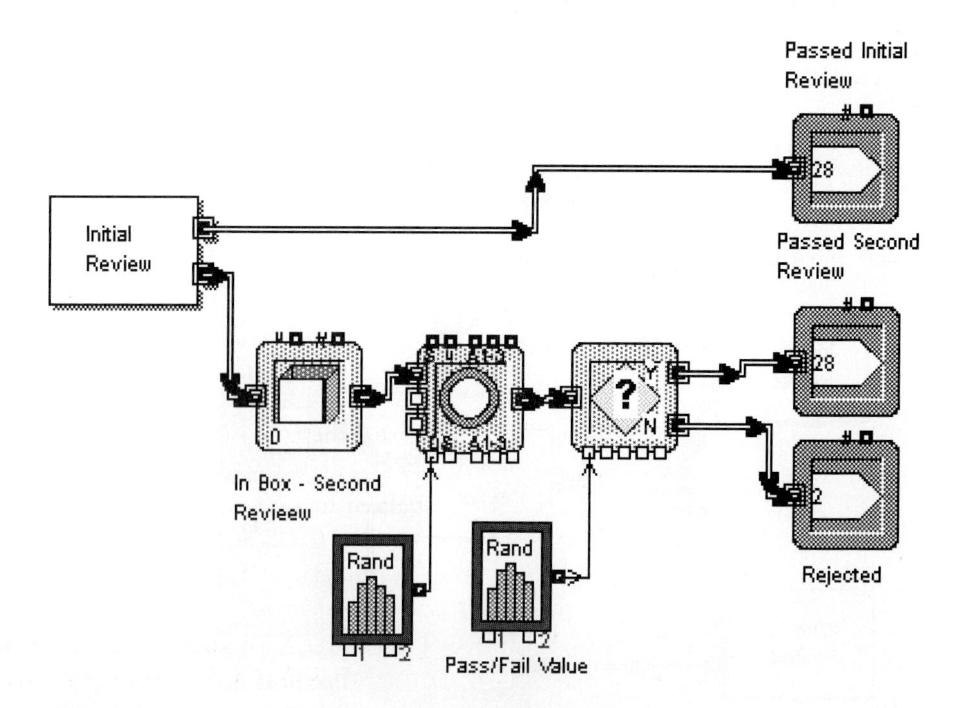

Figure 13.18 Each Processor Handles Applications to End.

This may appear to be an accurate model, but it is not. The problem is that a processor cannot be doing *both* the initial review and the additional review at the same time. In the model, the availability of the processor at the initial review must be accounted for, and that person must be made to be unavailable until either the initial review or the additional review is completed.

This is done in the following way: The processor is treated as an object in the simulation, and the operation is set up to require that a credit application *and* the processor must be available to perform the initial review. When either the initial review or additional review are completed, the processor is released and made available again for an initial review. This is accomplished by an Operation, Reverse block, which is used to break an item into its original components. In this case, the original components are the application and the processor (Figure 13.19).

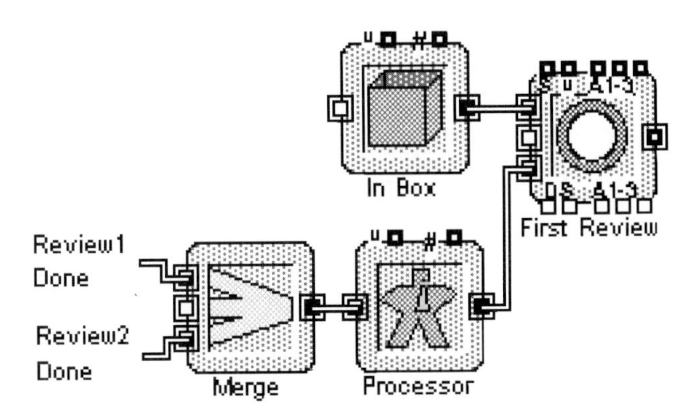

Figure 13.19 Two Inputs to Initial Review.

The model in Figure 13.19 now shows that there are two inputs required for the initial review operation, an item in the In Box and the Processor. A Labor Pool block is used to represent the processor, and initialized to a value of 1, showing that there is only one processor in the model. Note also the use of *named connectors*. This is a feature of Extend which allows a modeler to connect pieces of models that are physically separated.

As shown in Figure 13.20, the model has undergone some further refinement.

Two Operation, Reverse blocks have been used to release processors back to their respective initial reviews. The named connectors are the same as those entering into the Combine block in the previous picture. In addition, a Transaction block has replaced the Operation block for the second review. A Transaction block simulates multiple activities of the same type happening in parallel and, in the final model, we may have two processors performing a second review at the same time.

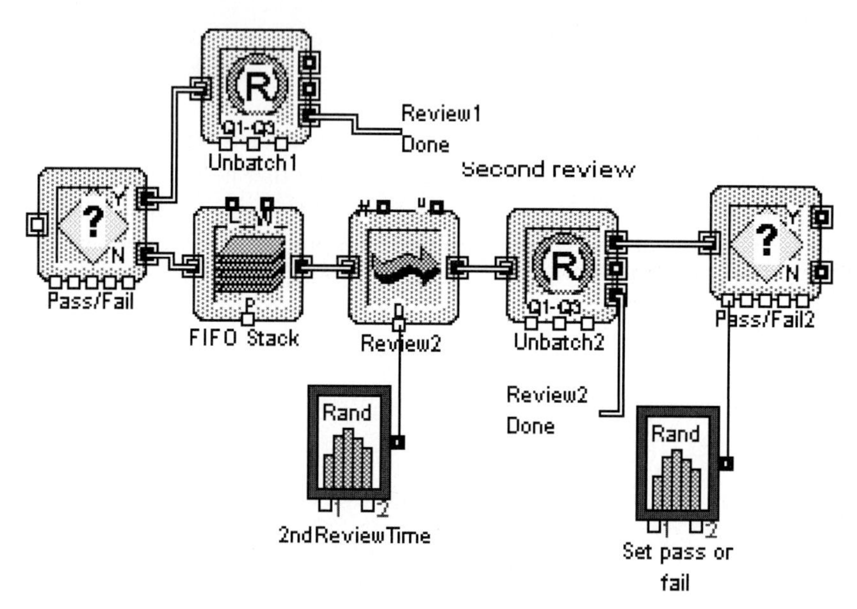

Figure 13.20 Use of Operation, Reverse Blocks to Free Processor.

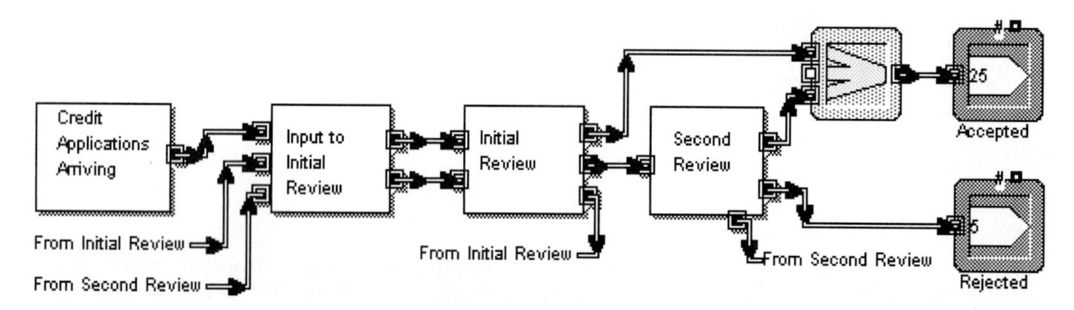

Figure 13.21 Results of Model After Elimination of Additional Reviewer.

Now the model will be cleaned up (Figure 13.21) by using hierarchical blocks and executed to see the effect of the elimination of one person.

Each of the hierarchical blocks shown in Figure 13.21 are shown in detail in Figures 13.22, 13.23, 13.24, and 13.25.

Note that each hierarchical block has named connections, such as "Labor Out." These refer to the connectors on the block and represent objects either entering or exiting the block. The ability to provide custom names to the connections enhances the communication of the model.

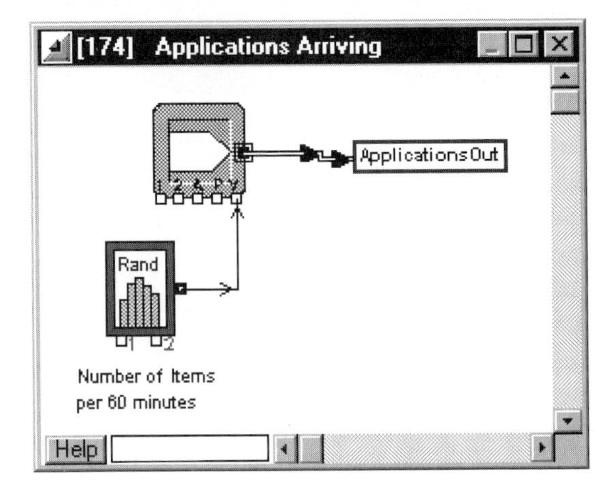

Figure 13.22 Hierarchical Block (Applications Arriving).

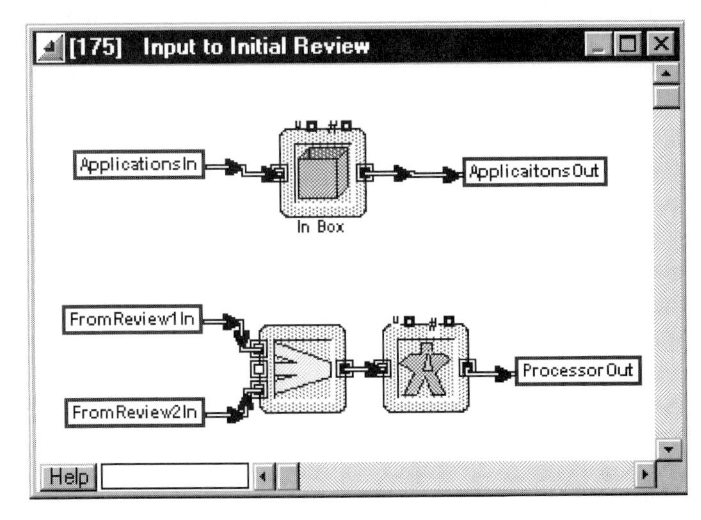

Figure 13.23 Hierarchical Block (Input to Initial Review).

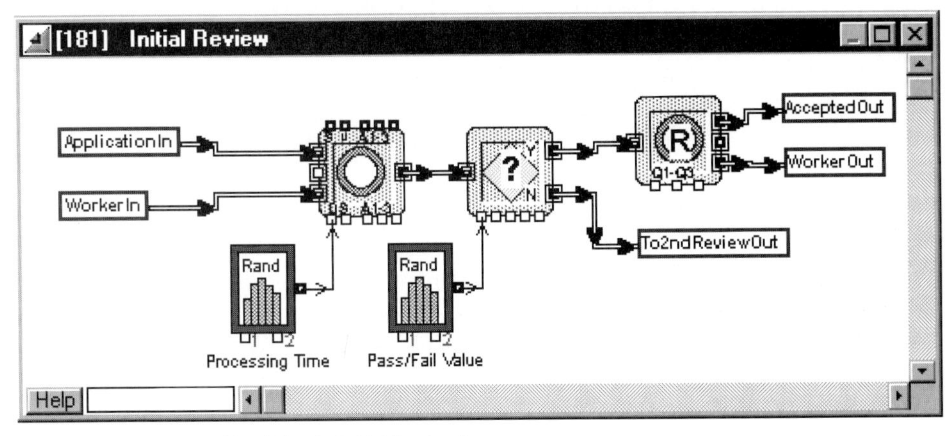

Figure 13.24 Hierarchical Block (Initial Review).

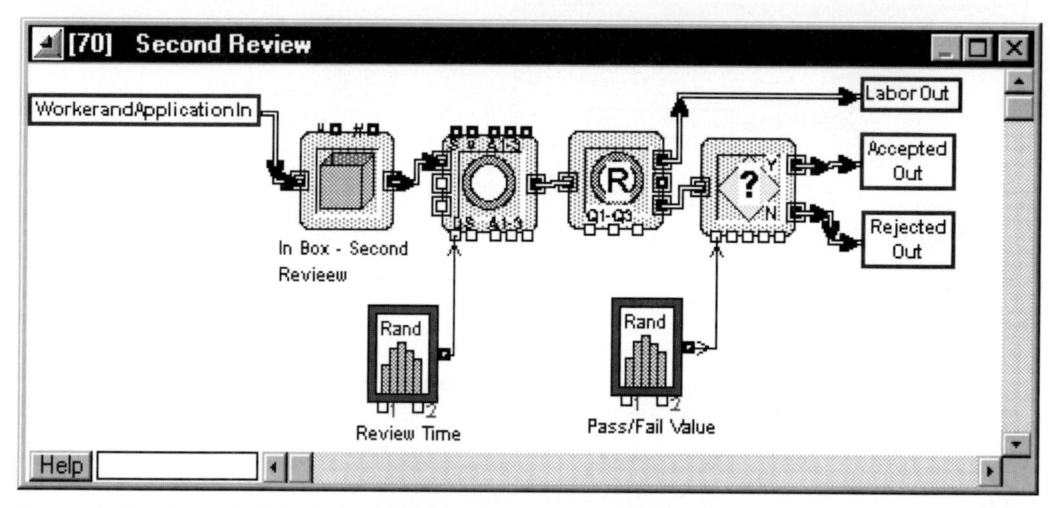

Figure 13.25 Hierarchical Block (Second Review).

The model in Figure 13.21 demonstrates that there is quite a drop in productivity when one processor performs both reviews. Instead of each application processor handling 24 applications per day *plus* an additional 5 or 6 that require further review (for a total of 53–54 applications processed), they now handle only about 15 each (about 30 total processed). This represents about a 40 percent *decrease* in productivity!

To predict the effect of downsizing on the process, the two models (original and downsized process) were run simultaneously. A comparison of the two processes in terms of productivity loss is shown in graphical form in Figure 13.26.

This graph reveals that, on average, the improved (downsized) process is about 55 percent as effective as the original process. In other words, if the combined number of applications handled per day by the three personnel in the original process were 55, the combined number of applications handled per day by the two personnel in the "improved" process would be 30. There is also a corresponding increase in cycle time, graphically shown in Figure 13.27.

This graph shows that the average cycle time for applications in the original process, including the time for applications that were rejected, was about 30 minutes. The cycle time in the new process, after 20 days, was almost 900 minutes, or two days!!

The assumption that there was sufficient idle time available to perform two tasks was wrong. This assumption was based on looking at the results of tasks that occurred in *parallel*. After the change, these tasks now occurred in a *serial* manner. The power of simulation allowed the modeler to account for that and to simulate the

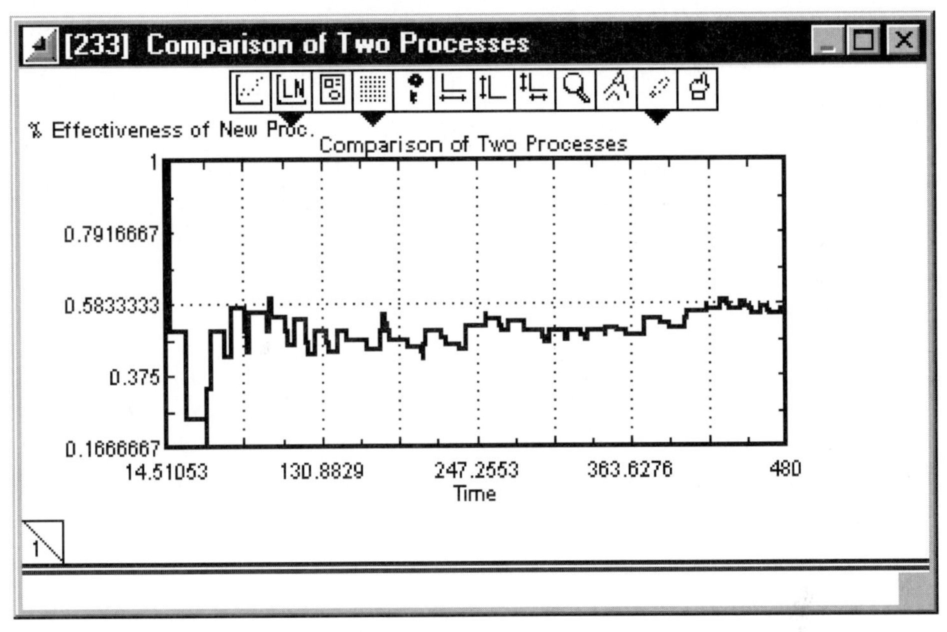

Figure 13.26 Effectiveness Comparison.

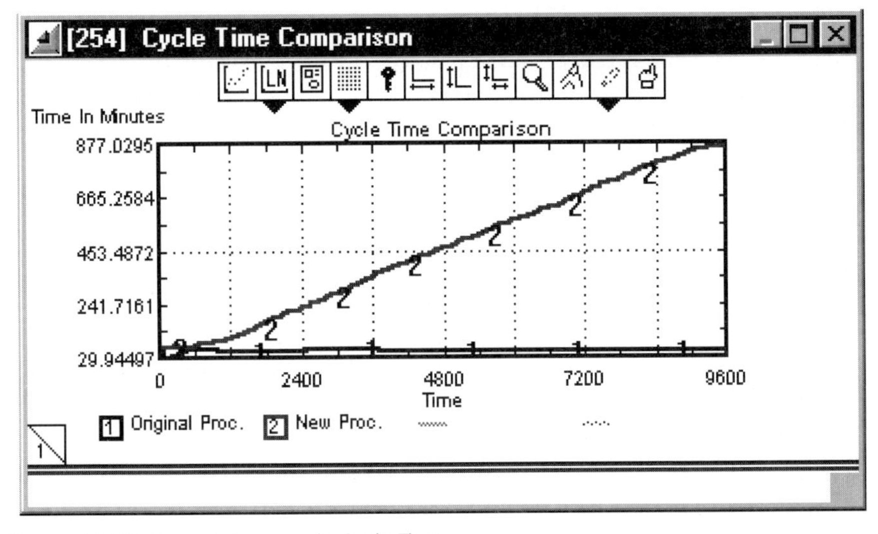

Figure 13.27 Average Increase in Cycle Time.

new process accordingly. The use of these techniques can prevent implementation of counterproductive changes as this would have been.

The question now is: What can be done to correct the problem? Management may choose to forgo the additional review and, if an application is questionable, return it to the applicant. Although this is a fictitious example, this scenario represents a trend in industry today—services that the public were accustomed to receiving without charge now require payment, or are just not being done.

Summary

There are many possible scenarios that can be tested using CAPRE technology. The scenarios described in the credit application example would be almost impossible to test using pen-and-pencil tools. This example, although presented for informational purposes only, represents the types of problems that management encounters every day, and the type of problem that can quickly and effectively be solved only with modeling and simulation.

14

Reengineering a Process Using Extend+BPR

Two words that have become popular in business circles are *downsizing* and *rightsizing*. Rightsizing, realistically, means the same thing as downsizing—it is just a little less harsh sounding.

Whatever word is used, the bottom line is that virtually all companies around the world, attempt to deal with cost problems by reducing staff. Although process reengineering does not prohibit the idea of adding staff, most published examples of successful process reengineering efforts have dealt with applications of downsizing to cost and productivity problems.

Staff reductions immediately affect the bottom line, which explains the popularity of process reengineering through downsizing. But is this simplistic approach the right approach? Is the bottom line an effective *process parameter*?

The impact on a process caused by adding or reducing staff is not easily analyzed using the manual practices of TQM and CPI. In fact, the impact on cost is not easily analyzed when cost is viewed as something other than the bottom line. Since staff size is a process parameter, a change in staff will result in a change to other process parameters. The relations of parameters to each other are complex and can only be effectively analyzed using computer aided process reengineering.

The following story is about a fictitious company. Although the company is fictitious, the scenario is based on reality. As in the Origami Process, the manufacturing process in this story consists of parallel subprocesses that merge. This story makes several points.

- Popular process reengineering approaches, no matter how well intended, are not always effective.
- Downsizing is not always the answer; it is just the "obvious" answer.
- Computer aided process reengineering facilitates the reengineering decision process and, in fact, eliminates the need for costly pilot programs.
- Computer aided process reengineering provides data to management that are essential to the decision-making process.
- Data collection, or process measurement, must be done carefully, or the incorrect decisions can result.

The company in this story is the Enlightened Amusement Company, Inc. It makes several entertainment items, one of which is the finest propeller hat in the world. (It has been suggested to me that only "propeller heads" will understand this book.) The manufacturing process is comprised of craftsmen, each specializing in a particular aspect of the hat production process. One craftsman makes the basic ("raw") hat, one makes the propeller, one assembles the hat by attaching the propeller to the raw hat, and one finishes the hat by sewing custom logos onto the brim. Each craftsman has his own office, remote from the offices of other workers, which provides a very comfortable working environment.

The company is, as the name implies, enlightened, and subscribes to all the most relevant TQM theories. It is a Theory Z company and has never laid off an employee. In fact, there is one employee who, before Deming's theories became popular, performed product inspections at each stage of the manufacturing process; however, once the company learned of Deming, it ceased depending on mass inspections but kept the employee on as an expediter. Because he was not a craftsman, however, he was paid less than the other employees. The company still employed an inspector at the end of the process, who sorted out defective hats for sale to a secondary market.

The Problem

Everything in this process worked smoothly, until one day the owner of the company visited the employees and said, "Now that NAFTA [the North American Free Trade Agreement] seems to be a reality, there are a lot of dynamics at work which will affect the way we do business in the future." The owner explained these dynamics by using the causal diagram shown in Figure 14.1.

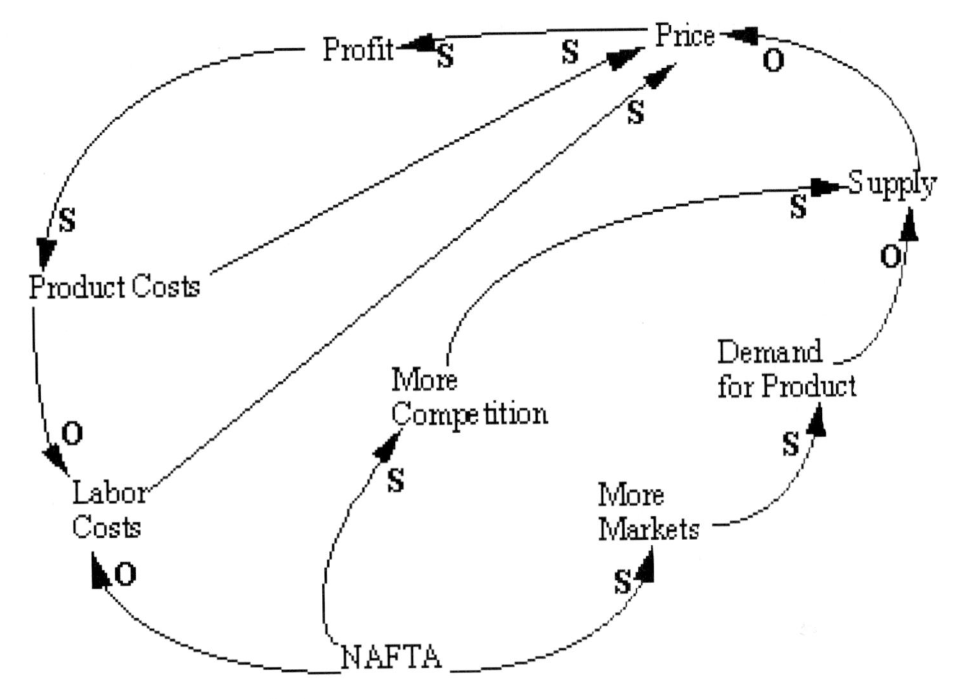

Figure 14.1 NAFTA Causal Diagram.

"With NAFTA," the owner explained, "we have more markets, which means that we have more demand, which means that our supply of product is decreased and we can raise our prices. When we raise our prices, we make more profit. When profits increase, we can afford higher labor costs. That's the good news."

The owner continued, "The bad news is that with NAFTA, we have more competition, which increases supply, which drives down prices and, therefore, profit. In addition, our competition is using cheap labor and piecework pay, which also drives down prices and profits. When profits drop, we must lower our costs to lower our prices. We must also increase output to make up for the lost revenues. In addition, we must eliminate defective products, since the downward trend in prices has eliminated our secondary market. Therefore, the company must find ways to lower costs, increase quality, and increase productivity. I am empowering you to reengineer our manufacturing process to meet these goals."

The employees were stunned—their comfortable way of life had been totally upset within a matter of minutes. Nonetheless, they were well versed in many TQM methodologies, and they convened a cross-functional team meeting to discuss cost-cutting ideas. The meeting consisted only of the crafts people, however. After all, they reasoned, they were the only ones involved in the actual manufacturing process.

Reengineering Change Number 1

The first action the cross-functional team took was to ask the inspector if he could categorize the types of problems he discovered during his inspections. The inspector explained that he could not, since there was no pattern to the problems that caused hats to be rejected. Moreover, without performing inspections at each stage of the process ("like we used to do," he added, pointedly), he could not determine where the problems were coming from.

After a long session, the hat maker and the propeller maker both spoke up: "We think the problem is with the expediter. He seems to always be around and, if there is nothing for him to do, he waits by our doors. This makes us nervous, and we probably make mistakes. We think that if we eliminated his job, we would reduce rejects and lower costs. We might even increase productivity, since he sometimes causes delays by not being around when he should be." These workers, experienced in systems thinking, used a causal diagram (Figure 14.2) to explain their assumptions.

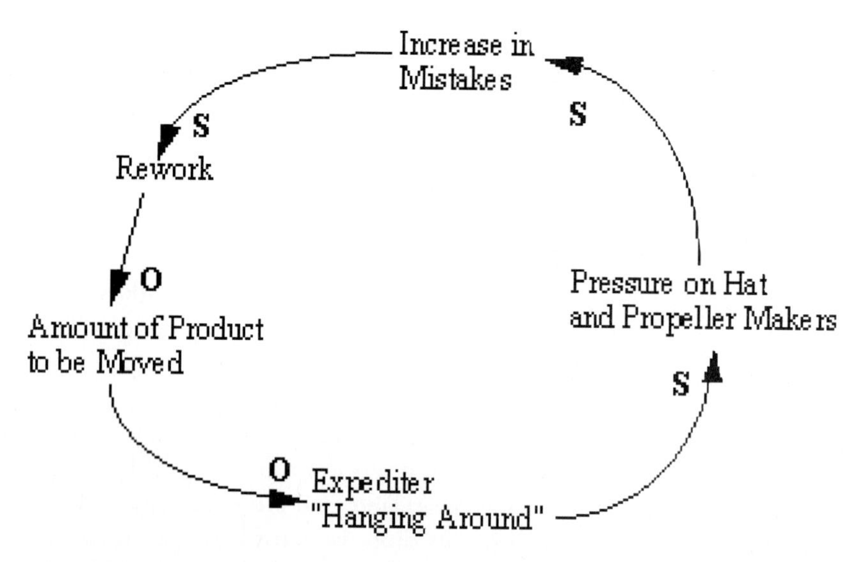

Figure 14.2 Causal Diagram of Pressure Caused by Expediter.

The team thought about this information and reached the following conclusion: If they moved their work areas close together, they could downsize by eliminating the expediter. With his salary removed, costs would drop. Also, if the hat and propeller maker were correct in their assumptions, the number of rejects would be reduced as well. This represented total involvement, since the remaining workers were sharing the burden. No one liked this idea, but a consensus agreement was reached and the matter was closed.

This was a bitter pill to swallow. After all, the company practiced Theory Z. On the other hand, that *is* just a theory, and theories change. With great regret, the expediter was laid off.

The result was not exactly what had been expected. Before removing the expediter, costs were about \$5.80 per hat. After removing the expediter, costs had dropped slightly to about \$4.90 per hat, representing a fairly significant cost reduction. On the other hand, rejects remained. Apparently, the assumptions of the hat and propeller maker about the causes of rejections were incorrect mental models.

The owner met with the craftspeople again and congratulated them on their effort; however, the owner reemphasized that the company could no longer tolerate the production of any hats that did not pass inspection. Rejected hats, no longer sold on a secondary market, represented a loss of revenues for the company. When the newly improved process was viewed from the perspective of cost per *acceptable* hat, costs were approximately \$6.00 per hat. (This number varied, since the percent of rejects changed daily.)

The owner of the Enlightened Amusement Company strongly suggested that the employees find out why hats were being rejected and fix the problem. If the employees could institute a zero-defects process, the cost of hats to market would be reduced. The owner also suggested that there was no time to conduct a lengthy statistical study, and that the problem had to be corrected quickly.

Reengineering Change Number 2

To address the quality problem, the employees convened a quality control circle meeting. An energetic brainstorming session took place, in which many ideas were considered. Finally, the craftsman who finished the hats (the finisher) explained that the work had to be done cautiously, since sewing logos on a hat with a propeller attached was cumbersome. The worker admitted to making mistakes, such as bending the propeller and scuffing the hat, and suggested that, if he were to finish the sewing *before* the hat were assembled, not only would rejects be reduced, but the work could also be done twice as fast.

The finisher explained his reasoning using a causal chain (Figure 14.3).

Changing the order of the manufacturing tasks to reduce the difficulty of assembly represented an opportunity to reduce rejects and increase productivity. The employees decided to rearrange the process as suggested by the hat finisher and then run a pilot program. The result was that rejects were eliminated and the cost per hat shipped was reduced by almost 50 percent to \$3.10 per hat.

Now *this* really was a success story. Cost was down, and rejects had been eliminated; however, the employees were troubled by one thought. If the hat finisher was working twice as fast, why didn't productivity increase?

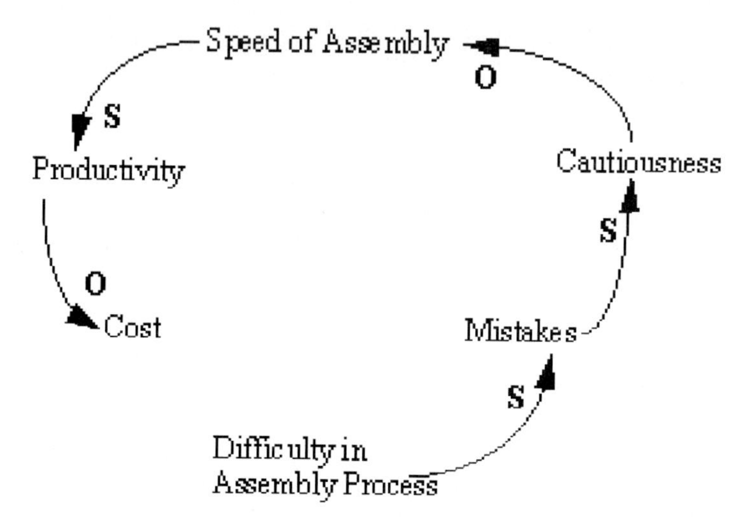

Figure 14.3 Hat Finisher's Causal Chain.

Reengineering Change Number 3

Another cross-functional team was convened, and the hat finisher explained that inventory was not available fast enough, so the finisher was idle much of the time. The employees, who wanted to increase productivity, considered this problem and arrived at a unique solution: Since the process was now free of rejects, the inspector was no longer necessary. Therefore, instead of inspecting products, he could participate in making hats and propellers. The craftsmen reasoned that the causal chain depicted in Figure 14.4 was true.

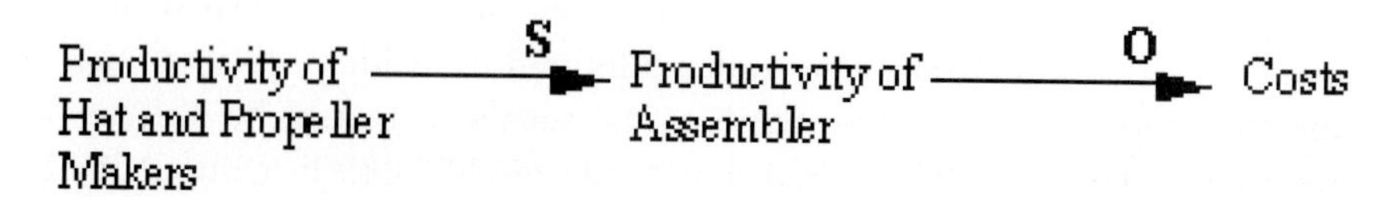

Figure 14.4 Assembler Productivity.

This required training for the inspector, but instituting training is one of Deming's fundamental theories, so the suggestion was accepted. When the inspector was trained to be a hat and propeller maker, delivery of inventory to the hat finisher increased.

A pilot program was tried, but productivity did not increase, nor did cost come down. It seemed that since the initial steps of the process were now being done more quickly, the just-in-time inventory was now "not in time." As a result, the workers found themselves with a great deal of idle time. A very simple causal loop (Figure 14.5) explained why this happened.

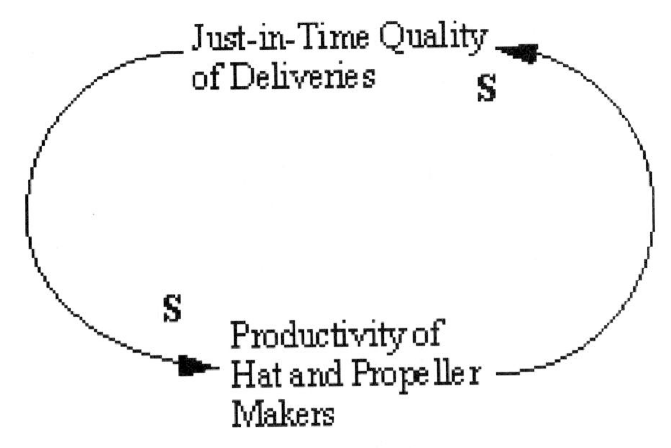

Figure 14.5 Just-in-Time Causal Chain.

It was easy to understand how this might have been overlooked. The pilot program was run under controlled conditions. One of the conditions that was controlled was the availability of raw materials, so it was not an issue. Another pilot program was run during which the productivity of the hat and propeller makers was measured. With the addition of the inspector, the workers determined that they could *triple* the delivery of raw materials.

Reengineering Change Number 4

Fortunately, a change in delivery schedule did not present a great problem to the Enlightened Amusement Company; after all, they practiced Continuous Improvement and had a long-term partnership arrangement with their supplier. The employees suggested to the owner that if their suppliers could deliver raw material at three times the current rate, the output of propeller hats would increase.

The owner met with the supplier and, even though the supplier had to juggle priorities and cut back deliveries to other customers, he agreed to triple his delivery schedule to the company. The owner, the employees, and the supplier were all quite pleased—this demonstrated the value of Continuous Improvement.

Once again, the pilot program failed to meet its goals. Raw hats were being produced faster, propellers were being produced faster, inventory was being delivered faster, the finishing touches were being applied faster, but the same number of hats was being produced every day. The owner, who practiced management by walking around, was walking around more than ever.

Reengineering Change Number 5

The workers held another brainstorming session in which they found a bottleneck. That is, the hat assembler was working at the usual speed, which was acceptable with the old flow of inventory but not acceptable in the improved process. The employees were faced with two options: either reduce productivity or add a worker to the assembly task.

Reducing productivity was not an acceptable option, so the workers concluded that the company should bring back the expediter and train that person to be a hat assembler. The former expediter knew the operation and would require a shorter learning curve than anyone else. The employees also decided that they could increase deliveries once again, even if slightly, since production all around would be increasing. Figure 14.6 shows the causal chain that was used to reach these decisions.

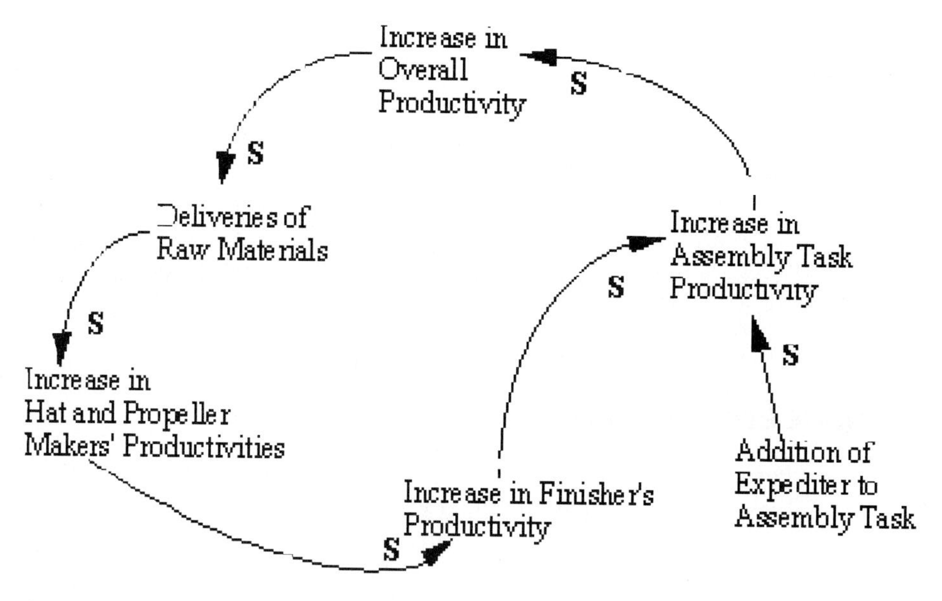

Figure 14.6 Effect of Adding the Expediter to Work Force.

The craftsmen had been burned once, but this time, after the previous pilot program, they were convinced that this change would work. The expediter agreed to return to work but demanded back pay and an increase in salary equal to that of the other craftsmen. The workers and owner were becoming desperate and had no choice but to accept the terms. A short pilot was run and productivity was up to almost 3000 hats per week and cost was down to about $1.90 per hat. And there were no rejects! Finally, a true success story.

Trying a Different Approach

Unfortunately, the change was not as successful as the pilot program suggested it would be. Shortly after the pilot, rejects from customers began appearing. This was a major problem, not simply because there were rejects, but also because each rejected hat had to be replaced with a new hat. The cost of *good* hats was once again up to almost $4.50 per hat. The owner of the company, who believed in coaching rather than judging, asked the employees if a process reengineering consultant should be hired. Seeing the opportunity to relieve themselves of this burden, the employees readily agreed.

Not happy with the results of the process reengineering efforts that were based on TQM and CPI approaches, the owner decided not to look for consultants who practiced traditional process reengineering methods. Instead, the owner chose a consultant who practiced computer aided process reengineering.

The consultant who was hired believed in the Rules of Process Reengineering and applied them in sequential order. First, the consultant met with all the employees and discussed their roles in the process. After documenting each worker's tasks, the consultant developed an overall map of the original process using Extend+BPR. When the workers agreed that the consultant understood the process, the consultant began to model the process.

After developing models of the original process and the subsequent reengineered processes, the consultant demonstrated that the results of the changes could have been predicted at every stage. The consultant also pointed out that pilot programs were always run under controlled conditions, and that the results of pilot programs should never be considered as completely reliable.

For example, after modeling the process after Reengineering Change Number 5, the consultant was able to demonstrate that inventories at the very beginning of the process were growing more rapidly than had been anticipated. When inventories became too large, workers hurried and mistakes were made.

The consultant considered what the owner had asked the workers to do: Increase output, increase quality, and lower costs. The key process parameter in the owner's mind was the cost of producing acceptable hats. Therefore, a process that had high productivity and rejects was acceptable, as long as the cost of each good hat produced was lower than the cost of good hats produced by a process that had no defects and lower productivity. The process parameter that the consultant focused on, therefore, was *cost per good hat.*

The consultant then modeled several other scenarios and performed some what-if analyses. In a short period of time, the consultant told the owner, "The solution is to add one person at the beginning of the process. In other words, upsize." Add a worker? "What a crazy notion," thought the owner; however, the consultant ran a model showing that by adding one worker productivity would rise to 4000 hats per day, and cost would drop to $1.75 per hat. Moreover, since inventory levels would be stabilized, there would be no rejects returned from customers.

The owner remained skeptical, so the consultant explained the approach that was used to model the process, how the model was used to provide information about the process, and how that information had been used as the basis for the conclusions.

Using Extend+BPR to Model the Original Process

First, using Extend+BPR, the consultant began by developing a model of the process as it was initially implemented (Figure 14.7).

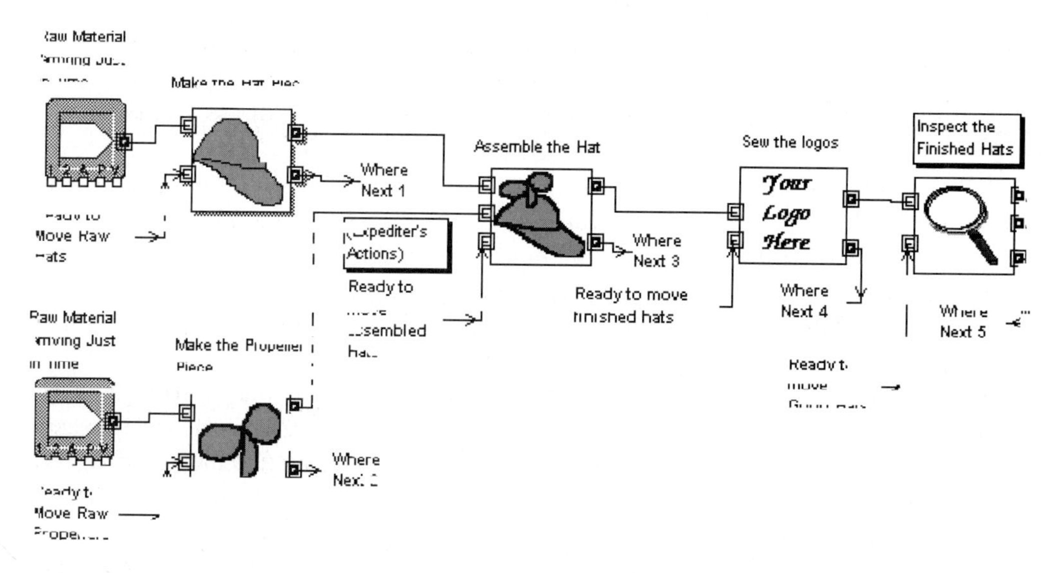

Figure 14.7 Extend+BPR Model of Original Process.

The owner and workers looked at the model and said, "This looks just like a process map. I thought you told us that maps alone can't give you the information required to reengineer a process."

"That's true," said the consultant, "but this is more than a map. I used hierarchical blocks to make the overall view of the model easy to understand and hid a lot of the detail in those blocks." The consultant then explained how the model had been developed.

The first element of the model to be considered was the delivery of raw materials, in this case, raw materials for hats and propellers. This just-in-time delivery process was represented in Extend+BPR by utilizing the Import block. Figure 14.8 is the dialog for that block.

The frequency of items generated for the simulation was set at once per eight simulation time units, and in this model, one time unit represented one hour. The dialog option "# of Items (V)" was set to 4, representing four batches of 50 items; therefore, the Import block simulated four batches of inventory, each of which was sufficient to make 50 hats, being delivered every eight hours. This is exactly the just-in-time schedule used by the company. In a similar manner, the same amount of inventory required to make propellers was being delivered every day.

The actual task of making raw hats was modeled as shown in Figure 14.9.

[1] Import

Generates items according to a distribution.

- Constant
- Erlang
- Exponential
- Integer, uniform
- Normal
- Real, uniform
- Triangular: [ ] is most likely.
- Weibull

(1) Constant = 8

(2) (unused) 1

OK

Cancel

No item at time zero

Set attribute, priority, or number of items (optional):

Attr. name =

Attr. value =

Priority =

# of items(V) = 4

Use block seed offset 0

Comments

Represents delivery of material for 4 bartches of 50 hats per day

Help

Figure 14.8 Import Block Dialog.

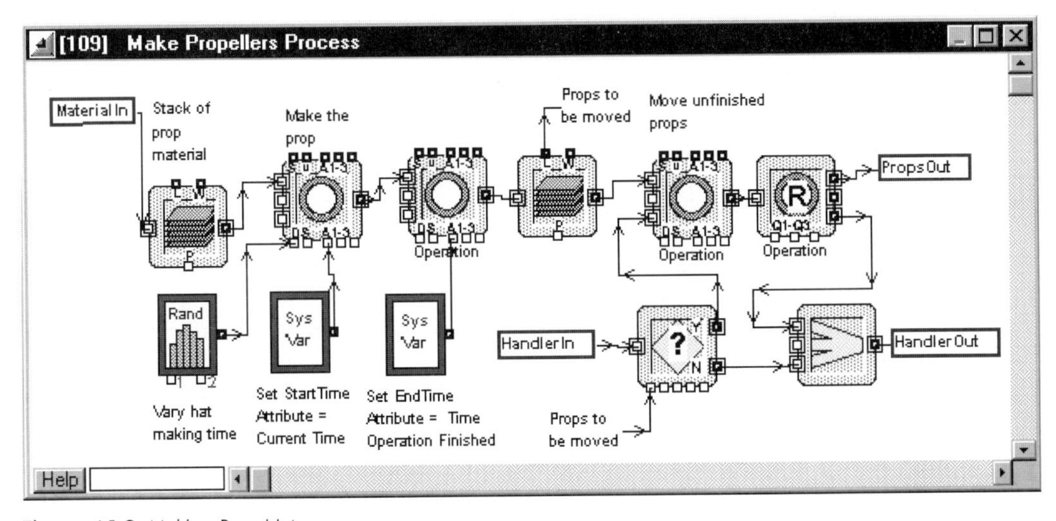

Figure 14.9 Making Raw Hats.

The connectors on the high-level model shown earlier, such as Ready to Move and Where to Next represent the movement of the expediter. The consultant included in the model a repository representing the number of batches of raw-hat inventory available to the hat-making workers and used an Operation block to represent the actual task of making hats. Based on information received from the workers, the consultant varied the amount of time required to make a batch of hats as a number between 0.9 hours and 1.1 hours. This was done using the Input Random Number block, the dialog of which is shown in Figure 14.10.

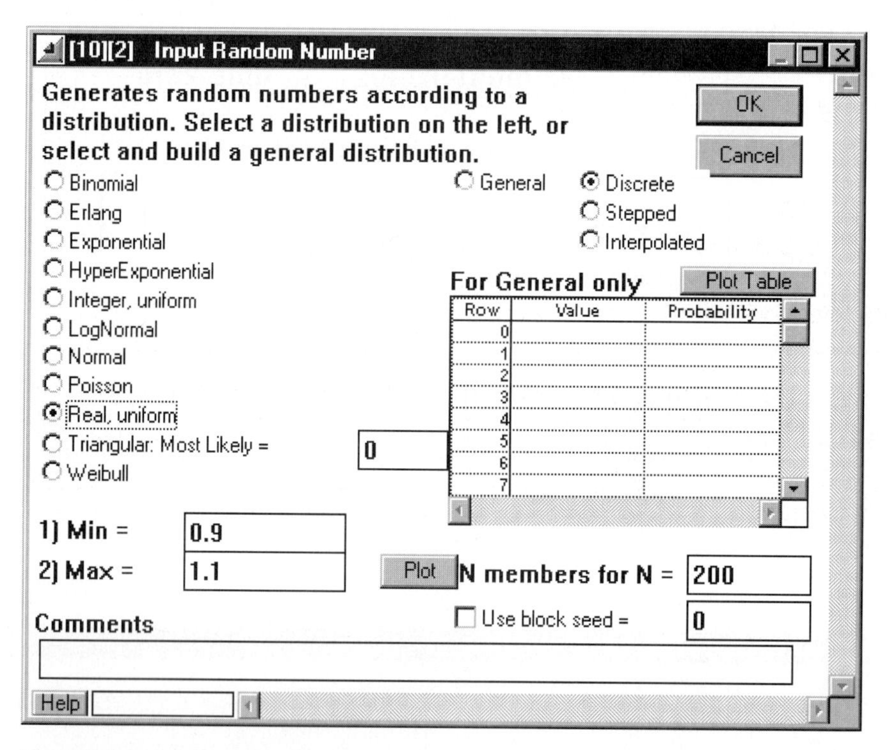

Figure 14.10 Dialog of Random Number Input Generator.

This immediately provided a clue about the process to the consultant: If it took *at most* 4.4 hours for the craftsman to make a batch of 50 hats, the worker must be idle some of the time. The consultant decided to wait to test this assumption until the model was complete.

Continuing the development of the process model, the consultant added the logic used by the expediter for moving batches of finished products. When the expediter arrived at a worker's office, the expediter looked at the amount of inventory to be moved. If there was not at least one batch of 50 items (in this case, raw hats), the expediter waited a bit. If enough material became available, it was moved; otherwise, the expediter decided what to do next. The decision logic of whether or not to move anything is contained in the dialog of the Decision(2) block used in the model (Figure 14.11).

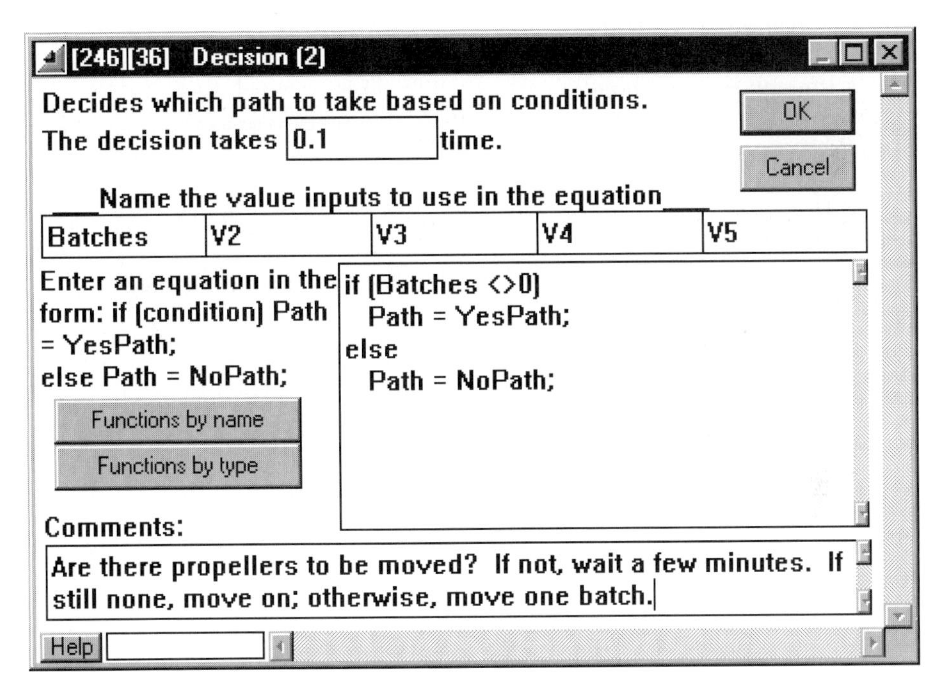

Figure 14.11 Dialog of Decision(2) Block.

Note "The decision takes...time" option. This allows the modeler to allow some time to pass before making a final decision, so that if material becomes available, the YesPath will be taken. When the YesPath is taken, the expediter becomes available for the Operation block that represents the movement of inventory. When the NoPath is taken, the expediter object is returned to a place in the model that determines where it goes next. Each set of logic will be explained below.

When inventory becomes available, an Operation block is executed, as shown earlier in Figure 14.9. This is a very complex set of logic captured in relatively few blocks. The model would proceed as follows.

1. The logic that determines if the expediter will move inventory is executed.
2. If material is to be moved, the expediter becomes available to the Operation block.
3. The expediter will move one batch of 50 hats, since that is all he can handle. This is reflected in the dialog of the Operation block (Figure 14.12) that represents the movement of hat inventory.

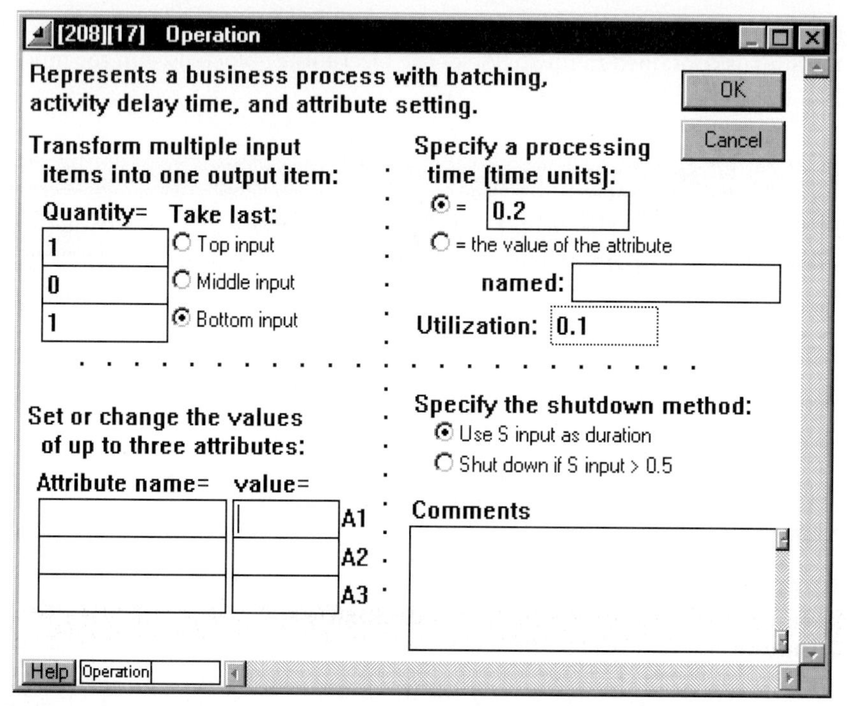

Figure 14.12 Dialog of Operation Block Used to Model Expediter's Activities.

Note that the "Bottom input" dialog option is set to 1, representing the expediter, and that "Take last" is checked. Since the block executes under an AND condition, it will execute only when items are available at both the top connector and bottom connector, but the Operation block has been set up so the expediter is not reserved by the block unless there is input at the top connector.

Note also that the time specified for the operation is set to 0.2, representing 0.2 hours of time, or 12 minutes.

The consultant then added an Operation, Reverse block to output the hats that were moved and to return the expediter to the model for use in other parts of the model.

Now, the actual movements of the expediter had to be modeled. The consultant asked the expediter how it was determined where he would go next after stopping at a worker's office. The expediter explained, "Well, we tried many different approaches to that, and the simplest was simply to go in a specific order." The consultant modeled these movements starting with a hierarchical block (Figure 14.13).

The Where Next connectors represent the expediter leaving a particular task, and the Ready to Move connectors represent the expediter arriving at a worker's office. The consultant then added the logic that defined the movement of the expediter (Figure 14.14).

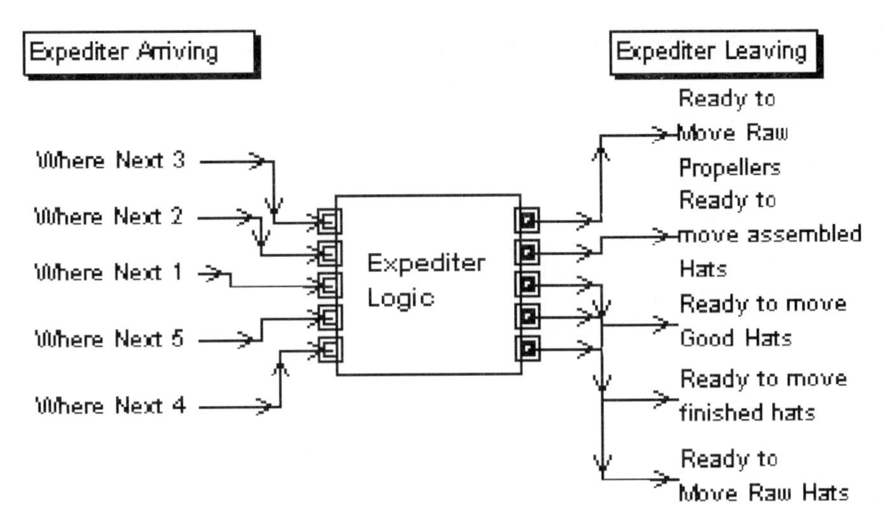

Figure 14.13 Hierarchical Block of Expediter's Routing Logic.

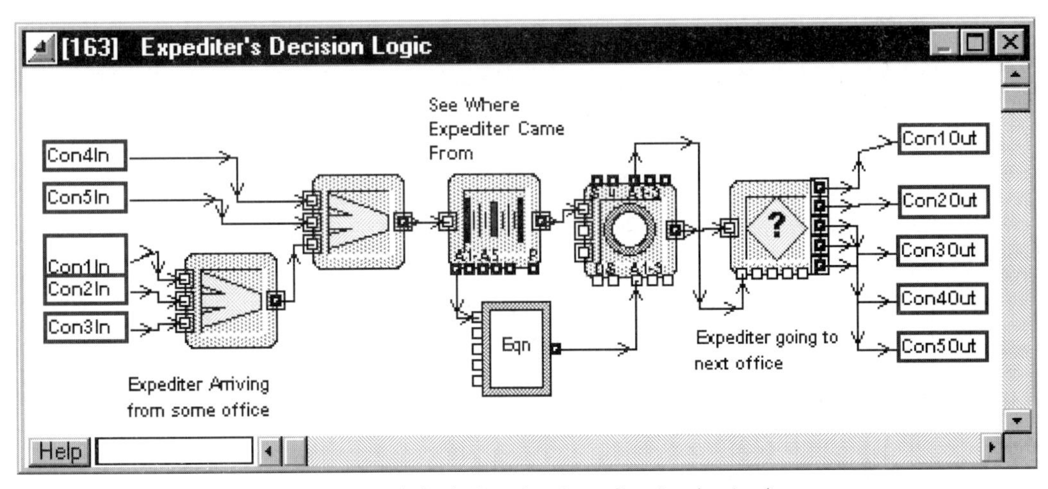

Figure 14.14 Decomposed Hierarchical Block Showing Expediter Routing Logic.

The consultant used two Merge blocks, since each block can merge only three paths, and the expediter could be arriving from one of five paths. To determine where the expediter should go next, the consultant assigned an attribute to the object representing the expediter. This attribute was a count from 1 to 5, representing the stops the expediter would make. A Measurement block read the attribute from the arriving object, which specified where the expediter had been, and then increased the attribute by one to specify where the expediter would go next. This is shown in the dialog of the Equation block (Figure 14.15).

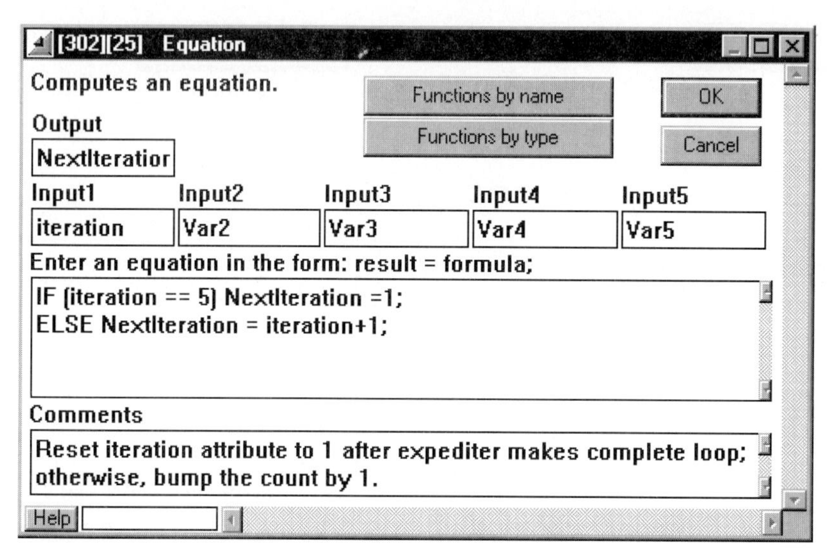

Figure 14.15 Dialog of Equation Block Used to Determine Expediter's Iteration Attribute.

The new value of the attribute was assigned to the expediter by using the Operation block shown in the model. Once that was done, a Decision(5) block was used to determine which of the five possible routes would be followed (Figure 14.16).

[314][37] Decision (5)

Decides which path to take based on conditions.

The decision takes 0 time.

OK

Cancel

__Name the value inputs to use in the equation__

iteration	V2	V3	V4	V5

Enter an equation in the form: if(condition) Path = PathName;

Functions by name

Functions by type

Name the paths

```
IF (iteration == 1) Path = Path1;
Else IF (iteration == 2) Path = Path2;
Else IF (iteration == 3) Path = Path3;
Else IF (iteration == 4) Path = Path4;
Else Path = Path5;
```

Path1
Path2
Path3
Path4
Path5

Comments:

Select path on iteration count.

Help

Figure 14.16 Path Selection Logic in Decision(5) Block.

The modeling technique used by the consultant was an object-oriented technique. The expediter was viewed as an object, and an attribute was assigned to that object to be used in logic that selected a path through the model.

The consultant had not yet finished modeling this task. The consultant wanted to test an assumption about the availability of idle time. Extend+BPR automatically determines the amount of time an object, in this case a labor "object," has been utilized in a simulation. To determine idle time, the consultant had only to look at the utilization of any worker and subtract that value from 1. Utilization is shown in Figure 14.17.

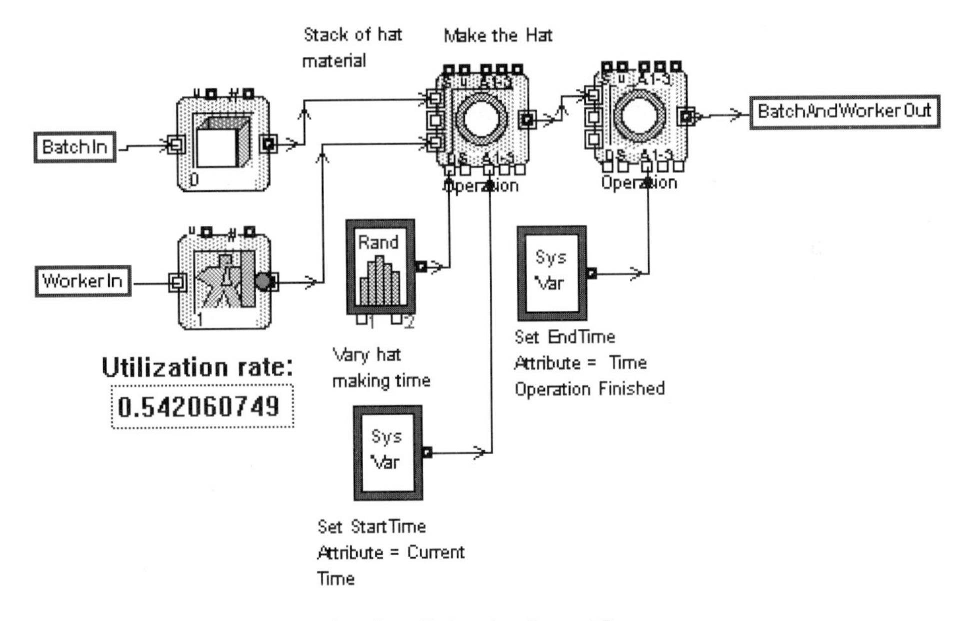

Figure 14.17 Using Inventory Levels to Determine Percent Busy.

The consultant discovered that the hat maker was busy between 50 percent and 60 percent of the time. The consultant asked the hat-making worker why it was never mentioned that the worker could have been doing more work. The worker simply shrugged and said, "No one ever complained about my output, and I was never asked to do more work."

This brings up an important point about TQM—team building, total involvement, continuous improvement, and so on, are theoretically fine concepts, but, if management expects employees to volunteer information, it may be greatly disappointed. It is important to constantly *ask* for input from employees to make these theories work.

The consultant measured the busy time of the other workers in the same manner and discovered that they were all busy about the same percentage of the time. This gave the consultant information about the productivity of the workers. It might have been simple to suggest that delivery of inventory should be increased, thereby increasing productivity, but the owner had asked the workers to look at two other factors in the model: the number of rejected hats and the cost of the hats. Since the goal of the reengineering effort was to reduce the cost of hats and reduce the number of rejects, the consultant defined a process parameter called *cost per good hat*. In this modeling effort, the consultant would measure cost in those terms only.

Modeling Reengineering Change Number 1

Modeling of the process after Reengineering Change Number 1 was a simple task—all references to the expediter were removed (Figure 14.18).

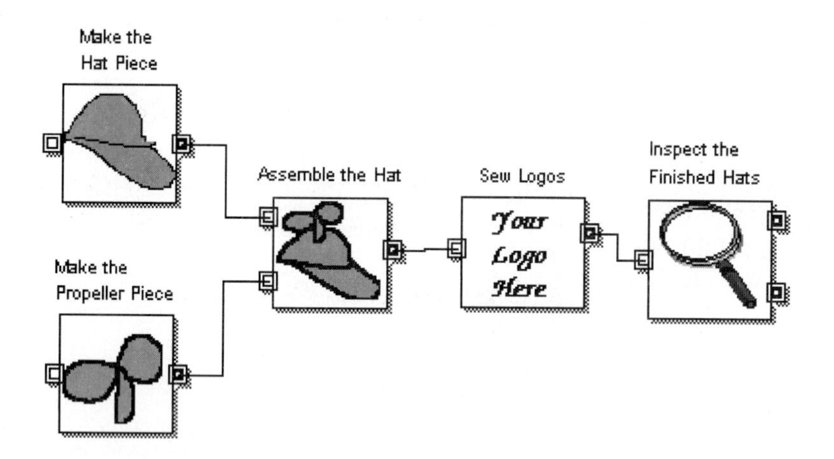

Figure 14.18 Propeller Hat Process Change 1.

The elimination of the expediter and the expediter's salary certainly lowered costs and, in fact, lowered the costs per good hat, but, this was misleading—the fundamental cause of the rejects had not been found, and the cost to manufacture a good hat was still too high. The consultant modeled the process without the material handler and demonstrated that cost would in fact be lowered, but nothing else, including productivity, would be changed. Therefore, elimination of the function of the expediter had no effect on the process, and the elimination of the person could have been postponed until more reengineering options had been considered.

The next step taken by the consultant was to add some information to the model, such as defining an attribute specifying the rejects created by the hat finisher.

The finisher estimated that this function created conditions for rejected hats about 20 percent of the time. The consultant modeled this as shown in Figure 14.19.

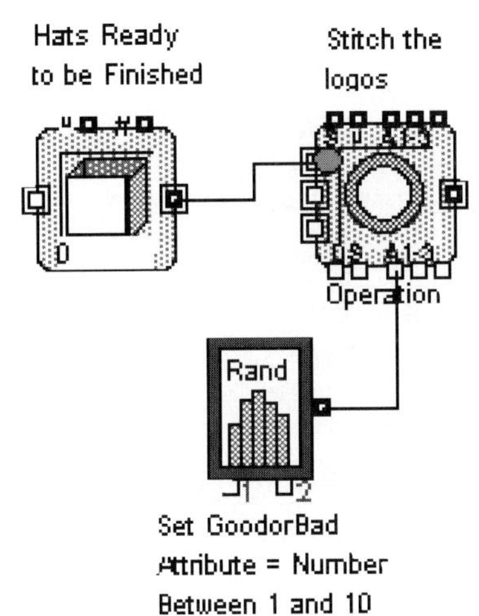

Figure 14.19 Set GoodorBad Attribute with Random Number Input Block.

He defined an attribute that represented the rejection percentage of the hat finisher and attached that attribute to every batch of hats worked on by the finisher. The definition of the attribute is shown in the dialog of the Operation block (Figure 14.20).

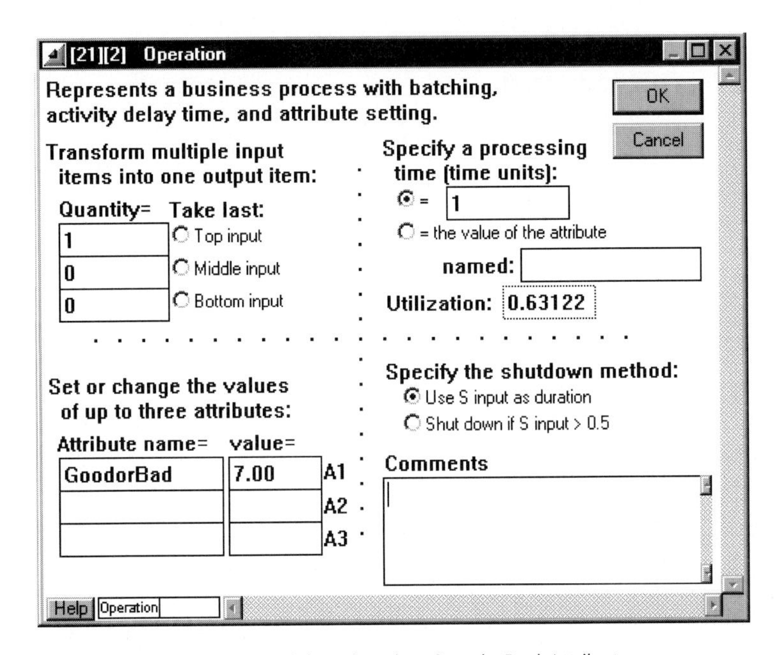

Figure 14.20 Operation Dialog showing GoodorBad Attribute.

The consultant then added logic that determined whether a batch of hats passed inspection or was rejected and calculated the cost per good hat (Figure 14.21).

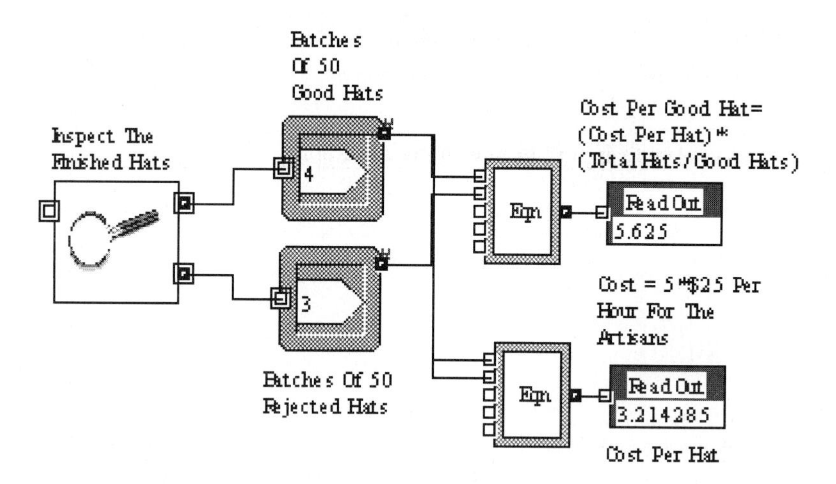

Figure 14.21 Calculating Cost per Good Hat.

The hierarchical block called Inspect the Finished Hats decomposed into its primary blocks, is shown in Figure 14.22.

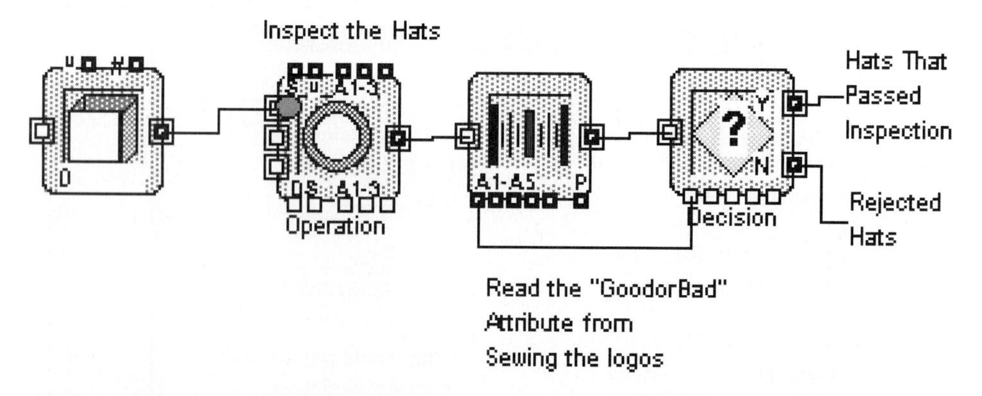

Figure 14.22 Details of Hat Inspection Task.

In this part of the model, the consultant interrogated the GoodorBad attribute that was set in the hat-finishing task. Based on that value, the batch of hats was either considered to be accepted or rejected. You might ask, Shouldn't hats be inspected individually? This is a valid question, but statistically, if 1 out of every 10 hats will be rejected, 1 out of every 10 batches of 50 (or 100, or 1000, and so on) hats will be rejected. This is a modeling technique that saves time and reduces complexity.

The logic for the inspection is shown in the dialog of the Decision(2) block (Figure 14.23).

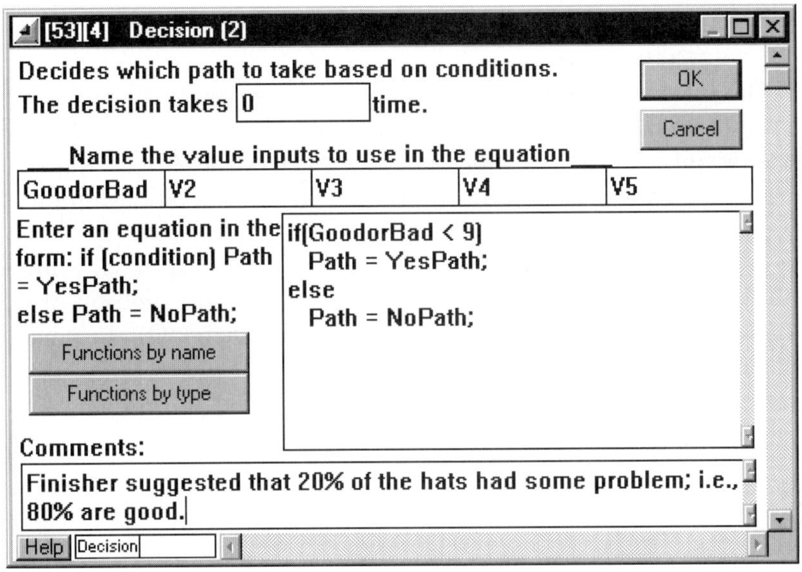

[53][4] Decision (2)

Decides which path to take based on conditions.

The decision takes 0 time.

OK

Cancel

Name the value inputs to use in the equation

GoodorBad	V2	V3	V4	V5

Enter an equation in the form: if (condition) Path = YesPath; else Path = NoPath;

```
if(GoodorBad < 9)
  Path = YesPath;
else
  Path = NoPath;
```

Functions by name

Functions by type

Comments:

Finisher suggested that 20% of the hats had some problem; i.e., 80% are good.

Help Decision

Figure 14.23 Hat Inspection Logic Contained in Decision(2) Block.

Finally, the calculation of the cost per good hat was performed in the Equation block (Figure 14.24).

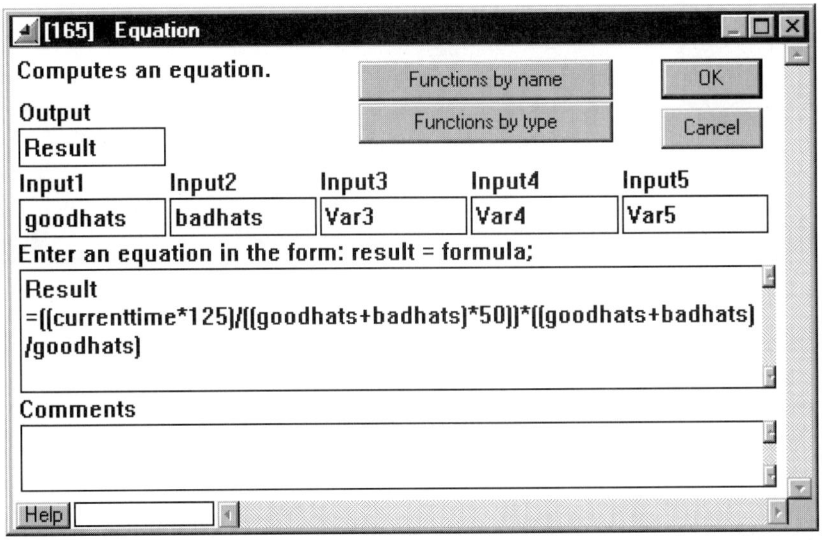

[165] Equation

Computes an equation.

Functions by name

Functions by type

OK

Cancel

Output

Result

Input1	Input2	Input3	Input4	Input5
goodhats	badhats	Var3	Var4	Var5

Enter an equation in the form: result = formula;

```
Result
=((currenttime*125)/((goodhats+badhats)*50))*((goodhats+badhats)
/goodhats)
```

Comments

Help

Figure 14.24 Calculating Cost per Good Hat.

This looks complex, but basically it is the cost per hat multiplied by the ratio of good hats to total hats produced. The consultant added another calculation (shown in Figure 14.21) that displayed the cost per hat, regardless of whether it was good or not. The difference is quite large and proved the impact of rejected products on the cost of accepted products.

This information showed that the elimination of the expediter did reduce costs, but cost per good hat would be reduced by a larger margin if rejects were eliminated. Elimination of the expediter function and the associated salary dropped the cost of each good hat about $.90, but the elimination of rejects in conjunction with the staff reduction would have reduced the cost an additional $1.30 per good hat. In other words, the elimination of rejects would have paid for the expediter! If the employees of the Enlightened Amusement Company had tackled the problem of rejects first, they would have found this important cost savings early on. On the other hand, one could reason that the elimination of rejects and the elimination of the expediter would reduce costs significantly.

Modeling Reengineering Change Number 2

Change Number 2, as implemented by the employees, did reduce costs slightly. The only change to the process, in terms of modeling, was a change in the order of the execution of the tasks (Figure 14.25).

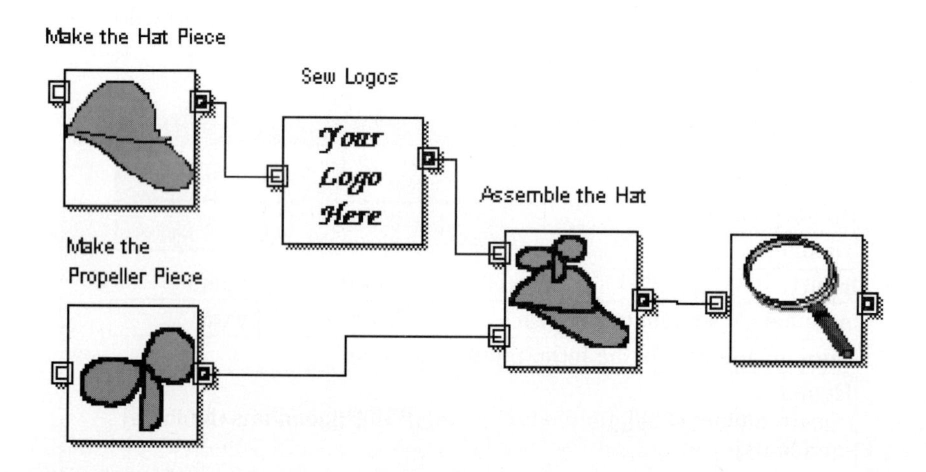

Figure 14.25 Propeller Hat Process Change 2.

The model demonstrated that no process parameters would have changed after this reengineering effort, except for the number of rejects and, therefore, the cost per good hat.

The consultant pointed out that the task of sewing logos was one that became necessary when the company decided to sell customized hats. The manufacturing process *evolved* and the task was put into place after the hats were assembled simply because no one gave much thought to the order in which tasks should have been executed. Like many business processes that evolve, the propeller hat manufacturing process was not as efficient as it could have been. Many of the problems the company faced might have been avoided if the process had been *engineered* before it became necessary to reengineer it.

Modeling Reengineering Change Number 3

When the consultant modeled the process after Change Number 3, it was assumed that all hats would be accepted, since that was the result of the pilot study. Therefore, the consultant eliminated the check for the reject attribute. Cost per good hat dropped to $2.50. This was a significant drop, representing a 66-percent reduction in cost from the original process (Figure 14.26).

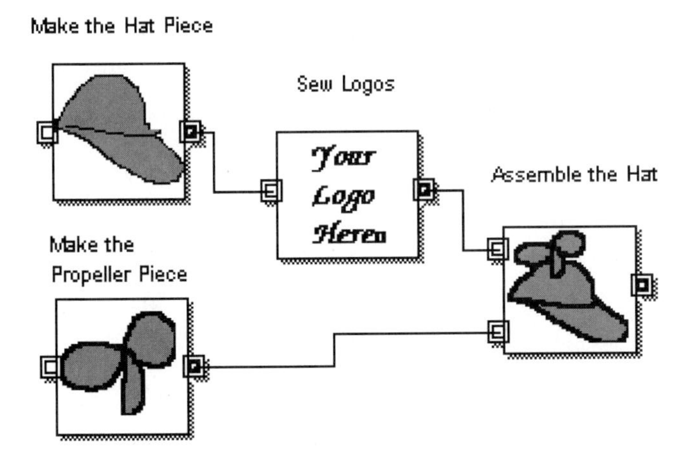

Figure 14.26 Propeller Hat Process Change 3.

This seems like a success, but remember that the owner of the company wanted to increase productivity as well. When the consultant examined the amount of time that the workers were busy, it was found that the hat and propeller makers were still busy only 50 to 60 percent of the time, the finisher was busy only 25 percent of the time, and the assembler was busy 50 percent of the time.

When the employees were reengineering the process, they had looked only at the first two tasks in the process. They saw that the hat and propeller makers had extra capacity, and that the finisher was idle almost 75 percent of the time. Given this information, they focused on increasing the supply of material to the finisher so that the finisher's productivity could be increased. Their reasoning went like this.

1. The hat makers could, by themselves, almost double their work load.
2. If the inspector were to participate, the output of hats could be increased threefold.
3. The finisher would still have spare capacity.

This is not an uncommon occurrence in reengineering efforts. It is often called *change in the small*, and change in the small can lead to suboptimization. This is exactly what happened when Change Number 4 was implemented.

Modeling Reengineering Change Number 4

Suddenly, the model was no longer simple. The craftsmen, without knowing it, had just added complexity to the process, a danger associated with "tinkering" with processes. Change Number 4 specified that inspections would no longer take place and that the inspector would be trained as a hat maker and as a propeller maker. The worker would alternate between the two tasks depending on which task had the greater amount of raw inventory available. That portion of the model is shown in Figure 14.27.

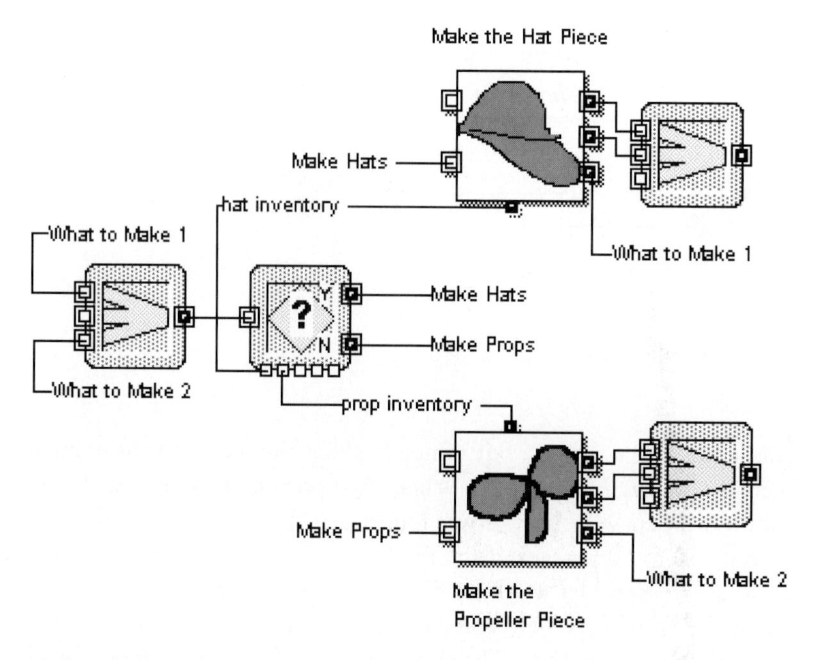

Figure 14.27 Inspector Alternating between Two Tasks Based on Inventory Levels.

The logic used to determine the task the inspector would perform is shown in the dialog of the Decision(2) block (Figure 14.28).

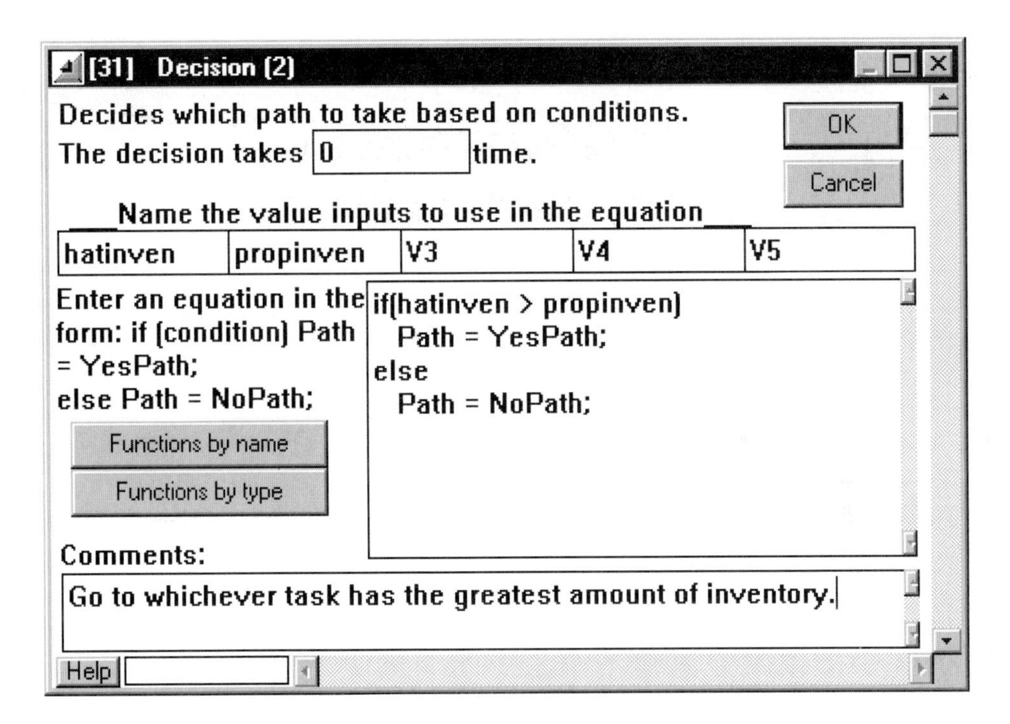

Figure 14.28 Logic Used to Determine Task to be Performed by Inspector.

Change Number 4 also called for the suppliers of raw materials to triple their deliveries when compared to the original process. Instead of four batches of material that could be used to make 50 hats, the Enlightened Amusement Company now was asking for twelve. In the model, this was a simple change to the Import blocks.

When the pilot program was run, the result was that the hat and propeller makers were busy almost 85 percent of the time, the finisher was busy 65 percent of the time, and the assembler was busy 85 percent of the time. Cost per good hat dropped to $2.95, the lowest it had been.

As is often the case with pilot programs, however, the results were misleading. Pilot programs are run under controlled conditions and for short periods of time. The data collected, if not interpreted correctly, can lead to a misunderstanding of the dynamics of the process. For example, while it was true that the assembler was busy only 85 percent of the time in the pilot program, a graph of this percentage of busy time could have been created using the Discrete Event Plotter from Extend's Plotter Library (Figure 14.29).

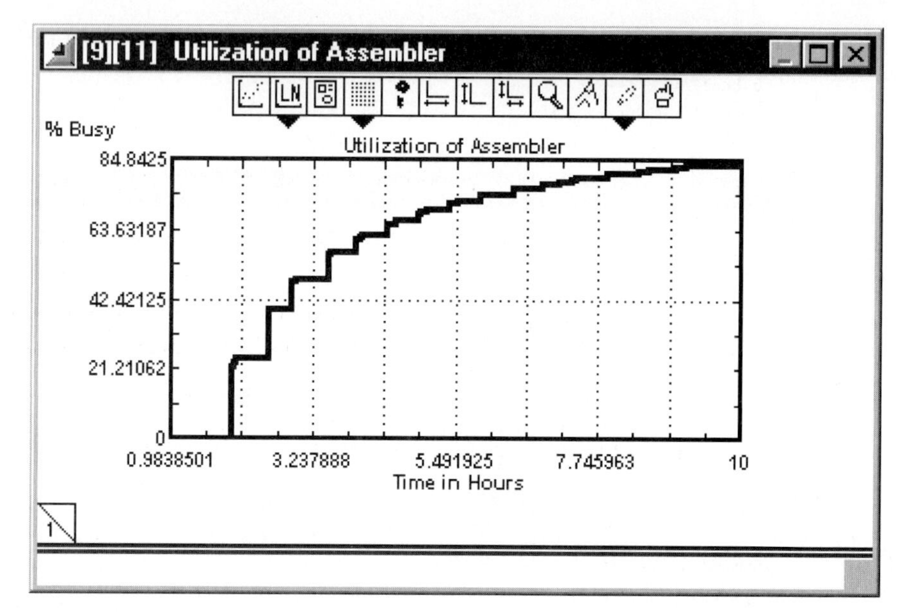

Figure 14.29 Percent Busy Trend Displayed in Graph.

This shows an *increasing* slope of busy time, indicating that as the process proceeded, the assembler would become busier since inventory would begin to stack up. Figure 14.30 is a graph of the inventory of the assembler.

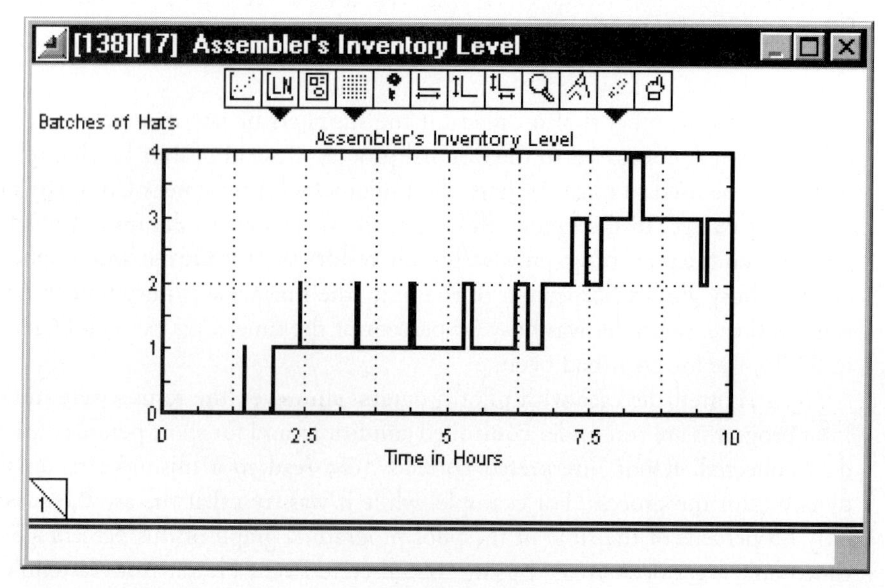

Figure 14.30 Graph of Assembler's Inventory Level.

Once again, this is an increasing slope, indicating that inventory to the assembler's task would continue to rise over time. In fact, both of these graphs revealed accurate information, as demonstrated by the consultant when a model of the pilot program was run for an extended period of time.

Figure 14.31 shows initial inventory when the model was executed for a longer period of time.

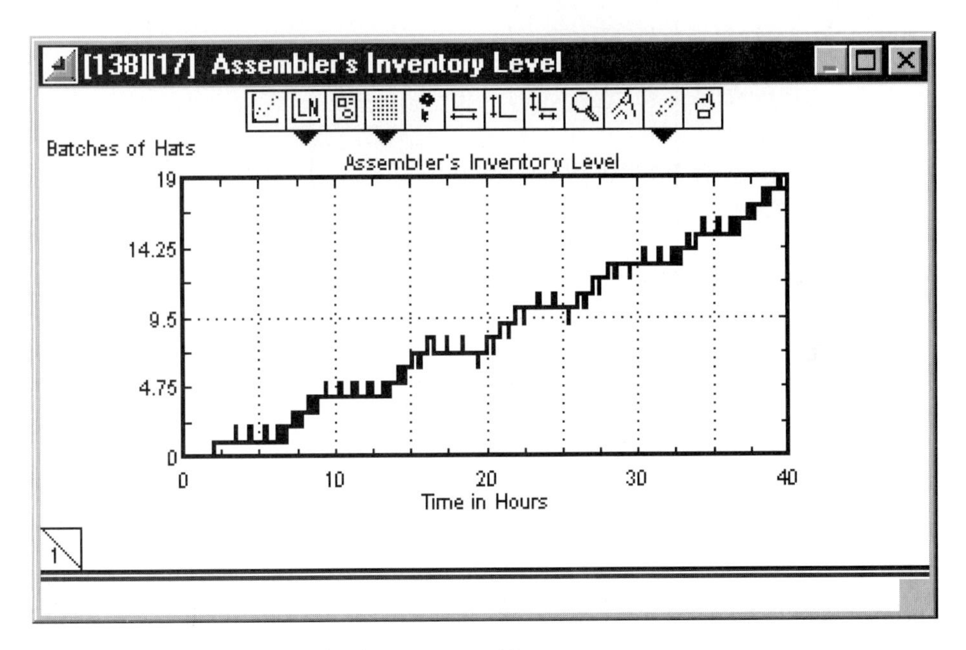

Figure 14.31 Graph of Beginning Inventory over Time.

This is a slowly inclining graph that even has some dips. It is unlikely that such a trend would be noticed in a pilot program. Given that there is at least one short downward slope, the data could mistakenly be viewed as positive. If the pilot program had run for a longer period of time, these trends *might* have been noticed, but there is no guarantee.

Modeling Reengineering Change Number 5

Change Number 5 resulted in the rehiring of the expediter as an assembler *and* an increase in deliveries of raw material. This decision was based on the fact that the assembler was a bottleneck in the process and was still performing at a 60 percent level. The consultant modeled the reengineered assembly task as shown in Figure 14.32.

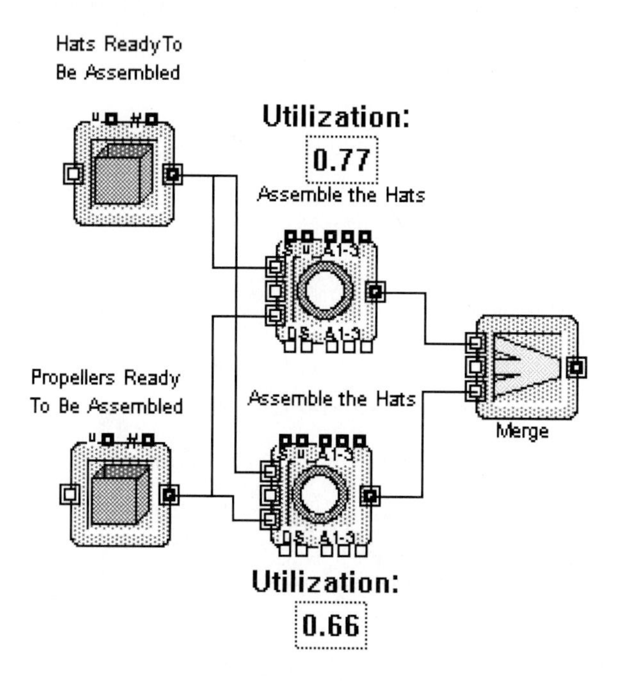

Figure 14.32 Model of Assembly Task with Two Assemblers.

In this portion of the model, there are two Repository blocks. One is a storehouse of finished hats, and another is a storehouse of propellers. There are two Operation blocks, each representing the activity of an assembler. Utilization of an Operation block (and, therefore, of a worker) is calculated automatically in Extend+BPR. The utilization of each worker is shown above each of the Operation blocks. Combined, the two workers are busy slightly more than 70 percent of the time.

The consultant then spoke with the hat and propeller makers about why they were making errors. They explained that when a stack of raw material grew too large, they felt under pressure and began to hurry their work. The size of the stack that triggered this problem was estimated to be about 16 batches of inventory.

The consultant captured this in two separate parts. The first part established an attribute called GoodorBad for each task. The value of this attribute was 1 if the size of a stack was less than 16 batches, or a random number between 1 and 10 if the size of the stack was greater than 16. Figure 14.33 shows how the value of the attribute was determined in the model.

First, the size of the input stack was measured every time a new task was initiated. This is shown in Figure 14.33 as the named connector Size of Stack. The value of Size of Stack was then input into an Equation block, as shown in Figure 14.34.

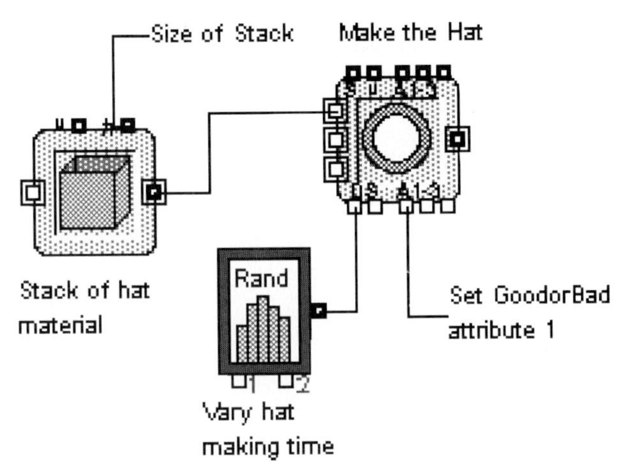

Figure 14.33 Determining GoodorBad Attribute When Inventory Is High.

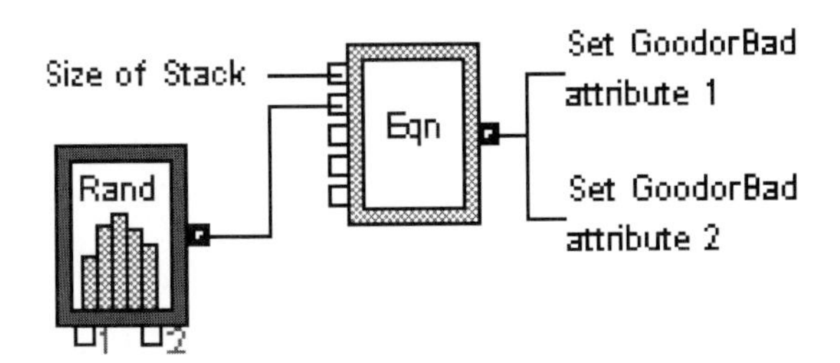

Figure 14.34 Size of Stack Value Read into Equation Block.

The result of the Equation block was then input to the Operation blocks that represented the activities of the workers. The logic of the Equation block is shown in Figure 14.35.

If the size of the stack were less than 16, the value of the Equation block would be 1, which would then be input to the attribute. Otherwise, it would be set to a random number, which would be interrogated later in the model, as explained earlier in this chapter.

This particular example points out the danger of focusing on only one aspect of a process. In the case of the workers, they focused on productivity, since they thought that the cause of rejects had been eliminated. It is true that they had eliminated one cause of rejects, but it was possible that there could be others. In fact, Reengineering Change Number 1 had been based on a *feeling* that the hat and propeller makers had. Although that feeling was related to the presence of the expediter, there was some indication that these workers occasionally felt pressured.

This is a case in which the "warm, fuzzy" aspects of TQM and CPI could have been useful. Knowing that the workers felt pressured under some condition, a model could have been analyzed to see if it could provide any information about what those

[45][28] Equation

Computes an equation.

Functions by name

Functions by type

OK

Cancel

Output

Result

Input1	Input2	Input3	Input4	Input5
stacksize	rejectfactr	Var3	Var4	Var5

Enter an equation in the form: result = formula;

```
IF (stacksize > 16) result = rejectfactr;
ELSE result = 1;
```

Comments

If either stack is greater than 16 batches, there is a chance that there will be rejects in the batch.

Help

Figure 14.35 Setting Rejection Factor.

factors might be. This particular model did provide some clue, such as the size of the input stack. Although the size of the stack might not have been the factor that created pressure, the fact that the size of the stacks was growing so large is important information in any event.

Running a simulation of the process after the change, even for a short period of time, would have predicted the effect of the change. Figure 14.36 uses a graph to show how inventory would have increased over time.

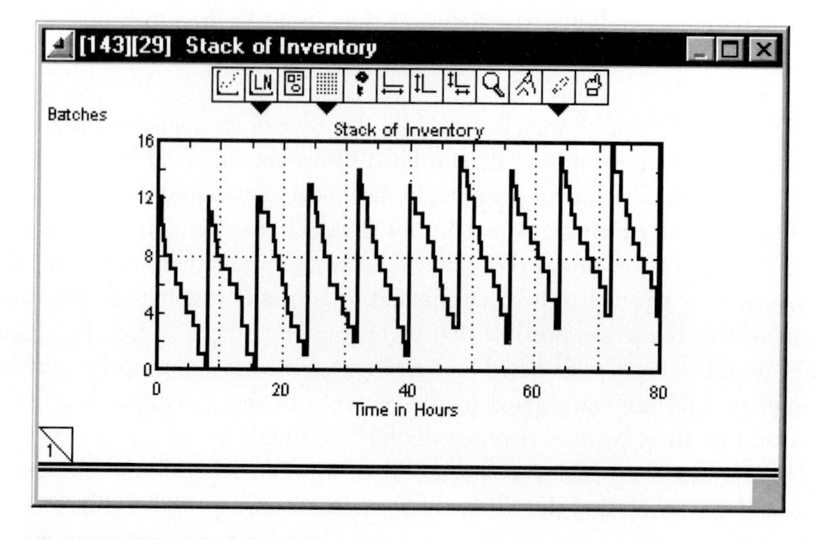

Figure 14.36 Graph of Inventory Levels.

This graph indicates a steadily increasing amount of inventory available to the hat makers and would have indicated that there was probably an assumption made that was not correct.

Arriving at the Decision to Upsize

The consultant looked at the history of the changes that had been made and noticed a pattern of shifting bottlenecks. It was true that some tasks had spare capacity and others were overloaded. He decided to do some what-if analysis and tried adding another worker at the beginning of the process where inventories were building. Instead of the former inspector moving from task to task, the consultant applied this worker to one task and added another worker to the other (Figure 14.37).

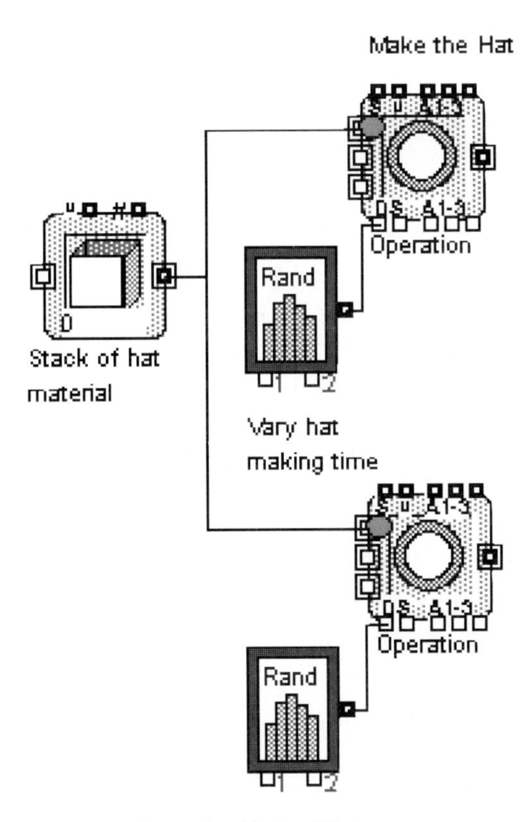

Figure 14.37 Hat Making Task with Two Workers.

The consultant calculated the cost per good hat, assuming that there would be no rejects (Figure 14.38).

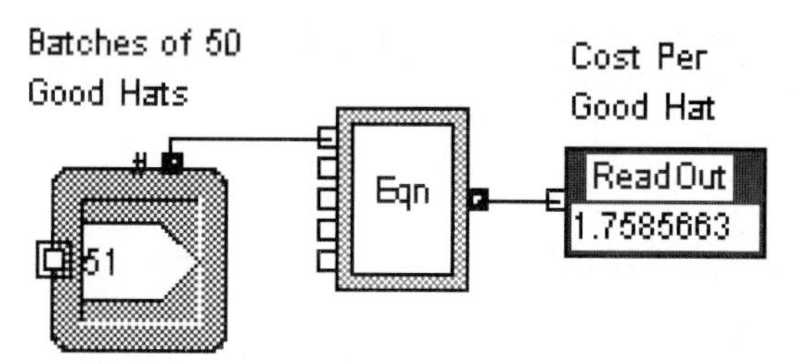

Figure 14.38 Cost per Good Hat in Upsized Process.

The productivity of the process was the highest it had ever been, almost 3800 hats per week, and the cost per good hat was the lowest it had been at $1.82. The consultant pointed out that predictions of rejections could only be based on the information given and that a careful review of the parameter most recently associated with rejections (size of inventory stack) showed that it would not be a problem. The consultant used the graph shown in Figure 14.39 to prove that point.

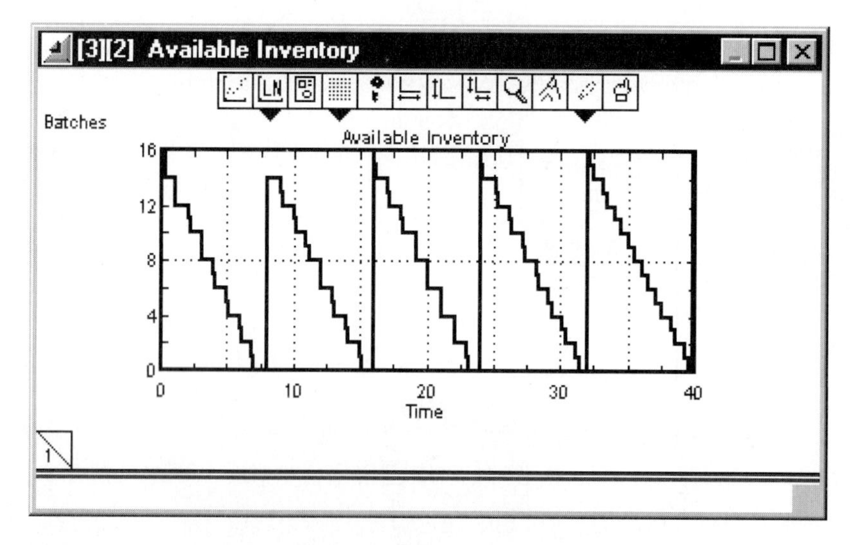

Figure 14.39 Graph of Just-in-Time Inventory.

The size of the input stack was greater than 16 for only very brief periods of time. The consultant and the owner met with the hat and propeller makers and assured them that, if this was the case, the problem was not theirs. Empowerment, explained the consultant, did not mean that someone had to assume responsibility for every aspect of a process. This graph also showed that the hat and propeller makers were working at almost exactly top capacity and the just-in-time delivery of raw materials was exactly that; however, the consultant was concerned about the work load on the assemblers, so the consultant graphed their percentage of time busy (Figure 14.40).

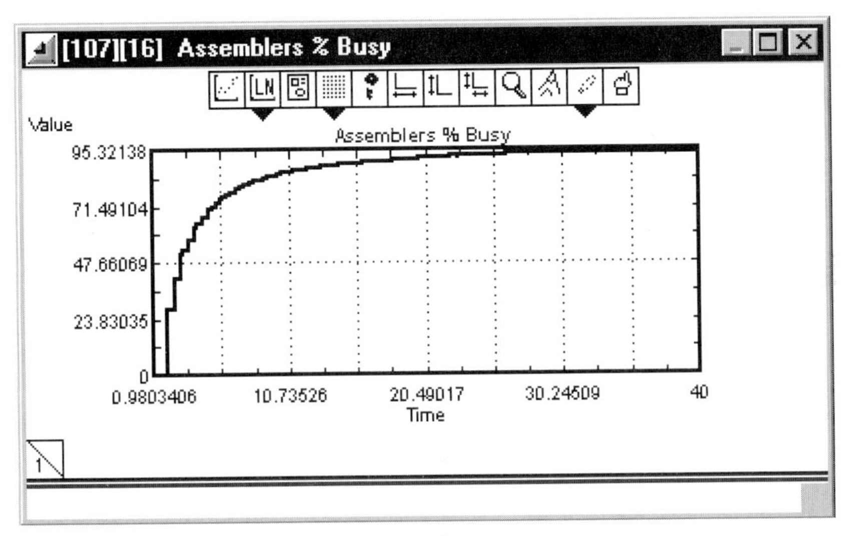

Figure 14.40 Assemblers Percent Busy Graph.

The assemblers leveled out at about 95 percent busy. The consultant suggested that the owner back off deliveries and build some slack into the process so that an event, like a worker becoming ill, would not have disastrous effects. This would raise costs slightly, but they would still be lower than they had been. Everyone agreed that this implementation of the process was the best, and it was a matter of fine-tuning it to get maximum value out of it, while at the same time avoiding quality problems.

Moral of the Story

It is very unlikely that any group of employees or managers would have independently reached a decision to increase staff. Computer aided process reengineering, on the other hand, made it possible to reach this decision painlessly and without the need for a pilot program. Moreover, it permitted continuing investigation of a process even after a supposed optimum process had been found. The primary moral of the story is that TQM and CPI, although important factors in process reengineering, cannot by themselves provide solutions to business problems.

Idle Time as a Process Measure

There is also a secondary moral. Idle time of workers is often viewed by management as an indication of the overall effectiveness of a business process. There is always the temptation to give a worker some other task to fill his or her idle time, thereby making him or her more productive. The fact is that idle time, in and of itself, is not an effective measure of an overall process, but only a measure of a part of a process. Moreover, process measures alone cannot determine the effectiveness of a process. That can only be done through analysis of the entire process.

Consider the following example. A worker is found to have 50 percent idle time. Management decides that the worker can perform an additional task that requires approximately the same level of effort as his primary task. Figure 14.41 shows how this would be modeled.

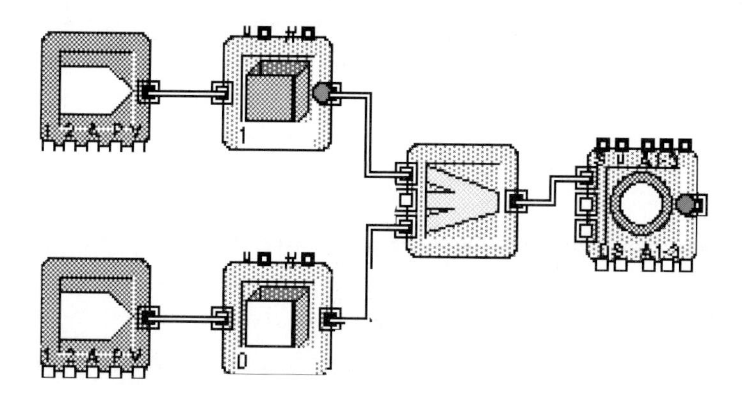

Figure 14.41 One Worker Performing Two Tasks.

The bubble in the Operation shows that there is one task in progress, and 1 in the Primary Task Repository shows that there is one task waiting to be performed. The worker performs well under the conditions that exist, since the amount there is an even, predictable flow of work to be performed.

However, change the parameters of the process slightly, and it begins to break down. The model in Figure 14.42 shows what happens when the worker's primary task load is increased 10 percent.

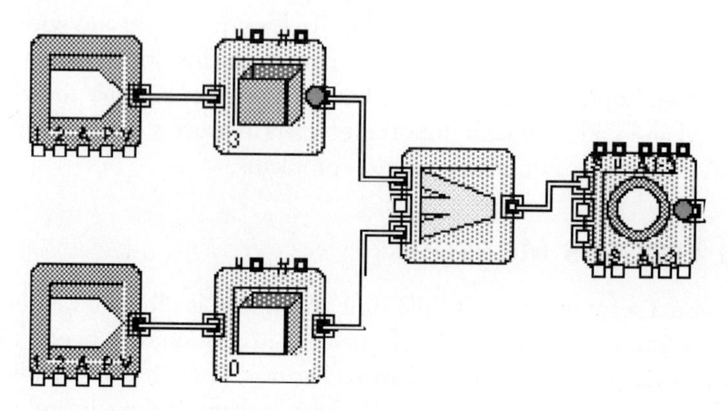

Figure 14.42 Effect of Increase in Work Load.

A backlog of work is beginning to build. When this happens, the typical management reaction is to view the worker as nonproductive when in fact the system, in which the worker is only a part, has changed. This is an example of Deming's theory it is the system, not individual skills, determines how workers perform.

15

Process Reengineering Case Studies Revisited

The Software Engineering Institute (SEI) at Carnegie-Mellon University has, on several occasions, been requested by the U.S. government to analyze and reengineer several software development and maintenance processes for the purpose of increasing the productivity of those processes. Two of those process reengineering studies are described in this chapter. In each of the studies, it was assumed that significant productivity gains had been achieved; however, the studies were done without the benefit of CAPRE technology, and later simulation of these processes with Extend+BPR demonstrates that these assumptions were not exactly accurate.

Case Study 1: Flight Software Maintenance

A branch of the Armed Forces contracted with the SEI to prepare a study of the software maintenance processes for two fighter jets, herein referred to as Jet A and Jet B. The customer's goal was to understand why the Jet A program released about 25 percent more software maintenance updates every quarter than the Jet B program. The SEI assembled a team of process reengineering facilitators to study the software maintenance process for Jet B, the first action of which was an attempt to eliminate some obvious reasons for the discrepancy in productivity. The team asked such questions as

- Were the personnel for the Jet A program more talented?
- Was the software for Jet A easier to maintain than Jet B's?
- Were there more personnel for the Jet A program than the Jet B program?

The customer claimed that the complexity of the software, the level of staff, the experience of the staff, and the budget were essentially the same for each program.

A team of four process reengineering facilitators met with government personnel to determine the purpose of the study. After meeting with the customer, the team established a goal: Determine the reason for the difference in *software maintenance changes per budgeted dollar* between the two programs. To learn more about the software maintenance processes, the team took the following process reengineering steps.

1 Personnel from each jet program were interviewed, and detailed information was collected concerning each task in the overall process.

2 The information collected about each task was documented, and the personnel interviewed were given the opportunity to validate the documentation. After several iterations of review, the documentation was accepted as valid.

3 Process maps were developed to provide overall views of each process.

The overall process for each jet program was basically the same, and the process maps (greatly simplified) for each program looked like the diagram in Figure 15.1.

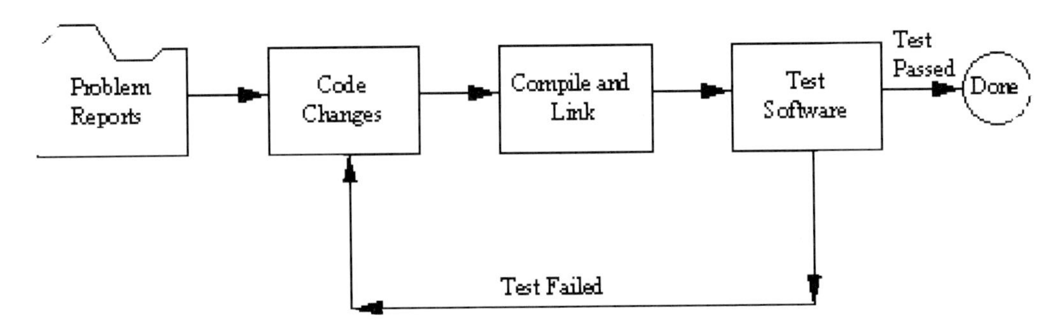

Figure 15.1 Process Map of Software Maintenance Process.

Each process consisted of the same tasks. In each case,

- A set of required software changes was established each quarter; the number was approximately the same for each jet program.
- Software personnel coded the changes, periodically submitting the code to be compiled and integrated into the overall aircraft system software.
- The software was then tested and either accepted or, if errors were found in the new code, returned to the coders for additional work.

However, there was one subtle difference. When the software maintenance personnel for the Jet A program discovered an error in code that had been changed, whether it was a preexisting error or an error that had been introduced in the coding task, it was logged as a new error and reintroduced into the set of changes that was to be made. The process map for the Jet A software maintenance process, therefore, looked like Figure 15.2.

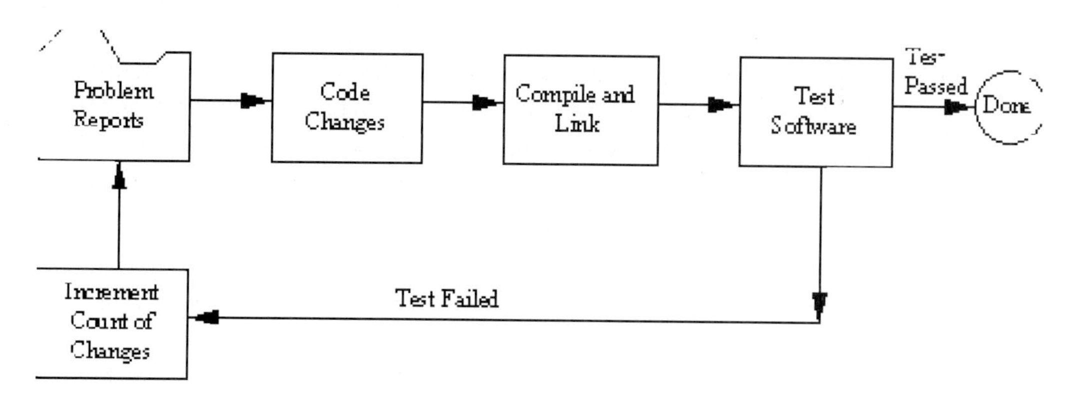

Figure 15.2 Jet A Software Maintenance Process Flowchart.

Both personnel in the Jet A and Jet B processes worked on returned software as soon as they received it; however, the personnel in Jet A increased their count of required changes to reflect rework. Depending on how this information was viewed, this was either an inflated count or an accurate count of code changes. The branch of service funding the study decided that the count was inflated, and the practice of logging errors introduced in the maintenance process was stopped.

A process parameter had not been accurately defined, and it was the interpretation of the parameter that gave the apparently different productivity numbers.

At this point, it appeared that the productivity discrepancy had been explained. To make sure the study had been thorough, the team decided to review both the Jet A and Jet B processes in more detail and measure other process productivity parameters, such as time to completion for each change, to determine whether differences existed between the processes.

Further investigation of both processes revealed that the compilation and system integration (hereafter referred to as "compilation") time for Jet B software was five days, whereas the time for compilation for Jet A software was less than one day. This seemed to make no sense, since compilation of software is normally a task that requires a few hours, at most. The personnel from the Jet B program explained they were using an obsolete version of a compiler that introduced errors, and those errors could only be corrected by hand. It was the effort of hand-correcting errors that resulted in the inordinately long compilation time.

Obviously, a big productivity inhibitor (bottleneck) in the process had been uncovered and there was potential for a significant increase in productivity. The process reengineering team recommended that the compiler be updated immediately and made an economic justification for doing so. The team explained that

- By eliminating the need to hand correct code, the customer was able to remove two software engineers from the software maintenance process.
- Those software engineers could be freed up to perform other tasks associated with the software, such as the development of new code.
- Based on the labor rates of the personnel and the cost of the compiler, the replacement of the compiler would pay for itself within one year.
- The elimination of the compilation task would increase productivity by reducing the overall time (cycle time) of the software maintenance process.

As in many other process reengineering studies, a change was made to a process *assuming* that obvious productivity gains would be made. There was little follow-up on the study; in fact, there was no question in any team member's mind that the study had been successful. Unfortunately, the team did not have CAPRE technology available to test its assumption. Had such technology been available, the team would have discovered that the estimations of cycle time reduction were not completely accurate.

The following model of the Jet B software maintenance process, developed with Extend+BPR, demonstrates that reengineering efforts that focus on obvious productivity bottlenecks do not always produce expected results.

The model of the original process is shown in Figure 15.3. The model operates under the following rules.

- A set of 100 required software changes is defined.
- Two software engineers work on the changes in parallel, and the amount of time required for a change is between one hour and one and one-half days.
- As changes are coded, the software is placed in a queue for use by the personnel who will perform the compilations, hand corrections, and integration.
- When either five days elapses since the last compilation, or when 20 changes are ready to be compiled, the compilation personnel begin their task.
- The integrated code is given to test personnel (in this case, pilots of the fighter plane) who load the software and perform flight tests. Testing is done at a rate of between one and five changes per day.

Notice that the model consists of hierarchical blocks, one of which is called Generate Event. This block is used to determine if 20 changes are available for compilation, or if five days have passed since the last compilation.

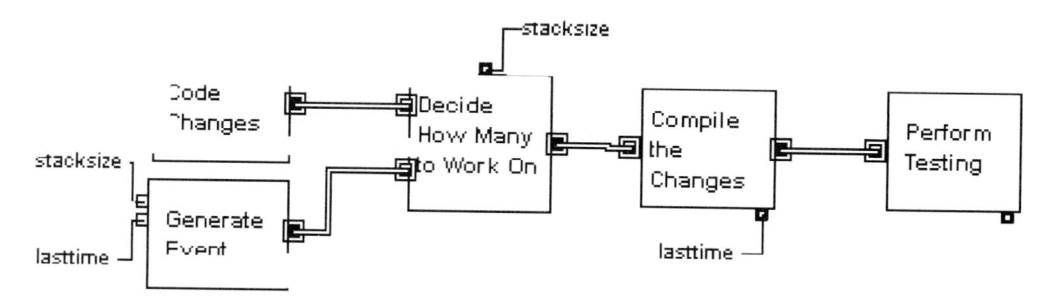

Figure 15.3 Overall View of Original Process.

Figure 15.4 shows the contents of the hierarchical block labeled Code Changes.

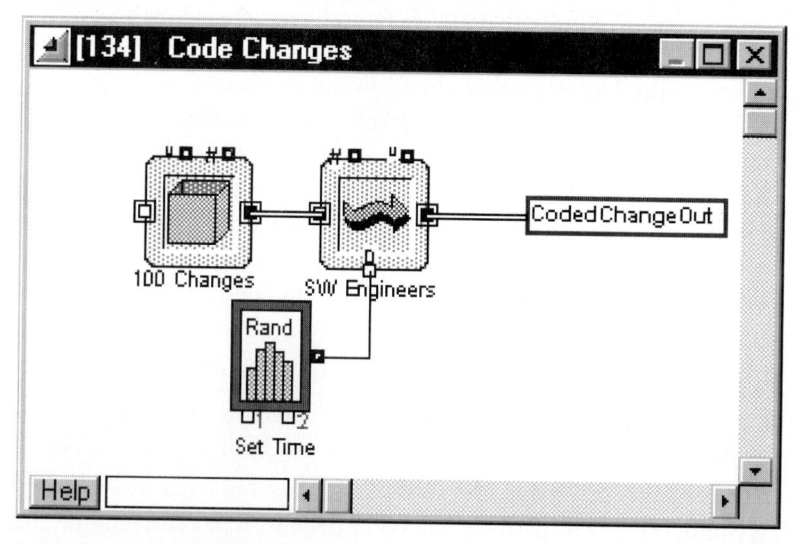

Figure 15.4 Code Changes Block.

This hierarchical block consists of three Extend+BPR iconic blocks:

- A Repository that contains the 100 changes that are to be made.
- A Transaction block that represents two software engineers working in parallel.
- An Input Random Number block used to specify the amount of time required to code a change.

The time required to perform the coding task is represented in terms of eight-hour work days, as shown in Figure 15.5. For example, a one-hour unit of time is represented as one-eighth, or 0.125, of an eight-hour day. One and one-half days is represented as twelve hours. The Input Random Number block represents a range of time from one hour to twelve hours with no particular distribution. Correct representation of time is an important consideration whenever one is developing a discrete event simulation.

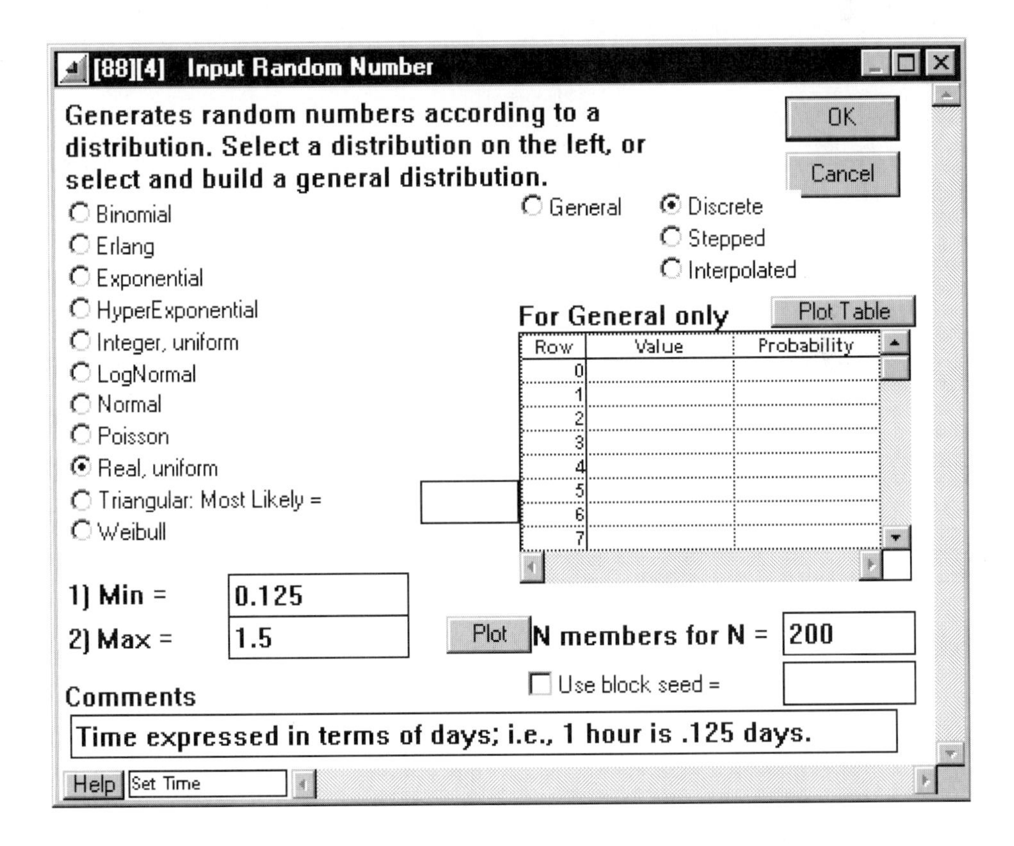

Figure 15.5 Random Number Expressed in Terms of Days.

Immediately below the Code Changes block shown in Figure 15.3 is a block labeled Generate Event. This hierarchical block determines if the compilation task will occur. If process conditions indicate that the compilation of coded changes should take place, then this block will generate an item that will represent an event. If compilation should not occur, then no item will be generated. The contents of this block are shown in Figure 15.6.

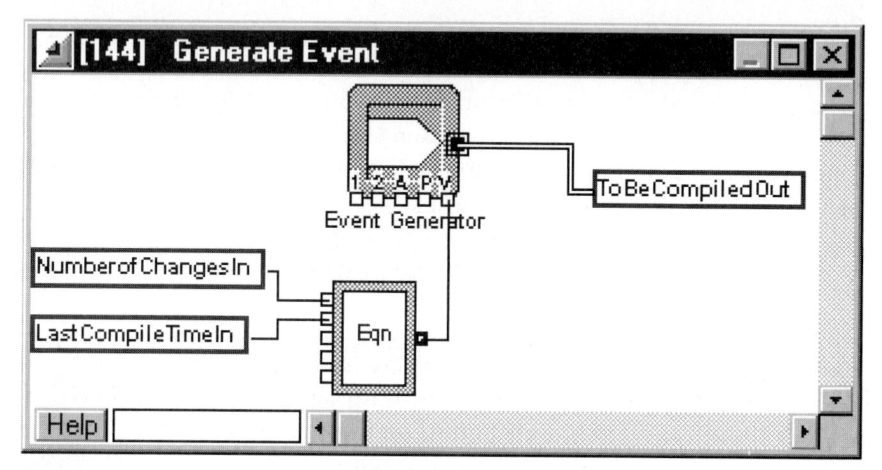

Figure 15.6 Contents of Generate Event Block.

The block consists of an Import block and an Equation block with two inputs. The input labeled NumberofChangesIn represents the number of changes that are available for compilation, and the input labeled LastCompileTimeIn represents the time at which the last compilation was completed. The logic used in the Equation block is shown in Figure 15.7.

[91][1] Equation

Computes an equation.

Functions by name | Functions by type | OK | Cancel

Output: Event

Input1	Input2	Input3	Input4	Input5
StackSize	LastTime	Var3	Var4	Var5

Enter an equation in the form: result = formula;

```
IF(StackSize >= 20) Event = 1;
ELSE IF (currenttime >= (Lasttime+5)) Event = 1;
else Event = 0;
```

Comments

If 20 changes are ready, or if 5 days have elapsed since last compilation, create an event; otherwise, no event.

Help

Figure 15.7 Logic within an Equation Block.

In everyday terms, the logic in this block is read as "If there are 20 or more changes ready, then begin the compilation task (generate an event); otherwise, if there are not 20 changes ready, but five days have elapsed since the last compilation, then begin the compilation task; otherwise, do not begin the compilation task." The output of the Equation block specifies the number of items the Import block should generate, that is, 0 or 1. The item generated by the Import block is an example of an "abstract" item, one that has no physical properties, but an item that can be used to control other events in a simulation.

Next, the hierarchical block labeled Decide How Many to Work On determines the number of coded changes that will be compiled. The details of this block are shown in Figure 15.8.

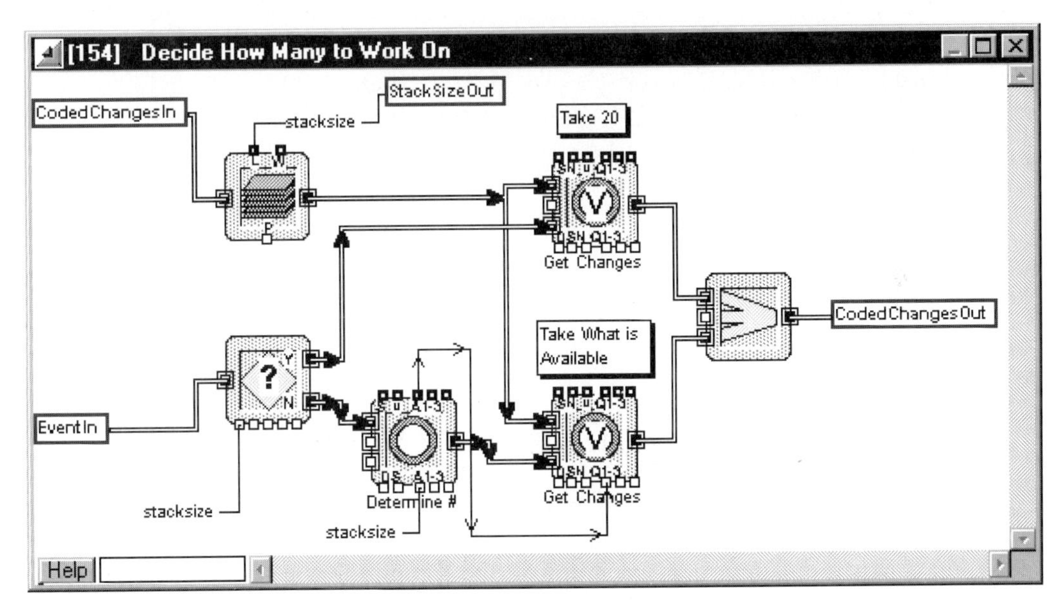

Figure 15.8 Details of Hierarchical Block Decide How Many to Work On.

Coded changes enter the hierarchical block and are held in the Stack block. When an event item is generated, it enters the Decision(2) block, where the number of coded changes in the stack is interrogated. If the number of coded changes is greater than or equal to 20, the event item is sent to the Operation, Variable block at the top of the block; otherwise, it is sent to the Operation block at the bottom of the model. The Operation, Variable block simulates an operation on a variable number of items, and the dialog of this block is shown in Figure 15.9.

The Q1–Q3 input connectors on the bottom Operation, Variable block are used to determine the number of items taken at each of the input connectors. In the case of the Operation, Variable block shown in Figure 15.9, the number of items at connector Q1 is 20, representing the coded changes. The number of items taken at Q3 is 1, representing the event. Notice also that two attributes, ToBeCompiled and

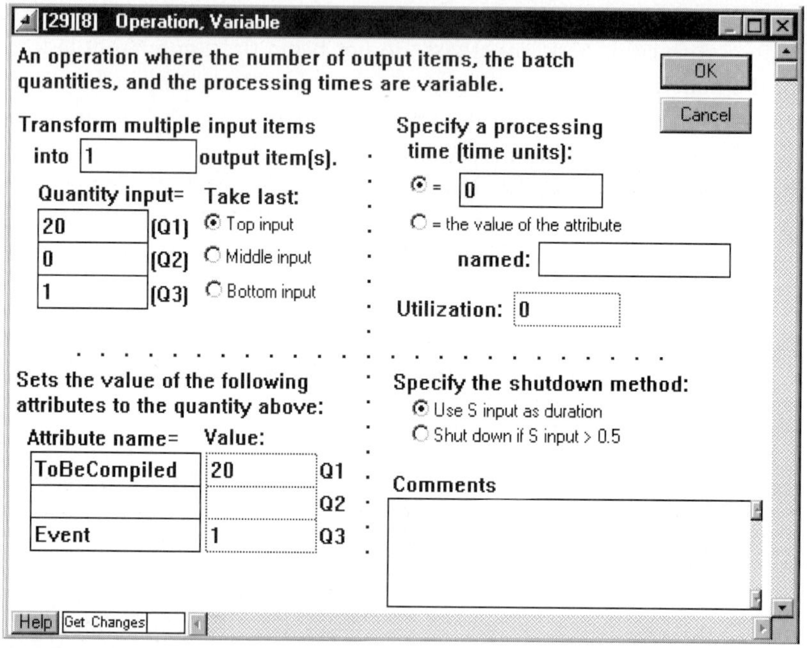

Figure 15.9 Dialog of an Operation, Variable Block.

Event, have been set, representing the number of each type of item taken. The ToBeCompiled attribute will be used later in the simulation. Also in this dialog, the Take last button is checked at connector Q1, meaning that the coded changes will not be taken by the block until the event is taken.

When the event item exits through the Y connector on the Decision(2) block, it is sent to the Operation, Variable block on top of the model and 20 coded changes are taken. When the event item exits through the N connector, it is sent to the Operation block on the bottom of the model. The Operation block determines how many items are in the stack and instructs the Operation, Variable block to take that number.

The items representing the coded changes are then sent, as one bundled (batched) item, to the hierarchical block labeled Compile the Changes. This hierarchical block simulates the actual five-day task of compiling the coded changes. The details of this hierarchical block are shown in Figure 15.10.

The batched coded changes enter the Operation block and are held for five time units, representing five days. Also, an attribute is set to the time at which the changes exit the block, and this value is sent to a block called the Retain block, which simply stores the time and outputs it through the hierarchical connector called FinishTimeOut. This is the value sent to the Equation block that controls the generation of events at the beginning of the simulation.

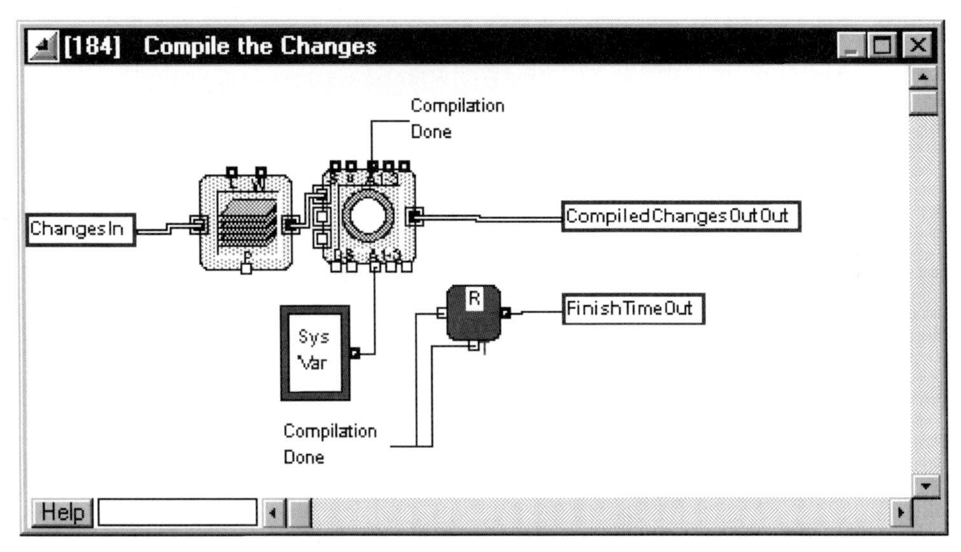

Figure 15.10 Details of Compile the Changes.

Finally, the changes are sent to the hierarchical block labeled Perform Testing which simulates the testing of the completed changes. When this simulation is executed, it reveals that, on average, the 100 changes are completed in 51 days.

Then to test the effect of technology insertion on the process, Extend+BPR was used to develop a model of the software change process with the compilation and hand-correction tasks removed. This model is shown in Figure 15.11.

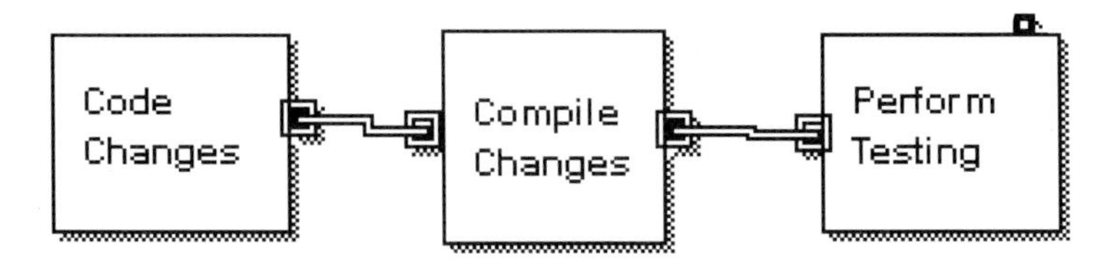

Figure 15.11 Simplified Process with Technology Insertion.

This model represents a simplified process in which software changes are coded and compiled and then sent directly to test personnel.

Both models were then executed simultaneously, defining the productivity for the software engineers and the testers to be the same values. Figure 15.12 is a graph showing the results of the simulations.

The lower graph, labeled "1," is the graph of the original process, and the upper graph, labeled "2," is the graph of the improved process. The difference in time to completion between the two processes varies by no more than five to eight days, no

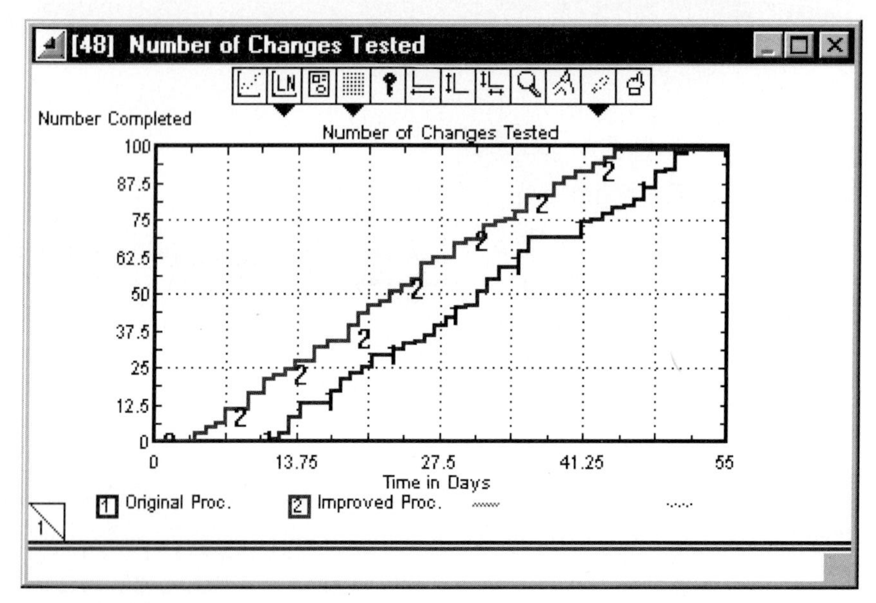

Figure 15.12 Graph Showing Time Required to Complete 100 Changes for Current and Improved Processes.

matter how many software changes have to be made. Moreover, the difference depends entirely on how quickly the first set of changes becomes available for the first instance of the compilation task.

The explanation for this is simple. Coded changes were collected and submitted in batch to the compilation task. While the compilation and hand corrections were being performed, more changes were being coded and tested in parallel. Because of this parallelism, the impact on process time was actually very minimal, and the five- to eight-day delay represents the delay in delivery of the *initial* set of changes to the testing group.

It is interesting to note that the process reengineering team never met its original primary goal, that is, an increase in *software maintenance changes per budgeted dollar*. In fact, after the practice of logging rework as software changes was ended, the number of changes per budgeted dollar dropped. To achieve the original goal, the team had to find a way to meet a secondary goal—an increase in productivity. Again, this goal had not been achieved, since cycle time had not been reduced significantly.

Nonetheless, the reengineering effort was considered a success. The replacement of the compiler was, however, beneficial in terms of other process parameters. For example,

- The confidence level associated with the software changes was increased.
- The number of errors introduced by coders was reduced.

- The personnel associated with the compilation task were freed up to do other work.

However, the cycle time of the software maintenance process actually changed very little after the insertion of the new compiler.

This proves the point that *without simulation of a proposed process, the effect of changes on that process cannot be accurately predicted.*

The process reengineering team did attempt to model the processes described utilizing a state-transition-oriented Computer Aided Software Engineering (CASE) tool. The CASE tool allowed the team to present the process as one in which software went through state changes (that is, coded, compiled, tested, and so on) and it had an animation feature which followed the execution of the process, including execution of tasks in parallel. Unfortunately, the CASE tool lacked an easily implemented simulation capability, so the team was unable (at the time) to predict what would happen when changes were introduced.

When this study had been completed, it was thought to be a major step in the state of the art of process analysis. The use of a CASE tool allowed the process reengineering facilitators to develop an *animated flowchart*, so that the overall view of the process reflected in the resulting activity diagram had more meaning and communicated more information to the process reengineering facilitators. In retrospect, however, the CASE tool was used as an advanced Level 3 tool, not a Level 5 tool.

Case Study 2: Technical Documentation Modification Process

The Software Engineering Institute was requested by the U.S. Air Force to investigate the process used to modify the technical documentation associated with the flight control software of the F-16 jet airplane. These documents, known as *Technical Orders*, are used to provide information about the operation of the F-16 to the pilots. The Air Force had determined that the process was too slow, and that cycle time could be reduced by applying technology to the process.

The SEI formed a team of process reengineering facilitators who studied the process in question and made recommendations for changes.[1] This team investigated the process using the following steps:

[1]Hansen and Kellner, "Software Process Modeling" (see chap. 6, n.1).

1. They interviewed as many process participants as possible to determine the tasks that were performed and the order in which those tasks were performed.
2. They documented the information gathered in the interviews and asked for validation of the documentation from the participants. This was an iterative process, and validation was eventually received.
3. They developed an overall view of the process using dataflow diagramming techniques (Figure 15.13).

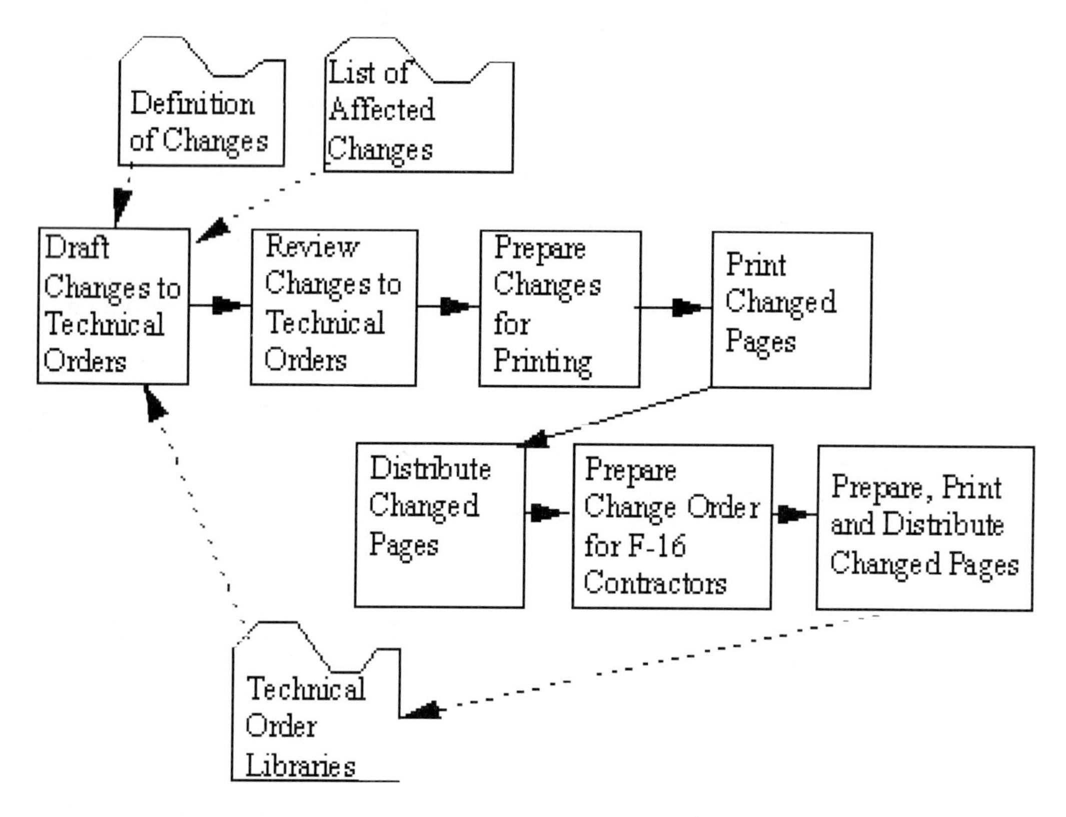

Figure 15.13 Technical Order Modification Process.

This simple seven-step process depicted in Figure 15.13 hides a great deal of detailed information; however, it is useful to present an encapsulation of the major tasks in the process. The process flowed as follows.

1. Software engineering personnel reviewed the Technical Orders that were affected by software changes and then modified the changes by *redlining* them. This was a completely manual process in which the personnel actually thumbed through documents and made changes in red pencil.

2. The modified pages were passed to a reviewer who either accepted or rejected the changes. The accepted pages were accumulated until notification was received that all software had been completely tested.
3. The modified pages were prepared for printing. This task included composing text, redrawing graphics, typesetting, and merging text and graphics into individual pages. This task was performed by both Air Force personnel and by contractors.
4. The prepared pages were then printed, usually by subcontractors.
5. The changed pages were distributed to the pilots and software engineers.
6. The changes were then passed to the prime contractors for the F-16, where the steps just outlined were basically repeated so that the contractors could include the modifications in the libraries they maintained.
7. The prime contractors reprinted the changed pages and distributed them to Air Force personnel.

This is clearly a process that suffers from redundancy and suboptimization and could be improved in many ways. The process reengineering team chose to focus on the first two steps in the process (redlining and review of changes), and the team applied sophisticated information management technology to those activities. An automated document management system was installed which allowed software engineers and document reviewers to perform their tasks electronically rather than manually.

For example, by using this technology,

- Software engineers could automatically find all references to text and graphics, eliminating the need for manual searches.
- Software engineers could redline documents electronically and insert both text and graphics in final form.
- Reviewers could modify the changes and electronically sign off on them.

Needless to say, productivity in the first two process tasks increased dramatically (between 200 percent and 300 percent).

The process reengineering team and Air Force personnel were satisfied with the results of this effort; however, it did not take very long to find out that the *overall* process time would not be changing. The only effect of technology insertion was that changed pages got to the other tasks more quickly; since the other tasks continued to operate at their current productivity levels, the overall process timing did not change. This can be demonstrated by looking at a model of the process built with Extend+BPR (Figure 15.14). This model uses custom graphics to add some intuitive meaning.

The redlining of the documentation was done by three software engineers. Because of the manual nature of the task, the time to complete each task was between two and three days, based on the complexity of the software affecting the documentation. Figure 15.15 shows how these parallel tasks were modeled using Extend+BPR.

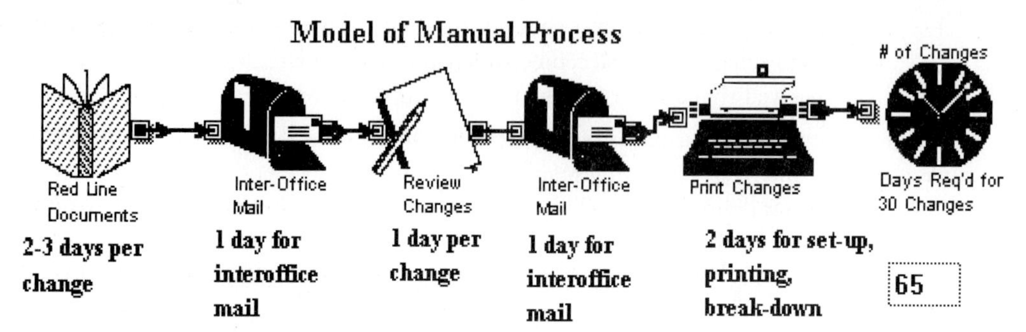

Figure 15.14 Model of Manual Process.

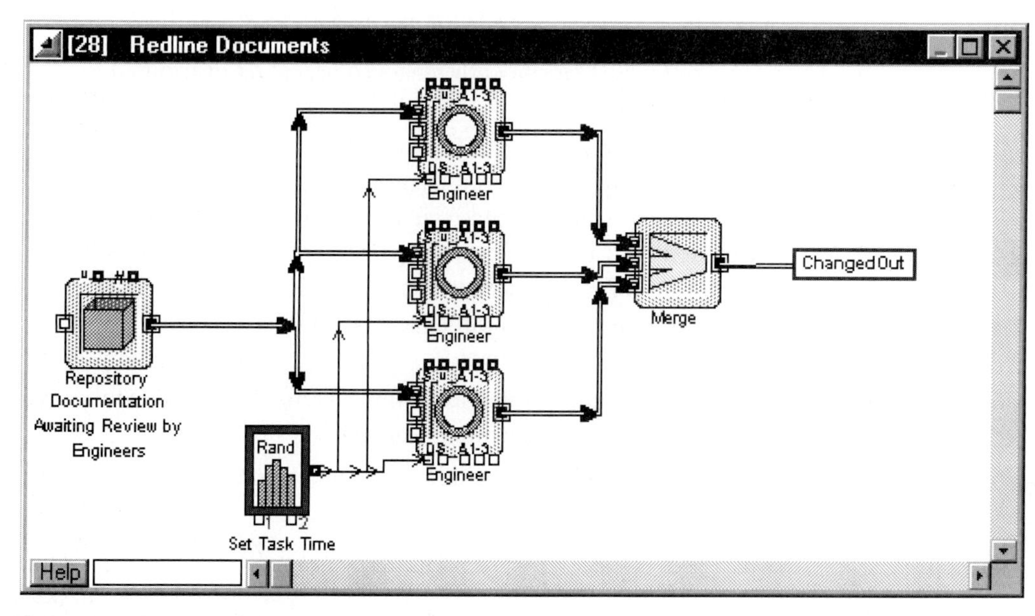

Figure 15.15 Model of Three Individuals Working in Parallel.

In the model shown in Figure 15.15, the three software engineering personnel work in parallel. The iconic blocks shown represent

- A Repository block, or the list of documentation changes to be made by the software engineers.
- An Operation block, or the actual task of making documentation changes.
- A Merge block, followed by a hierarchical connector, representing connection to another block.

In the model shown in Figure 15.14, there are 30 changes to be made, each requiring between two and three days of a software engineer's time. Each Operation block represents the activities of a software engineer, and each works independently of the others. Because of this independence, there is no relation between the Operation blocks regarding timing. The numbers used in the model are realistic, since a change could affect 50 pages of documentation contained in several different Technical Orders.

When a software engineer completed the documentation of a change, it was sent by interoffice mail to a person who would approve the changes. At this point, it can be determined that the time required to complete documenting and reviewing all 30 required changes is 33 days. Initially this seems surprising, but after some consideration, it makes sense. These are the parameters for the process:

1. The software engineers and the reviewer work in parallel.
2. Since there are three software engineers and 30 changes to be made, each software engineer will be responsible for an average of 10 changes.
3. Since each software engineer requires between 2 and 3 days to make a change, the average amount of time spent making changes will be between 20 and 30 days.
4. The interoffice mail happens *in parallel* with the software changes.
5. The reviewer, working at a rate of one change review per day, will require 30 days to perform his task.

Therefore, the total time required to (1) make 30 documentation changes, and (2) perform 30 reviews will be a maximum of 30 days (the reviewer's time) plus the amount of time required for the *first* change to reach the reviewer (two to three days plus one day for interoffice mail). Even with the facts just presented, this may be difficult to understand, so more explanation is provided.

- The software engineers will require between 20 and 30 days to make their changes. This means that, unless every change required 3 days, they would be done making changes in *less than* 30 days.
- Although interoffice mail requires 1 day, the net effect on the timing of the process is minimum, since the movement of changes by mail happens in parallel with all the other process tasks. While one set of documentation changes is in the mail, another set of changes is being started.

Therefore, the review process begins one day after the first change is completed. Since there are three software engineers working in parallel, it is likely that one would finish a set of documentation in a minimum of 2 days; therefore, the review process begins 3 days after the change process begins. Since the review process requires 30 days, the time required for changes to be made and reviews to be completed is 33 days.

The logic of all this is quite complex for a very straightforward process. The logic is complicated further by the fact that when a set of changes has been reviewed, the changes are forwarded by interoffice mail to the individuals who actually print the changed pages.

Those two tasks—*prepare for printing* and *print change pages*—will now be added to the model. There was one individual dedicated to the *prepare for printing* and *print change pages* tasks, and the two tasks combined required 2 days for each document being changed.

When these tasks are added to the model, the total time for the process becomes 65 days. Once again, this may seem surprising but, since all the activities are occurring in parallel, it is an accurate estimate of the timing of the process.

The process was then modeled to demonstrate the effect of technology insertion. For the sake of simplicity, the details of this process model will not be described. The model is basically the same as the model shown in Figure 15.11, except that the following changes were implemented.

1. The time required to redline a set of change pages was reduced from between 2 and 3 days to 1 day, representing a 200 percent–300 percent increase in productivity.
2. The time required for interoffice mail was reduced to 0 (zero) since electronic data interface technology was used. This took a one-day step in the process to no time—it can be viewed as an "infinite" increase in productivity.
3. The time required to review the changes was changed form 1 day to 1/2 day, representing another 100 percent increase in productivity.
4. The time for the second interoffice transfer was also set to zero.
5. No changes were made to the last task in the process, Print Changes, since no technology was applied to that process.

It is natural to assume that with all of the productivity increases in the process, cycle time would be greatly reduced; in fact, it seems *intuitively obvious*. When both processes are simulated in parallel, however, we see that the cycle time for the improved process is only 61 1/2 days, not very different from the original process! A model of both the original and improved processes, shown in Figure 15.16, demonstrates this fact.

This implies that the application of technology, which had the effect of reducing the amount of time to perform tasks *early in the process* and eliminated the delay caused by interoffice mail, actually had little effect on the *overall process* time. This reveals an important element of change that is often overlooked.

Improvements to a portion of a process may have little effect when the entire process is not improved to account for those changes.

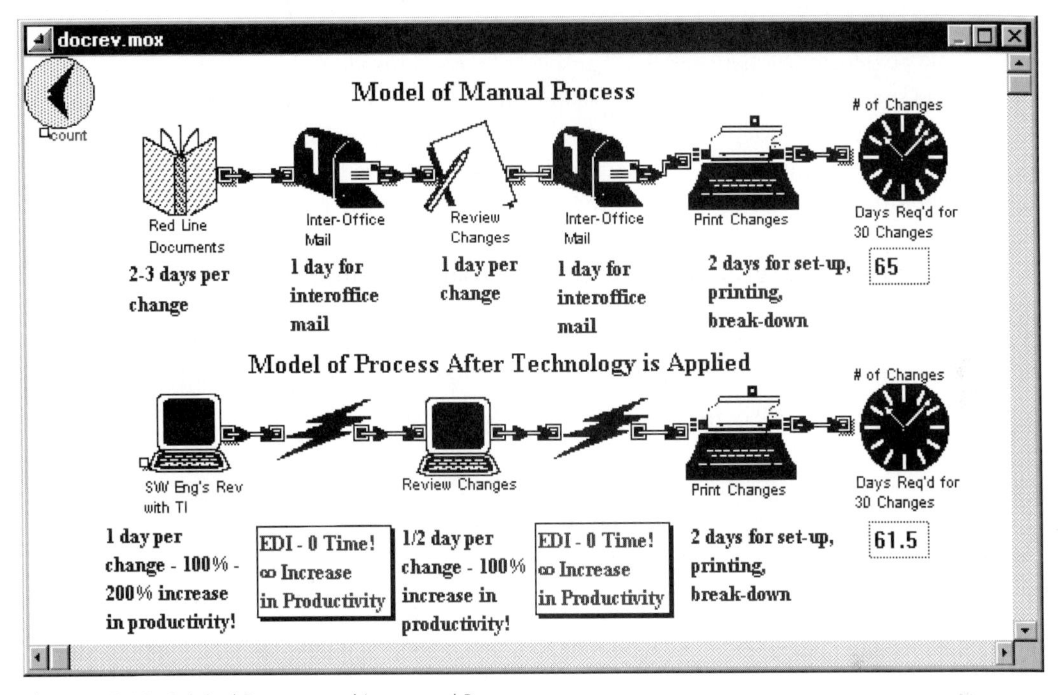

Figure 15.16 Original Process and Improved Process.

In other words, changes cannot be made in a vacuum, and suboptimization often results when changes are made to pieces of a process.

The original reasoning about the effect of technology insertion was based on the flowcharts of the process that had been developed. The process was viewed from the perspective of one change, and the cycle time of the process was seen as shown in Table 15.1.

Table 15.1 Perception of Documentation Process Timing

TASK	TIME	TASK	TIME
Write Changes	2–3 Days	Write Changes	1 Day
Interoffice Mail	1 Day		
Review Changes	1 Day	Review Changes	1 Day
Interoffice Mail	1 Day		
Print Changes	2 Days	Print Changes	2 Days
TOTAL	**7–8 Days**	**TOTAL**	**3 1/2 Days**

This table shows that the time required to process a *single change* was reduced by 50 percent, or between 3 1/2 and 4 1/2 days. This view of the timing of the process led the team to expect tremendous productivity gains. However, once technology was inserted, the error in the team's calculations was revealed. The time of the *overall process* was reduced by 3 1/2 and 4 1/2 days, since all tasks occurred in parallel, and the printing task now dictated the overall timing of the process.

Recognizing this, the team made recommendations to increase the speed of all the tasks in the process by changing the technology used in the latter stages of the process. There were many reasons for this rejection, collectively characterized as the *nontechnological barriers to technology transfer.* In this case the reasons could be summed up as politics, and the theory of *spheres of influence* directly applies. The spheres of influence that were prohibiting technology insertion are shown in Figure 15.17.

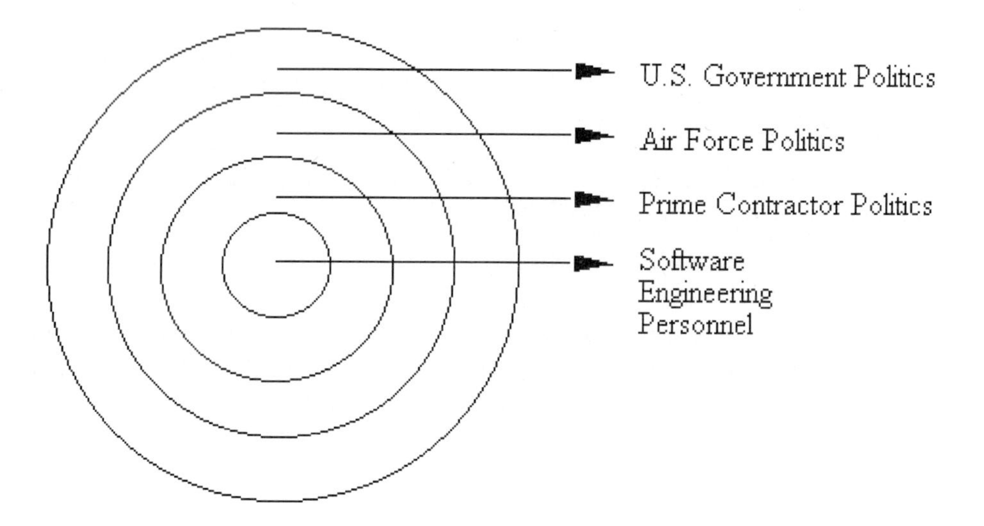

Figure 15.17 Spheres of Influence.

It is not important to discuss what issues were at stake that prevented an overhaul of the Technical Order modification process. It is only necessary to say that decisions were made in the past to invest in technology that would be rendered obsolete if the process recommendations were adopted. No one wanted to look foolish, so no changes were going to be made.

It is such spheres of influence that Deming has advocated eliminating and which Continuous Improvement theories, such as boundaryless organizations, also advocate eliminating. The reality of life is, however, that elimination of influence is a cultural change and will take a very long time to accomplish.

Spheres of influence can be attacked by facts, and computer aided process reengineering tools provide the facts that are required. In this study, the process reengineering team used a state-transition modeling CASE tool to analyze the Technical Order modification process. This tool provided an excellent diagramming and animation capability that allowed enhanced communication of the process to study participants; however, it did not give the team an easily implemented predictive capability. Therefore, the facts required to change the opinions of the management personnel in the outer spheres of influence were lacking.

Further simulation of the process using CAPRE technology reveals that literally millions of dollars could be saved in labor and cycle time reduced by orders of magnitude if the whole process were to be changed by the application of technology. This information, obtained through simulation, would have made the argument for change stronger and would have provided a greater opportunity to change the spheres of influence.

This case study also demonstrates that any process is a portion of a larger process, and that changes to a portion of a process may not result in the anticipated improvements. The changes implemented by the process reengineering team had the positive benefit of ensuring that all references to software documentation that had to be changed were, in fact, changed. The technology also improved tracing of changes to authors and accelerated the review process. In general, quality of the documentation increased.

On the other hand, expected productivity changes for the entire process did not improve. Time to completion is regulated by the slowest task in any process, so speeding up one or two tasks in a seven-task process had no effect at all on the overall process. This reinforces a major point of this book and Process Reengineering Rule 5—simulate the process. Simulation will reveal how a process will change, not how we hope it will change.

16

Applications of Computer Aided Process Reengineering

Although this book has presented the Rules of Process Reengineering sequentially, it is not always necessary to apply them in that order. It is highly recommended, however, that *all* the rules be applied when trying to reengineer a process. The following are actual examples of how effective process models were developed while varying the sequence in which the rules were invoked, and different ways CAPRE tools were utilized. These examples have been simplified for the sake of brevity and "sanitized" to conceal the identity of the organizations being discussed. Each reengineering effort was conducted by a process reengineering facilitator.

Example 1: Software Support Process

The manager of a software support organization that provided both off-line and on-line help to users wanted to use simulation to explain why the productivity of the group was less than he thought it should be. The goal was to reengineer the support processes so that time to completion (cycle time) for problem resolution was reduced. A facilitator working with the manager used this goal to define a process parameter by which the process could be measured.

> Establishing a goal early in a modeling effort allows a process reengineering facilitator to determine the attributes of the process that is to be investigated. This not only helps the modeling effort but also the task of interviewing process participants, since the facilitator can ask questions from a certain perspective.

Before interviewing the process participants, the facilitator asked for any documentation about the process that was available. He was given both written documentation and flowcharts. Therefore, Process Reengineering Rule 2 (document the process) and Process Reengineering Rule 3 (define the process) had already been applied to the process, and it was, to a certain degree, defined. After reading the documentation and reviewing the flowchart, the facilitator formulated some questions and began to interview software engineers (Process Reengineering Rule 1).

It did not take very long before one software support analyst vented his frustrations and explained that, in addition to reviewing and solving problems that had been submitted in writing, support personnel also had to answer telephone calls. In fact, answering the telephone, or real-time response, was given priority over problem solving. After discussing the process in a Continuous Process Improvement meeting, the support personnel had suggested to management that one person should handle all the phone calls while the others handled written problems. This request had been ignored, since management had concluded that (1) there was time to perform both activities, and (2) with three support personnel, if only one took phone calls, there would be a 33 percent reduction in problem-solving productivity.

The documentation of the process and the flowcharts treated off-line and on-line support as separate processes when, in fact, they were both part of one process. Unfortunately, the interaction of the two processes could not be captured either in text or in diagrams. Therefore, in an effort to understand better the interaction between the off-line software support process and the "phone call" process,

- The facilitator applied Process Reengineering Rule 4 and collected some data (measured process parameters) about each process.
- Using Extend+BPR, the facilitator began modeling and simulating both processes and the interaction between the processes (Process Reengineering Rule 5).

The data the facilitator collected included the following information:

- Written problem reports arrived at a rate of about four per hour, arriving either by mail, fax, or electronic mail.
- Problems required an *average* of 12 minutes each to resolve.
- Phone calls arrived at a rate of one per half-hour.
- The *majority* of phone calls required only eight minutes to resolve.

The facilitator used this information to develop models of both the off-line (written) problem resolution process and the on-line (phone call) problem resolution process. The facilitator first modeled the off-line process (Figure 16.1).

In this model, the facilitator used a Random Number block that was set to output values in a normal distribution with a mean of 12, representing an average of 12 minutes per problem. The output values were used to set an attribute for incoming problem reports, and this attribute was then used to establish the timing of the

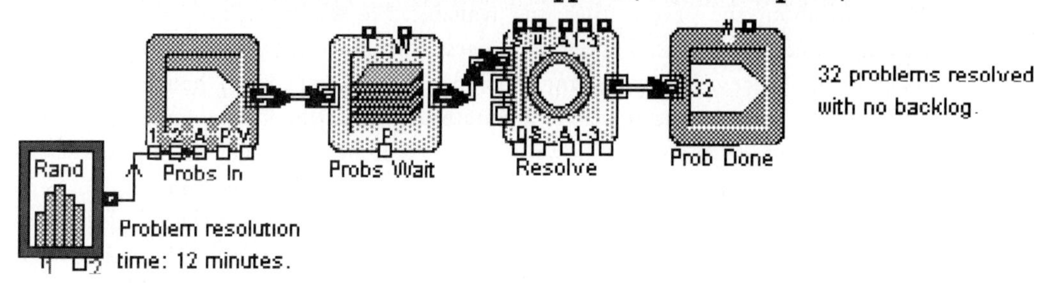

Figure 16.1 Off-Line Problem Resolution Process.

Software Engineer Task Operation block. At the rate of 12 minutes per problem, a software engineer could handle 5 problems per hour, or 40 in an eight-hour day; however, since problems only arrive at a rate of 4 per hour, the model revealed that a software engineer completes an average of 32 problems per day.

To determine the effect of having software engineers answer the phones and provide on-line support, the facilitator next modeled the on-line support process (Figure 16.2). In this model, the Random Number block was used to generate timing for the phone calls using the following probabilistic distribution: 50 percent (the majority) of phone calls required 8 minutes, 20 percent required 12 minutes, 20 percent required 17 minutes, and 10 percent required 20 minutes.

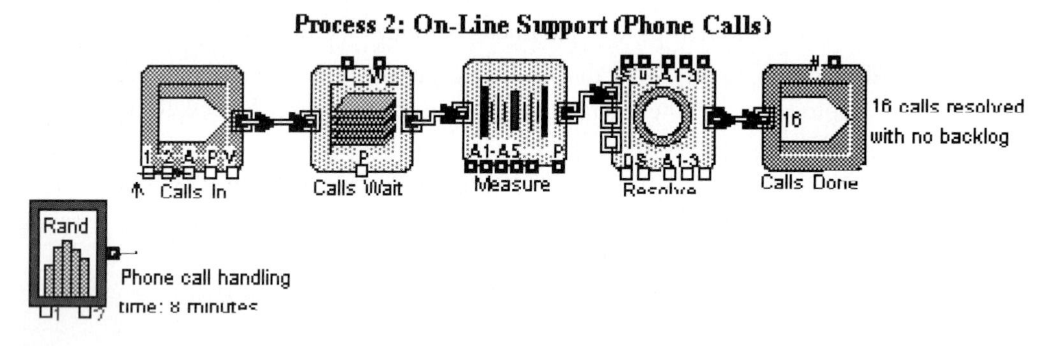

Figure 16.2 Model of Telephone Answering Process.

The dialog of the Input Random Number block (Figure 16.3) shows how a custom distribution can be created in Extend+BPR.

The results of a simulation of the on-line process representing one eight-hour day revealed that a software engineer could handle an average of 16 calls per day.

Given this information, the manager of the group concluded that he was correct in his assumption, that is, with a little extra effort, software support personnel

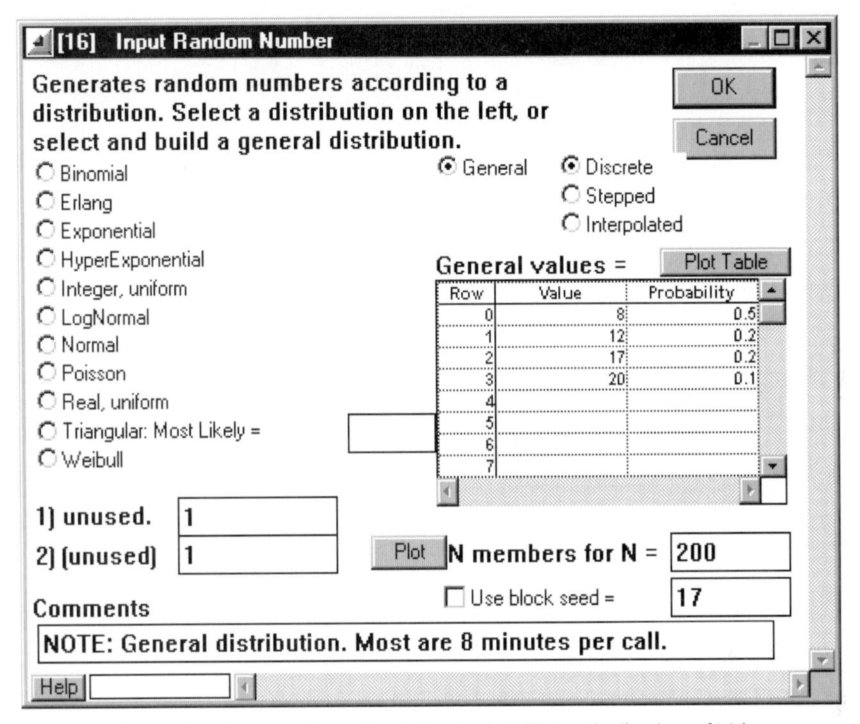

Figure 16.3 Random Number Block Providing Probabilistic Distribution of Values.

could handle both off-line and on-line support. His logic was the following: Thirty-two problems at 12 minutes each required 384 minutes, and 16 calls at an *average* of 8 minutes each required 128 minutes. That totaled 512 minutes, or slightly more than an eight-hour day. He reasoned that his personnel were professionals and would work the required extra 30 minutes. This math did not explain why productivity levels were not being met.

The manager, unfortunately, had made a fundamental mistake—he interpreted the fact that a *majority* of phone calls taking eight minutes meant that the *average* call was eight minutes. This was not the case. He also failed to realize that an individual could not handle both off-line and on-line problems *at the same time.*

To demonstrate the effect of handling calls on the off-line process, the facilitator made one simple modification to the model so that, when a phone call occurred, a support person would stop working on the written problem and handle the call. This is called *shutting down* an operation and is shown in Figure 16.4.

The facilitator used a modeling technique of connecting the phone call task to the software engineer's task and gave the phone call priority. In this model, a phone call would shut down the off-line support process for the period of time required to handle the call. This was accomplished using the Measurement block, which output

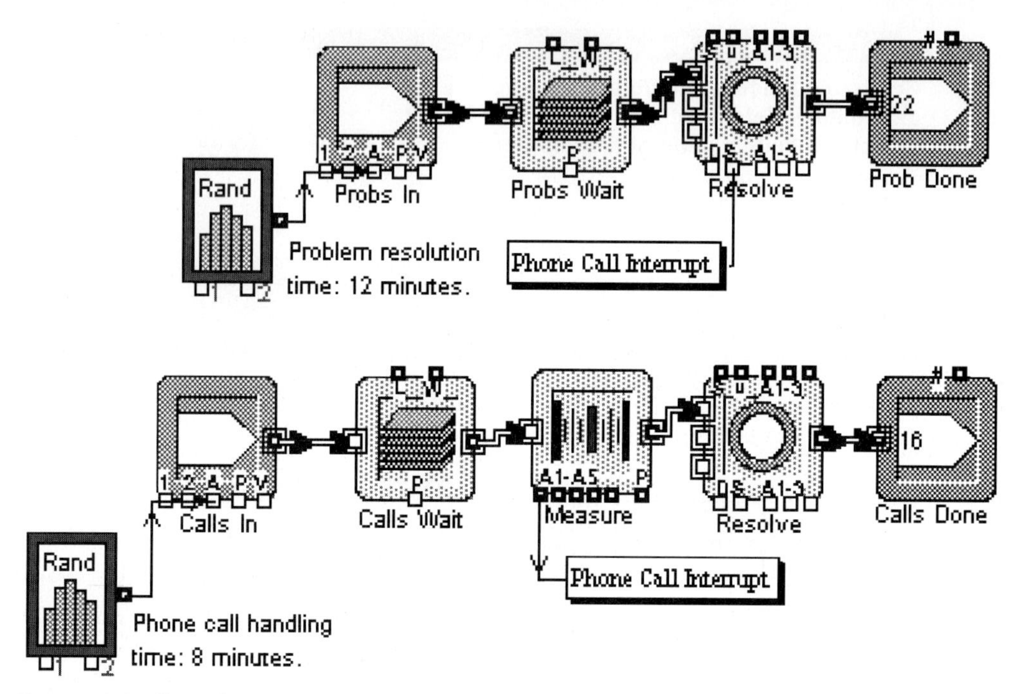

Figure 16.4 Effect of Answering Phones on the Problem Review Process.

the value of the attribute used to determine the length of the phone call. Figure 16.5 shows the interaction between the Measurement block and the Operation block depicting the software engineer's task.

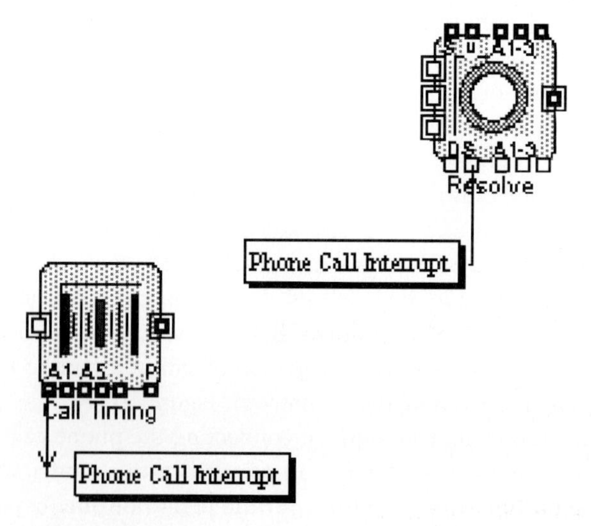

Figure 16.5 Measurement Block Used to Output Shutdown Value to Operation Block.

The simulation revealed that there was approximately a 36 percent reduction in the number of problem reports that were being resolved. This information was presented to the manager of the group and used to explain why his projections for problem resolution were being missed. The facilitator then explained that the support personnel's suggestion of having one person answer the phones might reduce productivity by 33 percent, but requiring each individual to provide both off-line and on-line support would result in a slightly greater loss in productivity.

As with any other good tool, there is more than one method of simulating a process in Extend+BPR. For example, let's assume we have learned more about this simple process over time. Specifically, let's assume we have learned that when a task is interrupted by a phone call, the time remaining to finish the task is increased by 10 percent. The rules for the simulation, then, are as follows:

1 When a phone call arrives, it must be answered; that is, phone calls are treated at a higher priority than ongoing tasks.
2 The interrupted task is resumed when the phone call is completed.
3 The time required to complete the ongoing task is whatever time remained plus 10 percent. In other words, if a task required ten minutes, and it was interrupted after five minutes, the time to complete would be the remaining five minutes plus 10 percent of the remaining five minutes, or five and one-half minutes in total.
4 In addition, the priority of the interrupted task will be increased by one to ensure that it is the next task begun after the phone call is interrupted. If this were not done, then the cycle time for interrupted tasks would increase tremendously.

To create this simulation, some new blocks from Extend+BPR will be introduced. The first is the Reneging Priority Stack, shown in Figure 16.6.

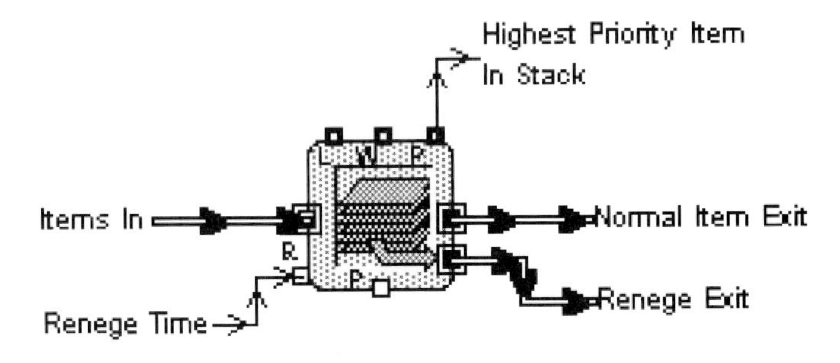

Figure 16.6 Reneging Priority Stack.

This block works in the following way. Items enter the Stack and are stamped with a "Renege time." This is the amount of time they will stay in the Stack if not processed.

If an item is processed before the renege time expires, it will exit normally; otherwise, it will exit through the renege exit. This block also provides a readout of the highest priority item in the Stack through the P connector at the top of the block. For this example, we will not use the renege feature, but we will use the priority readout.

The next block that will be used is the Transaction, Preemptive block, shown in Figure 16.7.

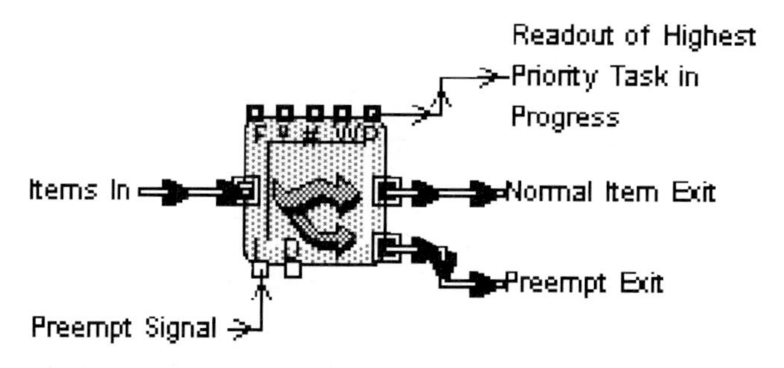

Figure 16.7 Transaction,Preemptive Block.

This block works in the following manner. Items (in this case, tasks or phone calls) enter the block and are held for a specified period of time. If an interrupt signal is received, and the conditions for an interrupt are met, the one item is preempted and exits through the preempt exit and another item enters the block for processing. The amount of time required for processing is set using an attribute and, when an item is preempted, the remaining time is stored in that same attribute. Therefore, when an item reenters the Stack, there is no need to explicitly determine if it has been preempted.

The two blocks just described work together as shown in Figure 16.8.

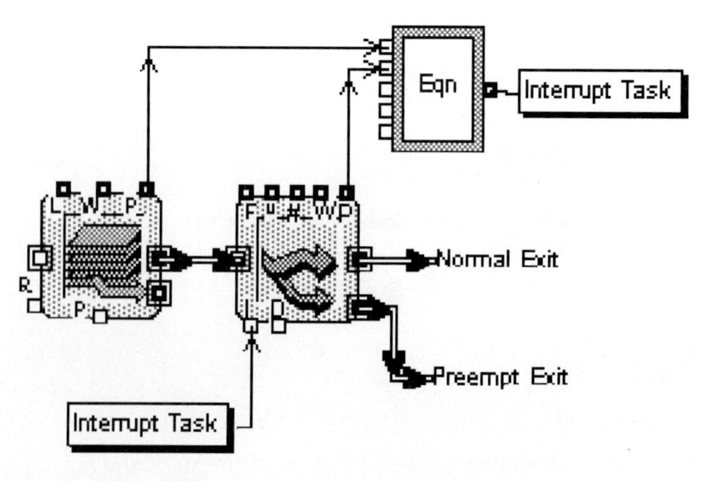

Figure 16.8 Reneging Stack and Transaction,Preemptive.

The logic of this simulation is as follows. When an item enters the Reneging Stack, its priority is compared to the priority of the item in the Transaction,Preemptive block. If the new item's priority is higher than that of the item being processed, it preempts that processing. When preemption occurs, the preempted item exits through the bottom connector and the higher priority item enters the Transaction,Preemptive block.

The model shown in Figure 16.4, when reconstructed using these two new blocks, is shown in Figure 16.9.

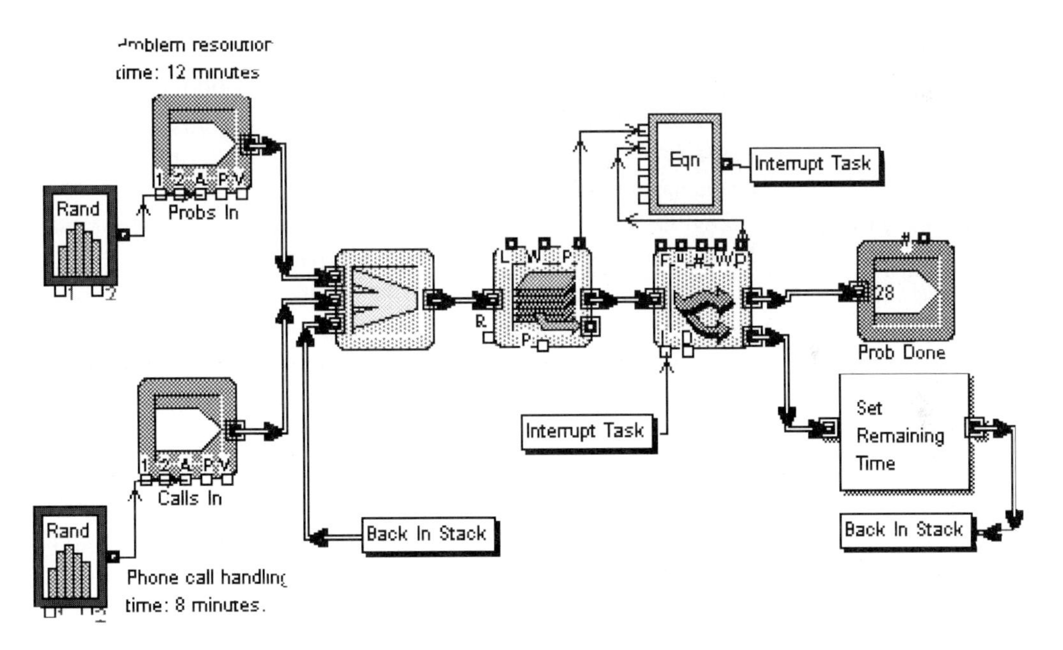

Figure 16.9 Model of Support Process with Preemption.

This is how the simulation works.

1. Tasks are generated with a priority of 3 and an attribute representing processing time as described earlier.
2. Phone calls are generated with a priority of 1 (the lower the number, the higher the priority) and an attribute representing processing time as described earlier.
3. When a phone call is generated, it enters the Stack as the highest priority. A comparison is made (in the Equation block) between the priority of the task in the Transaction,Preemptive block and the phone call. If the phone call has a higher priority, the task is preempted and the phone call is processed.

4 When a task is preempted, the amount of remaining processing time plus 10 percent is stored in the processing time attribute. In addition, the priority of a preempted item is increased, so that it is the next item of its type processed in the event there is more than one item in the Stack.

Note that, in this case, only 28 tasks and phone calls have been processed. This example demonstrates the flexibility of Extend+BPR and shows that, as details of a process become available, they can be added to a simulation to provide more understanding of the process.

This example was one in which both management and workers were right in their assumptions, and also a case where the philosophies of TQM and CPI were used effectively. The workers stated that having one individual answer phones would not reduce productivity any more than it would be reduced if they performed two tasks; management countered that either approach resulted in a productivity loss, so the workers should do it management's way.

There were intangibles that had to be considered, however, such as the feeling by the workers that they were either being ignored or unappreciated. The effect of these intangibles on worker morale could easily have resulted in additional losses in productivity. Ultimately, management accepted the support personnel's suggestions and reengineered the process. Software engineering personnel rotated the responsibility for phone coverage, and the productivity actually increased, perhaps because morale increased. This was a success story for everyone involved.

This example shows the power of using CAPRE tools to reinforce observations made about a process and to develop data that can be used to make informed decisions. It also shows that the Rules of Process Reengineering do not have to be applied in a particular order, but that all the rules should be applied in a process reengineering effort.

It also reinforces the validity of some of Deming's theories, namely "drive out fear" and "break down barriers between work areas." Employee suggestions were being systematically ignored, and management indicated, either directly or indirectly, that the employees were merely complaining about their work load. This had the effect of driving fear into the employees and lowering morale, so they kept their ideas to themselves.

This scenario is a particular type of vicious loop, "vicious loop," sometimes known as a *codependent loop*. Depicted in Figure 16.10, it raises the question: Who is in control, management or employees? If management viewed employees differently, perhaps the employees would communicate more. On the other hand, if employees mustered some courage, they might get management to change its view of them. Someone has to take action to break the loop, and this is what Deming is suggesting. If the situation could be reversed, the loop would become virtuous.

This example demonstrates how the philosophical approach of Continuous Improvement can be augmented by the analytical approach of computer aided

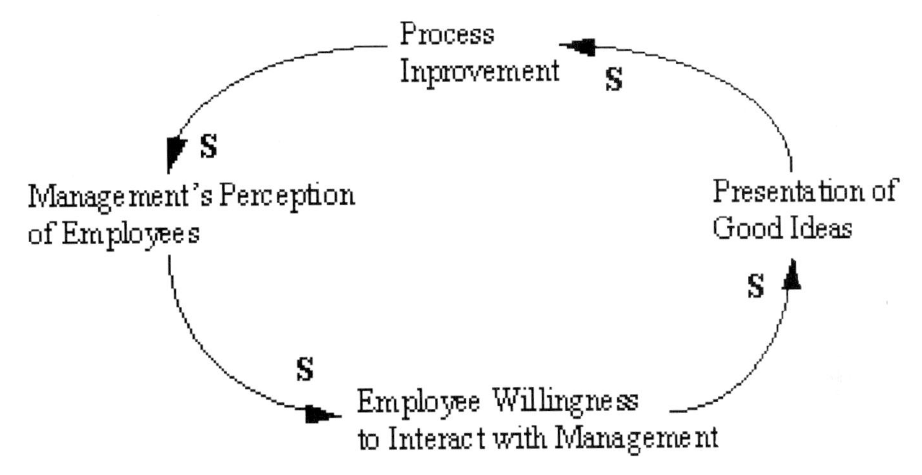

Figure 16.10 Codependent Vicious (or Virtuous) Loop.

process reengineering. It is important that management and employees have open communication. Although management should not blindly implement changes suggested by employees, it can test the validity of those suggestions using CAPRE technology. When a suggestion does not appear to be productive, management can explain why it is not being implemented. This serves several purposes: Communication remains open, the employee knows the suggestion was considered, and he or she has the opportunity to provide more information about the suggestion if clarification is required.

Example 2: Simulating Software Development Schedules

No matter how much effort is dedicated to the creation of software development schedules, it seems there is a continuing problem of trying to meet those schedules. There are a number of different techniques that have been popularized for determining software development schedules. Some companies use a *lines of code* estimate, some use *function point analysis*, some use historic data and a knowledge of their staff's capabilities, and so on. Whatever the case, software development organizations continue to have difficulty accurately predicting software schedules and delivering software systems on time.

One problem seems to be that, no matter how strongly emphasized by organizations such as the Software Engineering Institute (SEI), there seems to be little consideration given to the effect of the process on the schedule. This section will demonstrate that subtle changes to a software development process can have a major impact on the software development schedule. It will also show how CAPRE technology can help determine how the process used to develop software can affect a software schedule.

This simulation was done for a client that was beginning to implement a variation of the software development "waterfall process" called the "stage gate" process. The purpose of the simulation was to determine the level of effort required to manage the process, and this was directly related to the amount of rework that would occur at each stage of the process. Therefore, before beginning the development of the simulation, I asked the client what the percentage of rework was at each stage in the process.

When it came to the question of rework, there was at first general laughter and the suggestion that there was 100 percent rework at each stage of the process. However, when we defined rework as a substantial change to work that had already been done, the consensus was that rework at each stage of the process could be represented as shown in Table 16.1.

Table 16.1 Software Development Rework by Process Stage

Process Stage	Percent Rework(%)
Requirements Analysis	25
Preliminary Design	25
Design	25
Coding	10
Testing	10
Integration and System Testing	25

Whether you agree with these numbers or not, or, for that matter, whether they represent an industry average, is not important. This case study is only an example of how a process can affect schedule.

A Sample Process

The characteristics of the sample process to be used are as follows:

1. One hundred software modules will be developed.
2. There are three types of modules:
   - System modules, representing approximately 20 percent of the 100 modules
   - Application modules, representing approximately 40 percent of the 100 modules
   - GUI, Help, Reporting, and other types of modules, representing approximately 40 percent of the 100 modules

3 Each module requires a specific amount of time to be completed, depending on the process stage it is in. Type 2 modules, in general, require 65 percent of the time required for Type 1 modules, and Type 3 modules require, in general, 40 percent of the time required for Type 1 modules.

4 Each process stage has a dedicated staff. In other words, personnel doing design work do nothing but design work.

5 The first item—time—rework is required, the amount of time required to perform the specific process function is the same amount of time required to perform the function initially. After that, the amount of time required for rework is between two and four hours. This is based on the theory that after one iteration of rework, most changes are fixes, rather than fundamental, substantial changes.

6 Specifications are released to the requirements analysis function at the rate of 10 every two weeks (ten working days).

Given the times I used for each step in the process, the average cycle time for each type of module, assuming no rework at all, would be

- Type 1 Approximately 14.5 days
- Type 2 Approximately 9 days
- Type 3 Approximately 5.5 days

Modeling the Process

All of the above information, including the percentage of rework required, is necessary for exploring the dynamics of any process, including software development processes. To determine the effects of process changes, I developed a model of the waterfall process using Extend+BPR.

The initial model is shown in Figure 16.11.

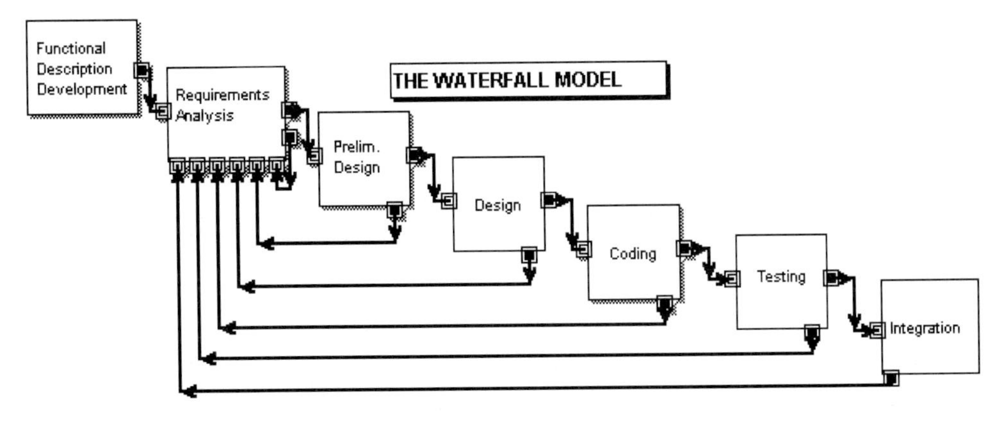

Figure 16.11 Initial Model of the Waterfall Process.

This may appear to be a simple flowchart, but it is actually a series of hierarchical blocks. For example, the block called Design, when expanded, appears as shown in Figure 16.12.

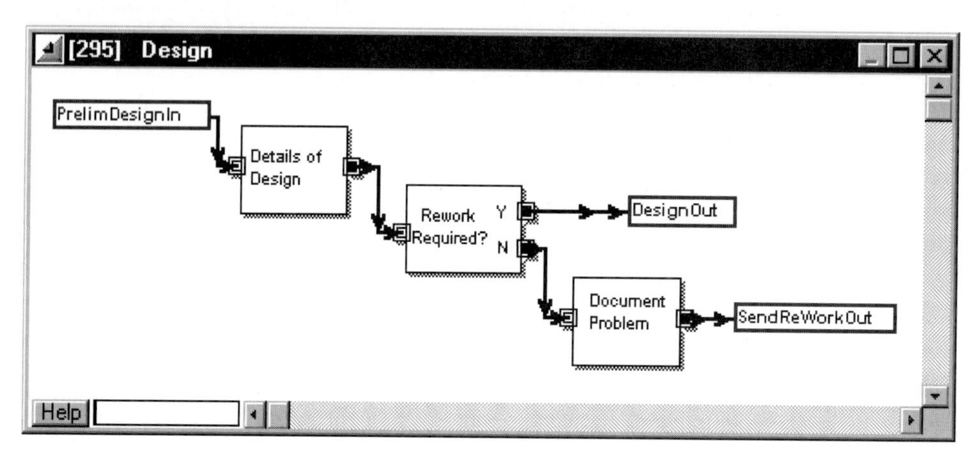

Figure 16.12 Contents of "Design" Hierarchical Block.

Again, each of these blocks is a hierarchical block containing more detail of the model. The block that is of most interest is the block entitled Details of Design. This block, and those like it representing other stages of the waterfall process, will change as the definition of the process changes. This block, when expanded, appears as shown in Figure 16.13.

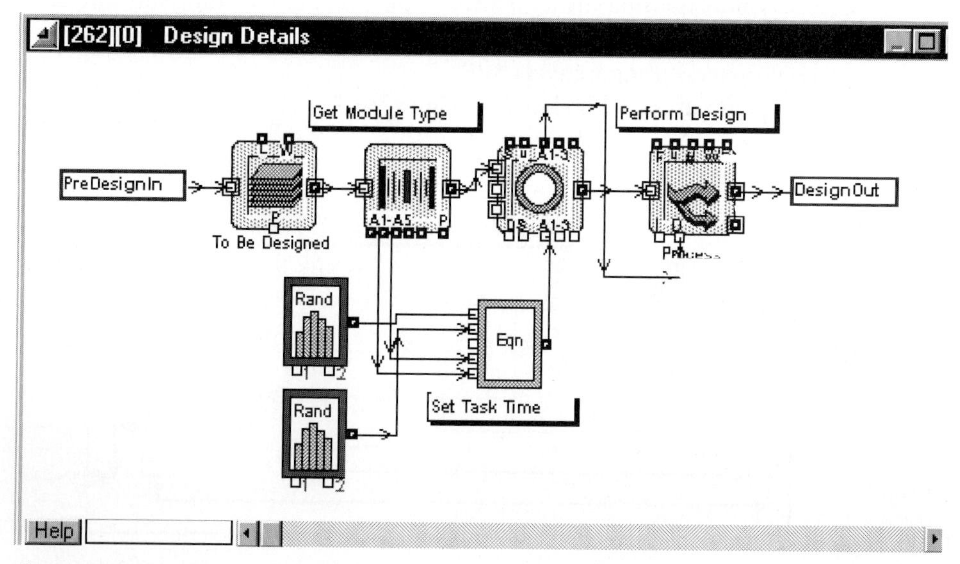

Figure 16.13 Expansion of the "Design Details" Hierarchical Block.

Since this is not necessarily a lesson on how to build models, the details of the Extend+BPR blocks will not be explained; rather, a general explanation will be given. Figure 16.13 represents a series of blocks that are used to determine the type of module to be designed, whether this is the initial design or rework and the number of rework iterations. This information is then used to determine the time required to perform the design.

Note that the model depicts rework from each step in the process being returned to the requirements analysis step, rather than being returned to the step preceding it. This is one variation of the waterfall process.

Example 1: No Prioritization

In the simplest process, workers in each task in the waterfall process operate on inputs to the task on a first-in-first-out (FIFO) manner. In other words, there is no prioritization of their work. When performed in this manner, the average cycle time for each type of module is as follows:

- Type 1 modules 40 days
- Type 2 modules 32 days
- Type 3 modules 39 days

These figures clearly show the impact of rework—cycle time for each type of module is greatly increased from the cycle time projected for a process with no rework.

Example 2: Prioritization of Work

In the next example, we can pretend that management has deemed Type 1 modules, the system modules, to be more important than Type 2 and Type 3 modules. Therefore, management has instructed workers at each stage in the process to perform their tasks on a priority basis, the assumption being that the cycle time for Type 1 modules will be reduced. In the simulation, I associated priority with the Type of module, so that Type 1 modules are the highest priority, Type 2 modules are the next highest, and Type 3 modules are the lowest priority. To simulate priority processing, I merely had to change the To Be Designed queue from a FIFO queue to a priority queue.

When this change has been made and the simulation executed, the cycle times for each type of module are as follows

- Type 1 23 days
- Type 2 35 days
- Type 3 35 days

These results are dramatically different than the results from the first process; at least for Type 1 modules. This seems to be an intuitively correct result.

Example 3: Preemptive Prioritization of Rework

Now we will pretend that management, having decided that Type 1 modules are the highest priority, has also decided that priority will be given to rework over new work. In other words, a Type 1 module that is to be reworked will have priority over new Type 1 work. Furthermore, if Type 1 rework is introduced into a process step, then it will cause work to stop on any other module of lower or equal priority.

Also, when work on a module is preempted, the priority of that module is increased. In other words, if a Type 2 module is preempted, it will be treated at the same priority as a Type 1 module.

The simulation of this process reveals that cycle time for each type of module is

- Type 1 41 days
- Type 2 29 days
- Type 3 25 days

This is a remarkable result and seems to be completely counterintuitive. It stands to reason that, if a priority-driven process yields better cycle times for a Type 1 module, shouldn't a preemptive priority-driven process yield even better results? These results can be explained by considering the nature of preempting. If a Type 2 module requires rework, its priority is increased, thereby taking on the same priority as a Type 1 module. In addition, there are more Type 2 and Type 3 modules to be developed than Type 1, so eventually, many modules will attain a high priority, diluting the effect of priority processing on Type 1 modules.

Therefore, from these examples, we can conclude that the priority process is the best process, and that a preemptive process is counterintuitive.

Is This Always the Case?

The answer is—*No*!

Despite our tendency to toss around the words "best practice." there is really no such thing. The best practice for a process is determined by many variable factors, such as prioritization of work, available resources, percentage of rework, and so on.

Table 16.2 shows the variation of cycle times for this process when there is rework, when there is no rework, when modules are introduced at a rate of ten every ten days, and when they are introduced at a rate of one every day.

This table clearly shows two facts:

1 Elimination of rework is essential to reducing cycle time.

Table 16.2 Cycle Time by Process Type

Type of Module	Interval	Rework (Yes/No)	Basic Process	Priority Process	Preemptive Process
Type 1	1 per day	Yes	31	29	30
	1 per day	No	14.8	12.7	14.1
	10 every 10 days	Yes	40	23	41
	10 every 10 days	No	14	14.6	14.5
Type 2	1 per day	Yes	24	21	31
	1 per day	No	8.8	9	9
	10 every 10 days	Yes	32	35	29
	10 every 10 days	No	11.8	10	9.8
Type 3	1 per day	Yes	14	28	23
	1 per day	No	5.3	5.2	5.5
	10 every 10 days	Yes	49	33	25
	10 every 10 days	No	9.3	10	9.8

2 It is better, in this case, to introduce work at the rate of one module per day, rather than releasing work in batches.

The concept of "best process" depends on the view you choose. For example, from the perspective of Type 3 modules, a preemptive process is clearly the best when modules are released once every ten days, but not as good as a priority process when modules are released once per day. From the perspective of a Type 1 module, a preemptive process is best when modules are released once per day.

Determining the Best Process

The examples given above were fairly uncomplicated and straightforward. There are many factors that can affect schedules. For example, consider the following.

- Software maintenance—What is the impact on new software development schedules if the same personnel handle maintenance questions and work (a very common situation for smaller companies)? How is maintenance prioritized?
- Technology—How does the insertion of technology, such as electronic data interface, affect schedules?
- Multitasking—What is the impact on schedule if staff performs more than one task, and how can you determine that impact?

Software processes, and processes in general, are too complex to be measured simply through the use of some lines of code or function point analysis technique. There are too many interrelationships between process elements to understand the effect of a change to one process element, not to mention the effects of changes to more than one element.

How Can You Use Modeling and Simulation to Determine Schedules?

The best way to determine a schedule is to mix empirical analysis, such as function point analysis, with process analysis. Use any technique you like to determine an estimate of time required to complete a software development project, preferably on a module or functional component basis. Then model your process, including all factors that can affect that process, such as interruptions, unanticipated work (requests for sales support, software maintenance activities, and so on). Then plug in the numbers and simulate the process. If you use the right tool, your simulation will develop a schedule for you.

Example 3: Simulation Model as Documentation of a Process

This simulation model was developed for a major utility (utility A) that had begun maintaining its equipment in-house, rather than paying equipment maintenance fees to the vendor. The utility's employees had undergone extensive training in the maintenance of the equipment, and the company had developed a great deal of documentation on the maintenance process. This decision seemed to be cost-effective until the managers at the utility discovered that another utility (utility B) had a higher Mean Time Between Failure (MTBF) on equipment it maintained in-house. Since the two companies were not competitors, utility B agreed to work with utility A in an effort to improve utility A's maintenance procedures.

A process reengineering facilitator hired to consult with the two utilities suggested that utility A develop a model of their process to present to utility B. The facilitator explained that a model of the process would convey more information than the printed documentation that had been developed. The management of utility A agreed to this approach and gave the facilitator documentation of one part of the maintenance process.

The part of the process that was modeled was a step in which mineral deposits were scrubbed from each turbine blade in the generator. Figure 16.14 shows the model as built with hierarchical blocks.

Contained within each of these blocks is a detailed submodel of a particular task in the process. Several layers of hierarchy can be contained within each other, so levels of detail concerning a process can be revealed as necessary.

The documentation that had been given to the facilitator did not contain enough detail to build a model. Apparently, the documentation for each process step

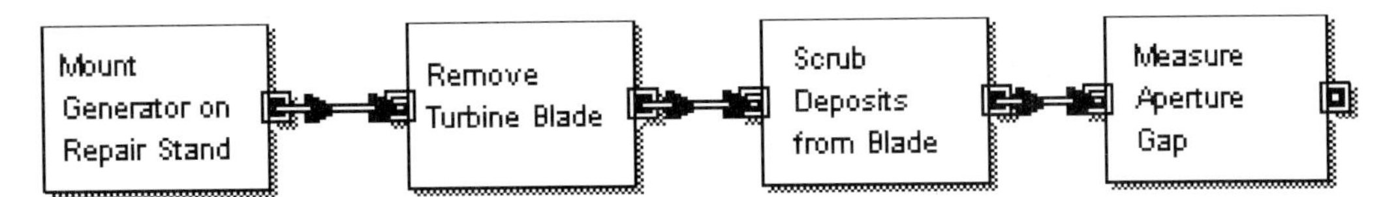

Figure 16.14 Generator Maintenance Process—Hierarchical View.

had been written by an individual who was intimately familiar with the process and, therefore, contained only sketchy information. For example, the "Scrub Deposits from Blade" documentation was essentially two sentences long, and said, "Grind the deposits from the turbine blade. Inspect the blade to make sure no deposits are left." These sentences provided no information about how the blades were scrubbed. They could have been scrubbed by machine, by sand blasting, and so on. The facilitator, when encountering documentation of this type, made notes for follow-up discussions. In this case his note was: "Find out how blades are scrubbed."

Another step, Measure Aperture Gap, was described as, "Measure the aperture gap. The gap should be no more that 0.002 inch. *Caution*—do not insert adjacent blades with side-by-side aperture gaps of more than 0.002 inch! Generator failure could occur." There are two things wrong with this description. For example,

- There is no indication of how an aperture gap is measured.
- There is no alternative to take if the condition in the caution statement occurs. This is an IF...THEN statement with no ELSE.

Millions of people were dependent on the smooth operation of this utility, but the process used to ensure that operation was at best a Level 2 process. Tasks were documented, but the documentation was so lacking in detail it was virtually worthless. The facilitator presented questions about these steps (along with many others) to the management of the utility and received the following answers.

- The grinding was done by hand, using a tool of a particular make and model.
- The aperture gap measurement was also done by hand using a micrometer.
- If the gap was too large, two new turbine blades (at a cost of $1800 each) had to be installed. Although no one actually admitted as much, this information was left out of the documentation as a subtle way of discouraging the installation of new blades.

By talking to maintenance personnel, the facilitator also discovered a very important, undocumented fact: There were two ways for a blade to be ground down too much: (1) normal decay due to use, and (2) technician error.

The utility obviously accepted blade decay, but when a technician made a mistake, it was noted on their record and, in some cases, they were reprimanded.

Therefore, technicians were reluctant to notify their manager when aperture gaps were too high. Since there was no way to determine which blades caused a failure, the technicians considered themselves safe from reprimand.

Of course, this attitude of workers and management was in direct opposition to Deming's philosophies. The causal loop in Figure 16.15 explains the problem.

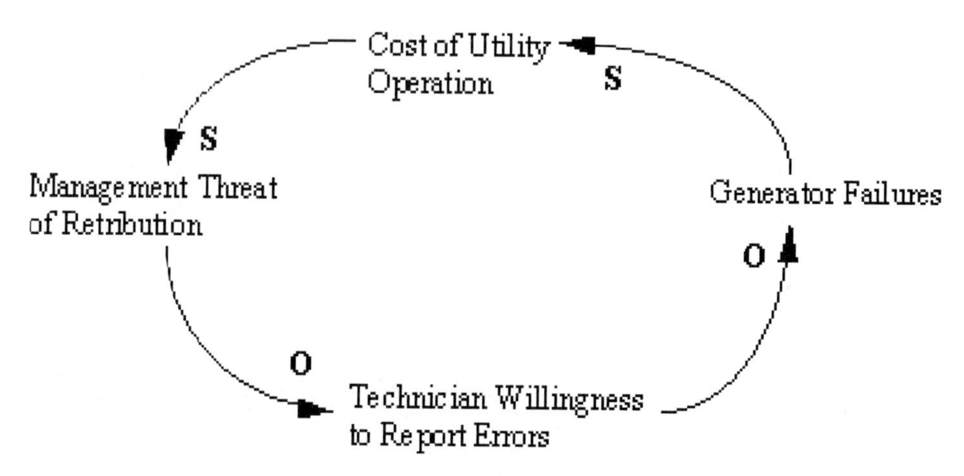

Figure 16.15 Effect of Threat of Retribution on Costs.

With this information in hand, the facilitator contacted the management of utility B to determine how they did things differently, if at all. He was given the following information:

- Technicians did not grind the blades by hand. Instead, the company had invested in technology that did the grinding. The grinding was done in small, iterative passes and took longer than a human, but rarely, if ever, did errors occur. A human was still involved in the inspection of the blade, and if problems occurred, it was usually due to a faulty inspection.
- Measurement of aperture gaps was done using an expensive imaging system. Although it was expensive, it was accurate to 0.0001 inch.
- When the gap limit was exceeded, new blades were inserted.

The facilitator also asked what happened to a technician who made a faulty inspection. The answer was interesting. Management of utility B said, "Obviously, if a technician continually makes errors, we will retrain him, or find another job for him; however, we want them to tell us when there is a problem. It is cheaper for us to fix problems when they occur rather than wait for a generator failure. No employee is ever punished for pointing out a problem, no matter whose fault it is." Here was a real example of Deming's "drive out fear" philosophy.

Utility A's management was quite surprised at this revelation. Their relationship with its employees had been adversarial for so long, it had no other way of thinking. Although it meant a radical change in corporate culture, utility A accepted utility B's approach. It asked the facilitator if the model could be used as a mechanism to convey the change in management's attitude.

The facilitator had never before been asked to use a model to convey a *philosophy*. After some thought, however, he arrived at a novel approach. The facilitator would capture the cost of each step in the process, and then model the process in two ways:

1. Introduce errors but not stop the process, force a generator failure, and add in the cost of a major repair job.
2. Introduce errors, catch them when they occur, add the expense of inserting new blades, and determine the cost of the process.

Figure 16.16 is an Extend+BPR model of this process.

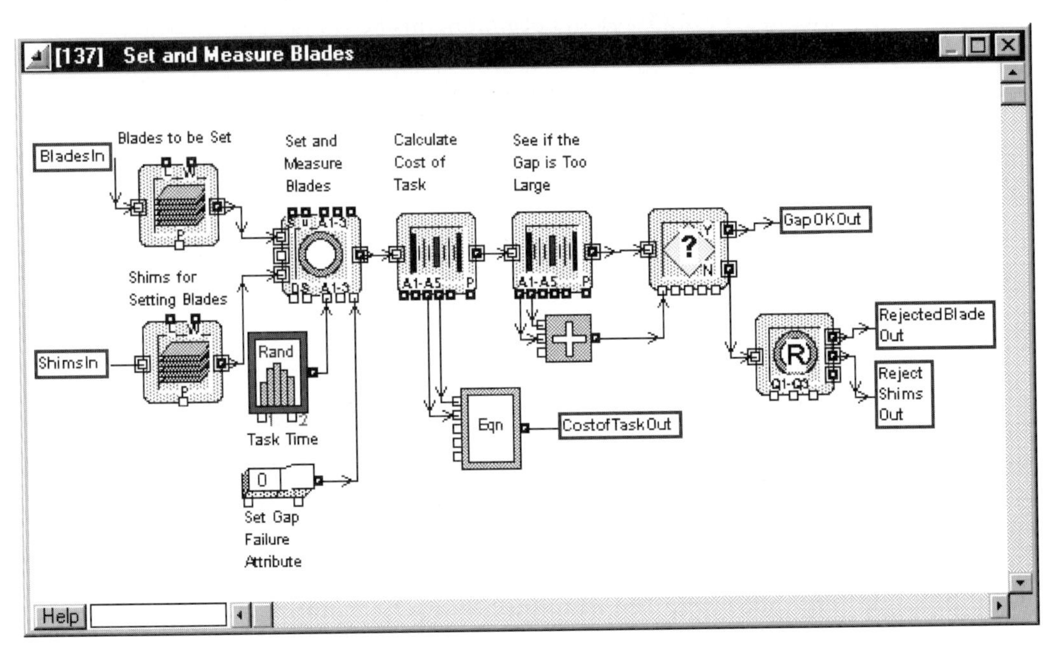

Figure 16.16 Blade Scrubbing/Testing Process.

The model shown in Figure 16.16 captures the insertion of one blade adjacent to a second blade, the insertion of shims between blades, the measurement of the aperture gap, and a decision of whether the aperture gap is too large. Each activity was modeled using an Operation block, and the cost of each activity (labor, parts, machine time, and so on) was defined as an attribute. An iconic block, called a Switch, was used to set an Aperture Gap Failure attribute that specified whether or not an aperture gap was too large. A Switch has either a value of 0 or 1 and can be set by the viewer of the model while a simulation is running. In this case, a 0 value meant the gap was within limits, and a 1 meant that the gap was out of limits.

The model was then executed in two iterations.

1 The first time through the model, no aperture gap failures were introduced into the model. However, at the end of the model, a generator test failure was forced, meaning that the entire model would be executed again. The Aperture Gap Failure attribute of the last blade through the model was set to 1, so the replacement of one blade with a new blade would be simulated. While the model executed, the costs of both passes were tabulated and summed.

   Setting the Aperture Gap Failure attribute and determining that the last blade was passing through the model required extensive use of the object-oriented features of Extend+BPR. Each blade object in the model was assigned a number at the beginning of the simulation, as well as an iteration attribute that was initialized to 1. When the last blade was encountered in the model, the Aperture Gap Failure attribute was set equal to 1, and the iteration attribute was increased by 1.

   This logic is captured in a Decision block and is stated as follows:

```
IF (ITERATION = 1 AND BLADE = 'LASTBLADE') PATH =
YESPATH;
ELSE PATH = NOPATH;
```

   When the YesPath is taken, the iteration attribute is increased by 1, and the Aperture Gap Failure attribute for the last blade is set to represent a failure. The whole simulation is then reexecuted, with the cost of installation of one new blade added to the cost of the second model iteration.

   When the NoPath is taken, the cumulative costs of the first and second iterations are summed and the simulation terminates.

2 The model was then reexecuted by the facilitator, who used the Switch to set the Aperture Gap Failure attribute to 1 for one blade in the model.

This resulted in one new blade being installed during the first pass through the model. At the end of the model, no generator test failure was forced. This technique could be used to force installation of two new blades, three new blades, and so on.

The object-oriented capabilities of Extend+BPR provided the mechanism to calculate costs as the model executed. When cost attributes were read by the Measurement block, their values were output to named connectors. The facilitator used the named connectors as input to iconic Extend+BPR blocks and he was able to accumulate costs while the model ran and displayed the results when the model finished (Figure 16.17).

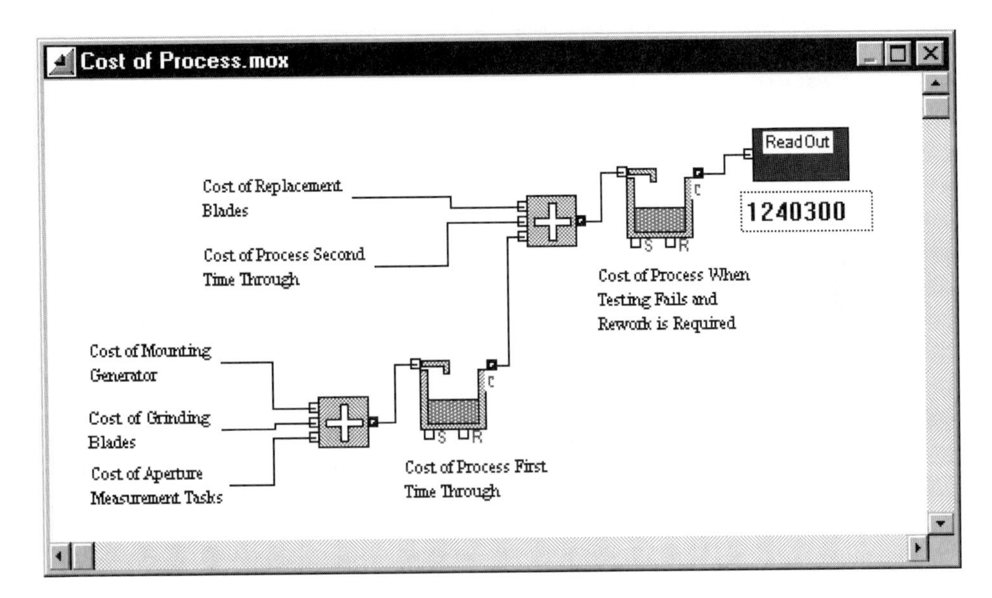

Figure 16.17 Capturing Costs in Extend+BPR.

After executing the model and simulating both processes, the facilitator was able to show both management and technicians the costs of the process as currently implemented, and the cost of the process incorporating management's new philosophy, including the cost of generator failures. The model proved to the technicians and to management that it was to everyone's benefit to report problems and solve them as they happened. It showed the *value* of doing so, instead of the cost. The numbers were the same—it was how one interpreted them that mattered.

To reinforce the change in management's philosophy, the facilitator then had to change the way management communicated with the technicians. Management was accustomed to saying, "Here's how much your errors cost us!" In the new philosophy, they had to say "Here's how much it is *worth to us* if you report problems when they occur. Your work is valuable to us and we appreciate your efforts to prevent small problems from becoming big problems."

This philosophical change was communicated through face-to-face meetings between workers and management, and by using the Help feature of Extend+BPR. The Help feature was used to add documentation to the hierarchical blocks in the model and to express the new philosophy in the documentation contained in the model. By clicking the HELP button associated with a model of a process step, a technician would see a window similar to that shown in Figure 16.18.

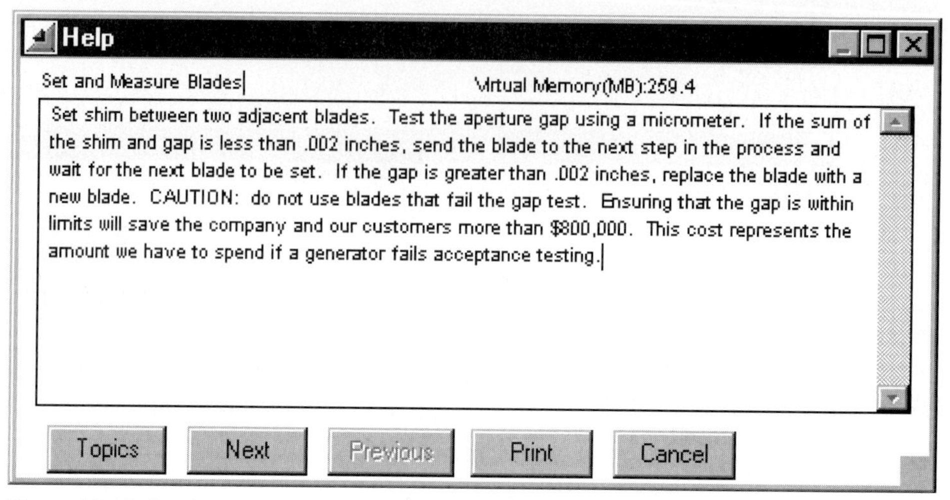

Figure 16.18 On-line Documentation as Part of a Model.

The documentation did not say, "Here's how much a mistake will cost," but rather, "Here is how much our customers and the company can save if a mistake is caught and addressed immediately." This is a subtle but important change in the interpretation of data. The Help feature of Extend+BPR has several other benefits.

1. It provides more detail on how the operation is performed.
2. It gives an alternative path in the event of a failure.
3. It explains the value of the operation in terms of real cost.

Eventually, the whole model was completed, and the on-line documentation replaced the written documentation.

This example provides a great deal of information about process reengineering and computer aided process reengineering:

- The fact that documentation about process steps exists does not necessarily make a process a Level 2 process; the documentation must be accurate and useful.
- The information that results from simulating a process can be used in different ways. In this example, the information could have been used in a confrontational or cooperative manner. This was a case in which the philosophies of TQM were very applicable.
- Process models are a valuable educational tool. This model allowed companies to share informational and "best practices" concepts much more easily than using written documentation, and the finished models served as a communication mechanism between workers and management.

> Modeling and simulation can be used to show the effects of changes in management philosophy, particularly if that philosophy affects the manner in which employees perform their tasks.

- Documentation developed when migrating from Level 1 to Level 2 does not have to be captured on paper. It can be attached to and become part of a model.
- Migration from Level 1 to Level 5 is iterative, and successive layers of detail can be supplied with each iteration. Using hierarchical decomposition enhances this iterative process.

The use of models as documentation is not new—we are all familiar with instructional video tapes. These tapes are, after all, a model of how to do something; however, they are one-directional, even those that are interactive. A user views or interacts with a tape, but cannot change what is represented on a tape.

Using tools such as Extend+BPR, however, users can actually modify the information presented or the model itself, so the model becomes a "living" document that is adapted to changing conditions in the workplace.

17

Case Study: Automating the Analysis of Customer Service Operations at Lexis-Nexis

Introduction

LEXIS-NEXIS[1] is the world's leading provider of enhanced information services and management tools. The company's mission is to help legal, business, and government professionals collect, manage, and use information more productively. Its products and services are marketed under the familiar brand names of LEXIS® and NEXIS® and those of Michie Butterworth, the state law publisher, and Jurisoft and the Folio Corporation, providers of information management software for the legal and business markets.

As part of its service, LEXIS-NEXIS provides on-line phone support to all of its customers, helping them find the most effective ways of using the system. This chapter discusses how modeling and simulation are being used at LEXIS-NEXIS to provide information about customer service operations and to determine how changes to LEXIS-NEXIS products and increases in the customer base will affect those operations.

[1] LEXIS-NEXIS is a division of Reed Elsevier Inc., and part of Reed Elsevier plc, one of the world's leading publishing and information businesses. Reed Elsevier has annual sales in excess of £3.06 billion ($5 billion) and 29,700 employees. It is owned equally by Reed Elsevier International plc (NYSE: RUK) and Elsevier NV (NYSE: ENL). The mailing address for LEXIS-NEXIS is P.O. Box 933, Dayton, OH 45401-0933.

Description of the Problem

LEXIS-NEXIS takes pride in the quality of its product offerings. The company's NEXIS service was named "Best News Service/Database" by the Information Industry Association and *ONLINE ACCESS* magazine in 1994. NEXIS offers 5800 sources of news and business information, of which 2400 appear in full text. In 1995, *American Demographics* magazine named the LEXIS-NEXIS services as one of the "Best 100 Sources for Marketing Information for 1995."

Due to increases in sales volume, the LEXIS-NEXIS customer base has grown in size. As a result, the magnitude of phone calls to its customer service operation has increased, demanding greater resources. Although the company's priority is to be a world-class leader in responsiveness to its customers, it must also provide those services in a cost effective manner. Maintaining high levels of customer satisfaction while lowering costs is a complex problem, and one that required a great deal of analysis.

Minimally, the factors that could affect both the responsiveness of customer service and the cost of customer service were

1. The number of Customer Service Representatives (CSRs) available at any time during the day.
2. The impact, both positive (reduction of calls) and negative (increase in calls), of new system software and information content releases.
3. The impact of training received by customers.
4. New technologies that could increase the effectiveness of CSRs.

The Need for Automated Assistance

Ann Beeson, Business Information Services Strategy Development Manager, was given the task of developing a LEXIS-NEXIS customer service strategy to anticipate and respond to changing business factors. According to Beeson,

> The numerous factors that affect customer service are complex and interrelated. For example, phone calls received by customer service can be broken out into more than 90 categories. Each type of call requires a different amount of processing time; some require a high percentage of follow-up work; and some are affected differently by system software changes.
>
> But that is just part of the complexity of developing a strategy. For example, CSRs must be trained when new system software enhancements are released, which impacts the availability of those CSRs. Therefore, just as software changes impact phone call volume, they also impact the number of available CSRs, since CSRs must be educated about the software changes in order to maintain a high level of customer responsiveness. I concluded very quickly that I needed some type of automated tool to assist me in the development of a strategy.

The "Aha" Experience

Steve Gabbard, director of business strategy, suggested that Beeson investigate the use of Extend+BPR, a graphical, object-oriented dynamic modeling and simulation tool specifically targeted for business process analysis and improvement. Based on Gabbard's recommendation, Beeson contacted Greg Hansen, president of Computer Aided Process Improvement (CAPI) and co-developer of the BPR library, to request a demonstration of the tool.

Hansen explains, that he "ran through a series of business process examples, all of which address business scenarios that are very common in today's environment, such as task elimination, downsizing, insertion of technology, rework, etc. The premise of the examples is that—no matter how simple a process may seem—the results of a change to the process are rarely what you expect and, in fact, can be completely opposite of what you expect."

"I was particularly impressed by one example Greg showed me," Beeson said after the demonstration. "It consisted of two side-by-side process models, one being the 'as is' process, and one being the process as enhanced through the use of information technology. I was surprised to see that, although information technology resulted in a 400 percent increase in productivity in some process steps, the increase in productivity for the overall process was minimal.

Greg also demonstrated how he could tie together all of the parameters that affect a process, and how I could test changes to the process all at once, one at a time, or some at a time. I immediately realized how the use of this technology would allow me to control the complexity of Customer Service process analysis. This is what I call my 'aha' experience."

Using Modeling and Simulation

Prior to creating process models, Beeson convened "focus groups" consisting of CSRs to develop customer service process descriptions, to determine the types of tasks performed during a typical day, to find out the amount of time spent performing each type of task, and so on. She also investigated what data were available to help in the analysis of the process, and found that, "We knew, almost to the second, how much time a CSR spent on the telephone, the types of calls that were coming in, the arrival rate of calls for different periods of the day, and so on."

Beeson then contracted CAPI to develop the initial models of the customer service operations that supported business, journalist, and corporate legal clients. "Because LEXIS-NEXIS had so much data available regarding these operations," Hansen noted, "the development of the models was enhanced. The data concerning arrival rates of calls, types of calls, etc., is what we call 'hard data' and it can be used to test the validity of the process descriptions developed earlier. This is known as 'as-is' process modeling."

"As Is" Process Modeling

Before beginning the development of the "to be," models, Hansen asked Beeson to establish process parameters by which process changes could be measured. She determined that, "We wanted to make sure that calls were answered within a preset service level target, that the number of customers hanging up (or abandoning) was kept to a minimum, and that the CSRs were not over worked." Beeson concluded: "It was important that any process model be able to predict these parameters when changes to the process were being made."

The process models were developed over a period of three weeks. The information collected at the focus meetings was incorporated into the process models, and ultimately the "as is" process model was accepted as being correct. The completed model for the News Customer Service operation is shown in Figure 17.1.

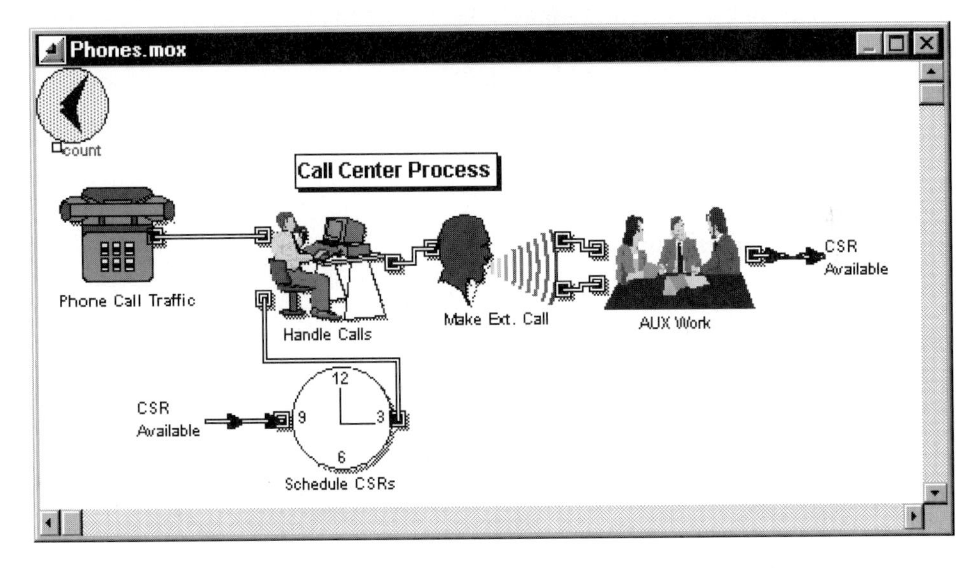

Figure 17.1 Overall Call Center Model.

If this model seems overly simplistic, it is because it consists of *hierarchical blocks*. Hierarchical blocks contain more information than is actually shown. For example, the block labeled Phone Call Traffic, when expanded, would look like Figure 17.2.

Notice that there are additional hierarchical blocks, so the model is beginning to take on complexity. According to Hansen, the use of hierarchical decomposition is important for two reasons: (1) Hierarchy reduces the visual complexity of a model, so that even very large models can be displayed in a few blocks. (2) In addition, anyone interested in seeing details of a model can do so by 'peeling back layers' of the model, exposing complexity a little bit at a time."

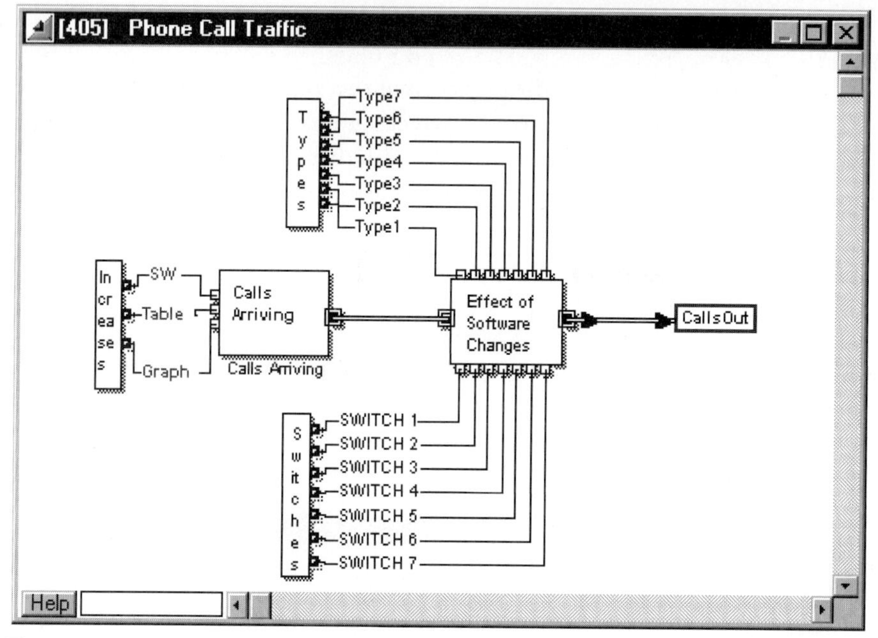

Figure 17.2 Details of Phone Call Traffic Block.

The hierarchical block in Figure 17.2 labeled Calls Arriving is the block that controls the arrival of phone calls into the simulation. This block is shown in more detail in Figure 17.3.

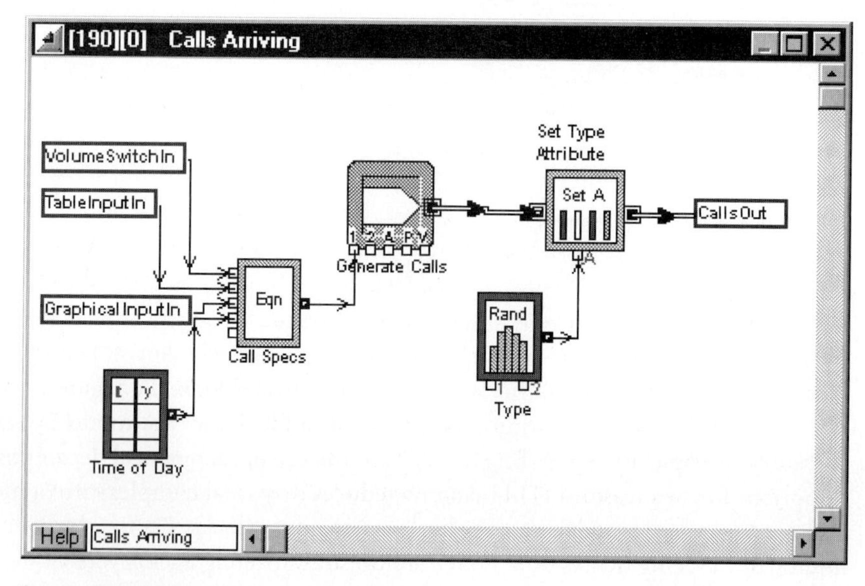

Figure 17.3 Details of Calls Arriving Block.

Notice there are a number of inputs to an Equation block (labeled Call Specs) which in turn provide input to an Import block (labeled Generate Calls). The Import block creates phone call "objects" for use in the simulation according to user-specified distributions. One of the inputs to the Equation block is a table that specifies the arrival of phone calls, or more accurately, it specifies the time interval between phone calls arriving by time of day. This table uses current data and is shown in Figure 17.4.

Row	Time (Time of Day (0 = 6 A.M.))	Y Output (Interval Between Calls)
0	0	8.571429
1	30	2.941176
2	60	1.973684
3	90	1.485149
4	120	1.421801
5	150	1.595745
6	180	1.376147
7	210	1.470588
8	240	1.554404
9	270	1.666667
10	300	1.973684

Figure 17.4 Table Containing Details of Phone Call Traffic.

Another input to the Equation block is from the hierarchical block labeled Increases (shown in Figure 17.2). This block determines whether the model will analyze the impact of an increase in call volume, and whether tabular or graphical input will be used in that analysis. The Equation block uses all of this information to specify the arrival rates of calls by time of day.

Once a call is generated, it is assigned an attribute, or numerical descriptor, that specifies its type (such as Request for Information, Search Logic Assistance, Hard Copy Help, etc.) This attribute is used later in the model to simulate the length of a call and determine if follow-up action is required.

Other existing data were used in the model as well, such as the number of CSRs available. LEXIS-NEXIS had this information broken out into half-hour increments, by time of day. These data, as was the case with phone call arrivals, could be captured in tabular form, as shown in Figure 17.5.

The table in Figure 17.5 specifies that at the start of business (time 0, or 8 A.M.), there are two CSRs available. At 8:30 there are four more (six total), and so on. With all the data available, Hansen was able to build models of each customer service operation that reflected actual behavior.

	Time of Day (Minutes)	Inc./Dec. in CSR Staff
Row	Output Time	Value
0	0	2
1	30	4
2	60	4
3	90	2
4	120	0
5	150	0
6	180	0
7	210	-1
8	240	-1
9	270	-1

Figure 17.5 Table of Available Staff.

"To Be" Process Modeling

In addition to predicting the effects of increases in phone call volume that resulted from an increase in the customer base, Beeson also wanted to determine the impact of proposed software changes on call volume. The impact of the software changes could be either an increase in calls or decrease in calls, and the number of changes to be tested could be as few as one or as many as seven.

"This type of testing demonstrates the real power of modeling and simulation," Hansen says. "If there are, for example, seven possible software changes, you can test the effect of change 1 and no others, of changes 1 and 2 and no others, of changes 1 and 3 and no others, and so on. In fact, there are 256 possible change combinations that can be tested. This is far too many to be tested using paper and pencil techniques."

To facilitate this analysis, Hansen set up a number of "switches" that could be used to implement a change. The switch would also specify whether the impact would be an increase or a decrease in phone calls. He then set up a matrix in Extend+BPR's executive interface (or notebook) that allows the user to set those switches and enables the user to determine the type of call that would be affected and the percentage of increase or decrease in call volume that could be expected. That matrix is shown in Figure 17.6.

"If we wanted to test more than seven possible changes, we could easily modify the model; but for now, seven is sufficient," Beeson stated. "The fact that we can test changes in combination is very powerful and very useful."

SWITCHES: 0 MEANS NOT IMPLEMENTED, 1 MEANS DECREASE, 2 MEANS INCREASE

Effect of Software Enhancements

Switch		Type of Call Affected	Percent Reduction
0	Change 1	21	1
0	Change 2	21	2
0	Change 3	4	25
0	Change 4	9	5
0	Change 5	4	1
0	Change 6	21	0.55
0	Change 7	22	10

Figure 17.6 List of Switches.

Results of Process Modeling

Extend+BPR automatically calculates the percent busy of simulated labor, the amount of time items (in this case, phone calls) stay in a queue (on hold), the number of items that abandon a queue (hang-ups), and so on. All this information was reported in the notebook for easy viewing, as shown in Figure 17.7. (Some numbers have been modified due to the sensitive nature of this data).

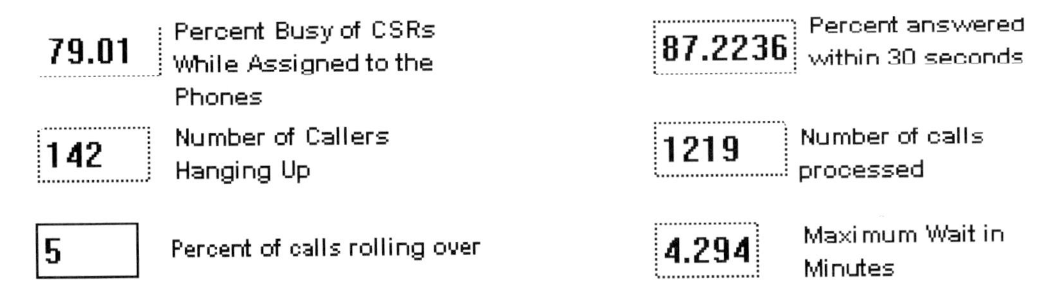

Figure 17.7 Results of Simulation.

All the parameters defined by Beeson are reflected in the notebook, and some others are included as well. When a number of scenarios were run, some interesting results were noted. "We saw that, as call volume increased, the number of abandons increased as well," said Beeson.

"Because the model predicted that callers would hang up, the number of calls answered in 30 seconds would remain high and the percent increase in workload would not be in proportion to the increase in calls. This information told us that we could not view the customer service operation in terms of any one parameter alone, and that all three had to be assessed together."

Future Use of Modeling

According to Gabbard,

> We are just scratching the surface of our analysis. We have to tie all of our operations to determine their relationships. For example, we have seen that software changes can affect call volume, but we also know that software changes affect the need for training, which will impact available staff. In addition to the analysis already done, we want to determine the impacts of marketing programs, policy changes, productivity tools for customer service representatives, and so on. This analysis is too complex to be done without automated assistance, and it is important to LEXIS-NEXIS that the analysis be done as scientifically as possible.
>
> Business process reengineering is not easy either. Automated tools facilitate reengineering and enable a user to view the process from a number of perspectives. In this case, for example, we were able to assess the impact on call volume, CSR utilization, callers hanging up, and so on.

"The big advantage of using a tool like Extend+BPR," Beeson claims, "is that not only can we determine the impact of process changes before they are made, but we can make the right decisions about process changes—those with customer positive impact. This effort will help LEXIS-NEXIS continue its high level of customer satisfaction."

18

Case Study: High-Volume Manufacturer Order Fulfillment Processes

Introduction

The business in which this particular manufacturer is engaged is extremely competitive, with margins being reduced every day. In addition to price pressures, the company was under increasing pressure from customers to deliver finished goods faster (reduce order fulfillment cycle time). After careful study, it was determined that the areas of purchasing, receiving, and accounts payable provided opportunities for lasting cost reductions, and also opportunities for reducing the cycle time. Specifically, these processes were manual in nature, burdened with a great deal of paperwork, and time-consuming. A number of attempts were made to improve and streamline these processes with little success. The vice president of finance decided on another method of reducing costs but, before proceeding, wanted to predict the effects of process changes before investing in the technology required to implement those changes.

Original Process Description

The first order fulfillment process to be addressed was the purchase order process. Purchase order processing is an important element in order fulfillment—delays in obtaining raw materials lead to production delays. The process itself was very simple: Designated employees, typically managers, would issue purchase orders and the purchase orders were sent to the Purchasing Department for review. On occasion, multiple signatures would be required. When this was the case, the purchase orders were placed into interoffice mail and, after being signed, returned to the Purchasing Department. Also, if the dollar amount of the purchase was very high, upper-level management signatures were required. In this case, a purchasing agent would walk the purchase order through the process. Ultimately, a purchase order would be sent to a vendor and a copy filed.

On occasion, mistakes would be made and the purchase order would be returned by the vendor for correction. A purchasing agent would contact the originator, resolve the problem, and resubmit the purchase order to the vendor. A model of the initial review and processing process is shown in Figure 18.1.

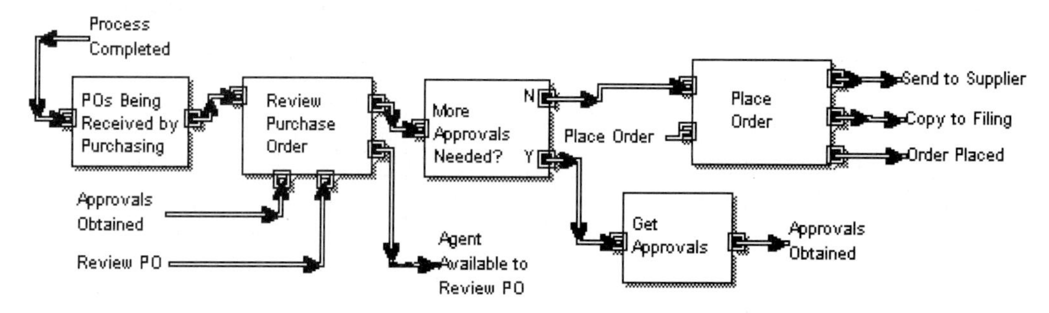

Figure 18.1 Model of Initial Review Process.

A continuation of this model, depicting the tasks of filing the purchase orders and resolving problems, is shown in Figure 18.2.

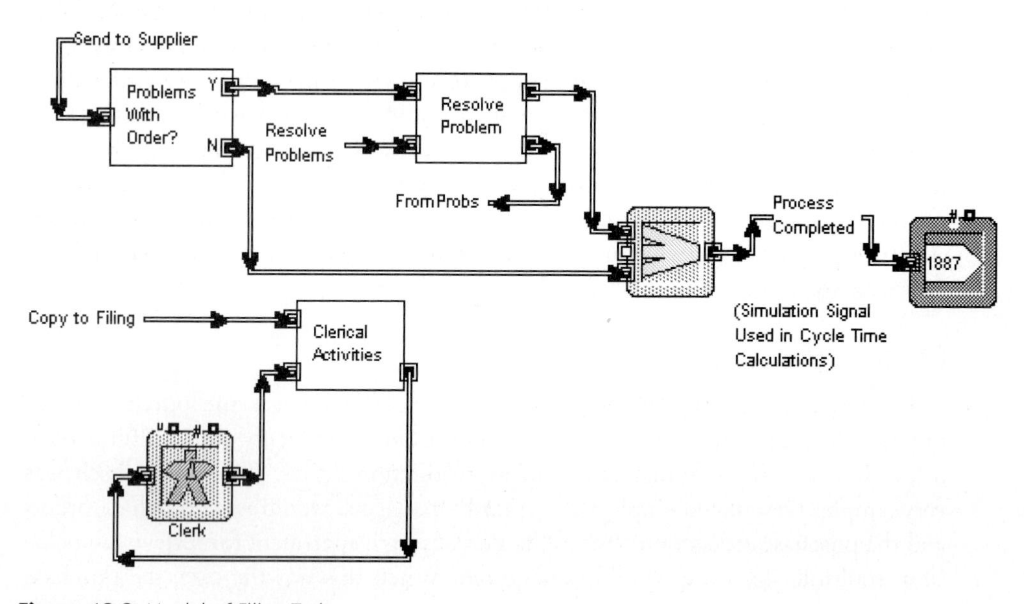

Figure 18.2 Model of Filing Task.

The Purchasing Department consisted of one full-time agent and one part-time agent. The main activity of the agents was negotiation with vendors, followed by reviewing and placing orders. Problem resolution, because it happened infrequently, was given a lower priority. The activities of the purchasing agents were modeled as shown in Figure 18.3.

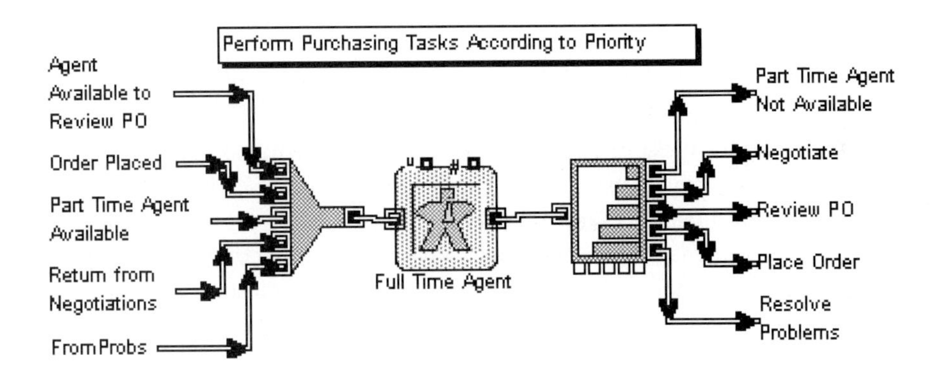

Figure 18.3 Prioritizing Purchasing Agents' Tasks.

The phrases on the left-hand side of the figure represent paths taken by the purchasing agent simulation "objects" as they return from tasks, and the phrases on the right represent the paths taken to areas of the model where work would be done. Note the phrase "Part Time Agent Not Available." The simulation was set up to remove the part time worker from availability for periods of time, representing the fact that he had other work to do.

In the original process, the average cycle time required to process a purchase order, including the time required to resolve problems, was 14 working hours, or almost two 8-hour days, and as long as 40 hours, or 5 working days. As this was considered to be excessive, causing delays in the procurement of essential materials, an implementation of process changes was begun.

First Attempt at Reengineering

The vice-president of finance of the company reviewed the process and determined that there were two main problems associated with the purchasing process:

1. The time required to complete a purchase order was too long.
2. The time required by purchasing agents to handle purchase orders was excessive, leaving little time for their most important task—negotiations.

The VP of finance recognized that more time had to be allotted to negotiating, a purchasing agent's main function, if costs were to be reduced. He set out to change the process so the agents would have less involvement in the clerical aspects of purchase order processing. He reasoned that there were two causes for the problems listed above:

1. All purchase orders, regardless of amount, were being sent to the Purchasing Department for review.
2. Purchase order originators relied upon the Purchasing Department to catch errors; therefore, they were careless when providing essential information, leading to problems and rework.

To resolve these issues, the VP of finance determined that all purchases under $1000 would be made using Small Purchase Order Forms (SPOFs). SPOFs would be sent directly to vendors, and no action on the part of the Purchasing Department would be required. To place an order, any employee, regardless of managerial status, could simply fill out a SPOF. The employee would be responsible for ensuring that the order was correct and that when the order was received, the accompanying invoice would be sent to the Accounts Payable Department for processing.

The VP of finance used the causal reasoning shown in Figure 18.4 to justify the SPOF system.

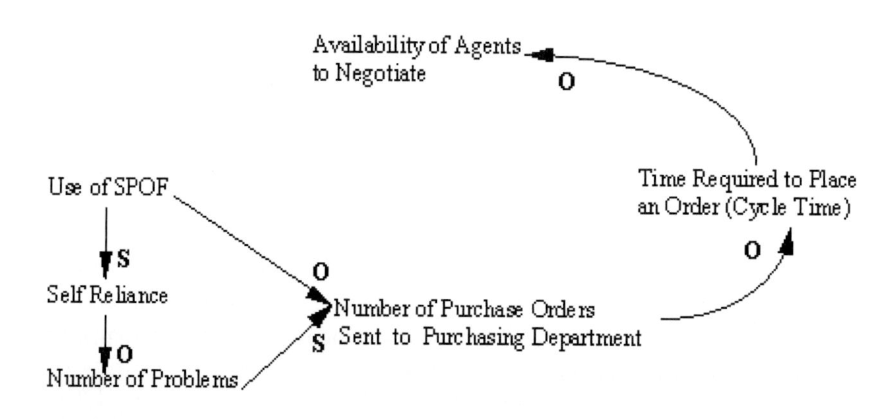

Figure 18.4 Justifying the SPOF System.

This logic seems to be simple and straightforward. Management was relying on the fact that empowered employees, given power and responsibility, would do their best to make the system work. Unfortunately, the end result was not what had been hoped. Figure 18.5 presents a causal chain that demonstrates what actually happened when the changes were implemented.

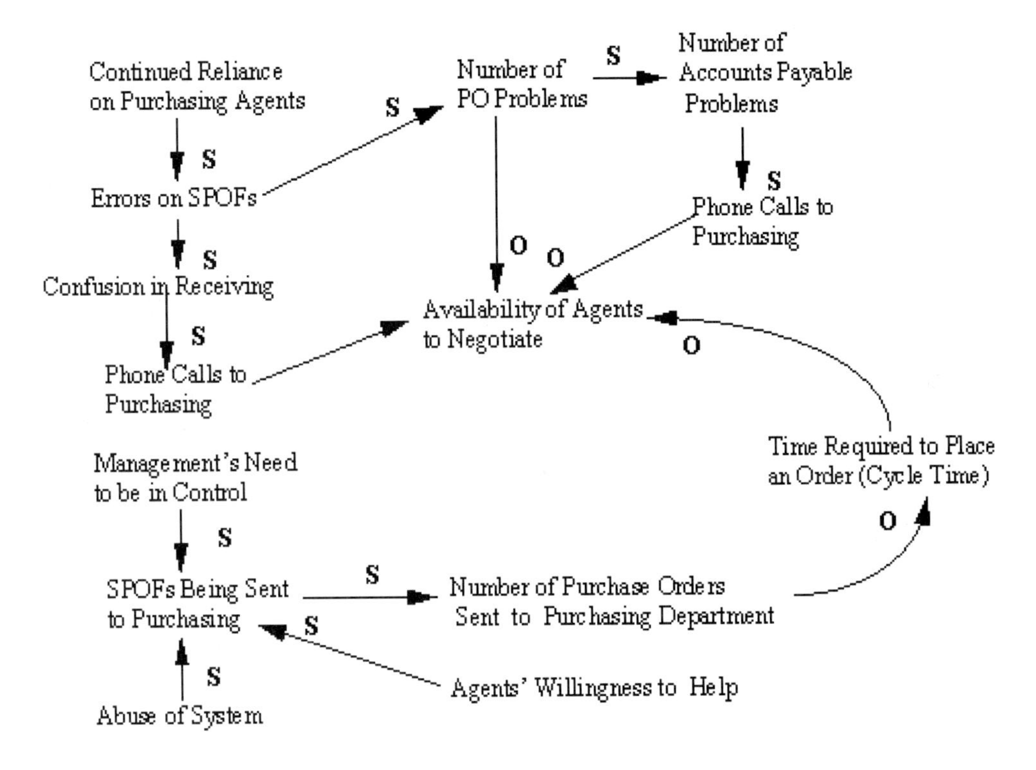

Figure 18.5 Causal Chain After Implementing the SPOF System.

Once the new process was implemented, a whole series of problems began to emerge.

1 Some mid-level managers were reluctant to grant even a limited amount of autonomy to their staff. Moreover, they were concerned that their staff would make a mistake and they, the managers, would be held responsible for the mistake. Therefore, some managers insisted that their staff pass all SPOFs to them for signature and then pass the SPOF to the Purchasing Department. This was a complete violation of the intent of the SPOF and caused delays in the processing of purchase orders.

2 Purchasing agents, also accustomed to being in a controlling position, readily accepted the SPOFs, although the agents could reasonably have refused to accept them.

3 Originators of SPOFs, seeing that Purchasing would participate in low-dollar purchases, neglected to assume responsibility for their orders and made more errors than in the past.

4. The Receiving Department could not determine where to deliver orders when they were received, since originators often neglected to put this information on the SPOF.
5. Accounts Payable began receiving calls from irate vendors who had not been paid and were at a loss to help the vendors, since they had never received invoices. Moreover, they had no idea who had placed the order.
6. Workers quickly learned that they could break orders for $5000 into five $1000 orders. This was abuse of the system and resulted in more paperwork than before, and more problems than before.

Ultimately, cycle time for purchase orders increased, and there were bad feelings among all the participants. Moreover, the main task for the Purchasing Agents became problem resolution, leaving less time for negotiating with vendors. Cycle time increased from an average of 14 working hours to an average of more than 50 working hours! Clearly, this was unacceptable and counterproductive.

Lessons Learned

Many other companies might have abandoned their attempt at reengineering, blaming their employees for the failure of the experiment. This company, however, chose to learn from this mistake. It realized that purchasing was not an isolated process and, in fact, what happened in Purchasing affected other departments, such as Receiving and Accounts Payable. Had the process remained in place much longer, it is quite likely that impacts would have been felt in other areas as well, including the manufacturing floor.

Company management decided not to abandon its reengineering efforts but decided instead to look at the larger process picture. Specifically, it decided to investigate technology that would link Purchasing, Receiving, and Accounts Payable and to investigate mechanisms for working more closely with its vendors.

Before investigating available technology, however, management decided to address the problems that had arisen from the first reengineering effort.

1. Certain managers wanted to maintain control over employees and demanded that SPOFs be signed by them and then sent to Purchasing. This presented both a cultural issue and process issue. The VP of finance met with all mid-level managers and explained how their actions hurt the company, and also that he would reconsider the level of purchasing authority in the new process.
2. Some employees took advantage of the new process by breaking larger-dollar purchases into a series of smaller purchases. This simply demonstrated that empowerment, which is based on the idea that employees will do their best job, does not always work. Again, this was a cultural issue that had to be addressed.

3 The purchasing agents wanted to remain part of the process for all purchases and also wanted to help resolve issues. These goals are counter to each other, and this was another cultural issue that had to be resolved.

It is clear that the culture of the company had to be changed before the processes could be changed. The question was, how could that be best achieved?

The Use of Simulation

The VP of finance decided to simulate proposed process changes and contracted with Computer-Aided Process Improvement (CAPI) to assist in that effort. The functional areas that were going to be addressed were Purchasing, Receiving, and Accounts Payable. CAPI first constructed "as-is" models of these functional areas, measuring cycle time, cost, and staff utilization. An overview of the model of all these processes is shown in Figure 18.6.

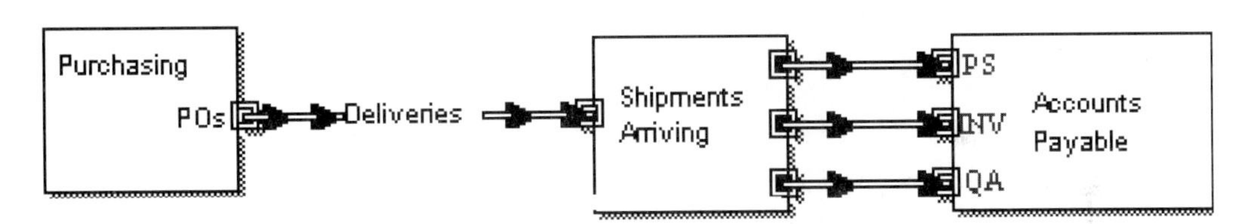

Figure 18.6 Overall Process Model.

The details of these models are too large to be presented in this format; therefore, a textual description of the processes is provided.

1 Orders were placed with vendors and, after some time, shipments were received at the receiving dock. Vendors, upon shipping, would send invoices to the Accounts Payable Department.

2 When a shipment was received, a packing slip was removed and sent to the Accounts Payable Department, and the material was placed in a holding area for Quality Assurance (QA) processing.

3 The QA Department then inspected the incoming material and, if acceptable, issued a QA slip to Accounts Payable.

4 The material was then moved from the QA area to the manufacturing warehouse.

5 Accounts Payable would only pay a vendor when it could match invoices to packing slips and QA slips. In addition, if the item received was a large dollar amount item, then Accounts Payable would also have to match the invoice to a purchase order.

6 Since backlogs tended to grow in Accounts Payable, invoices were paid at the end of each month, even if all the accompanying paperwork was missing. This was done to maintain good relations with vendors, although it did tend to create problems later on.

Although this is a stripped-down description of the process, it demonstrates that there was a great deal of paperwork and movement of material. Since management was interested in how staff could be used more effectively, one measure of the process, determined via simulation, was the percent utilization of the Accounts Payable staff, which consisted of four people. The simulation also measured the backlog of work in Accounts Payable. The results of the simulation of Accounts Payable are shown graphically in Figure 18.7.

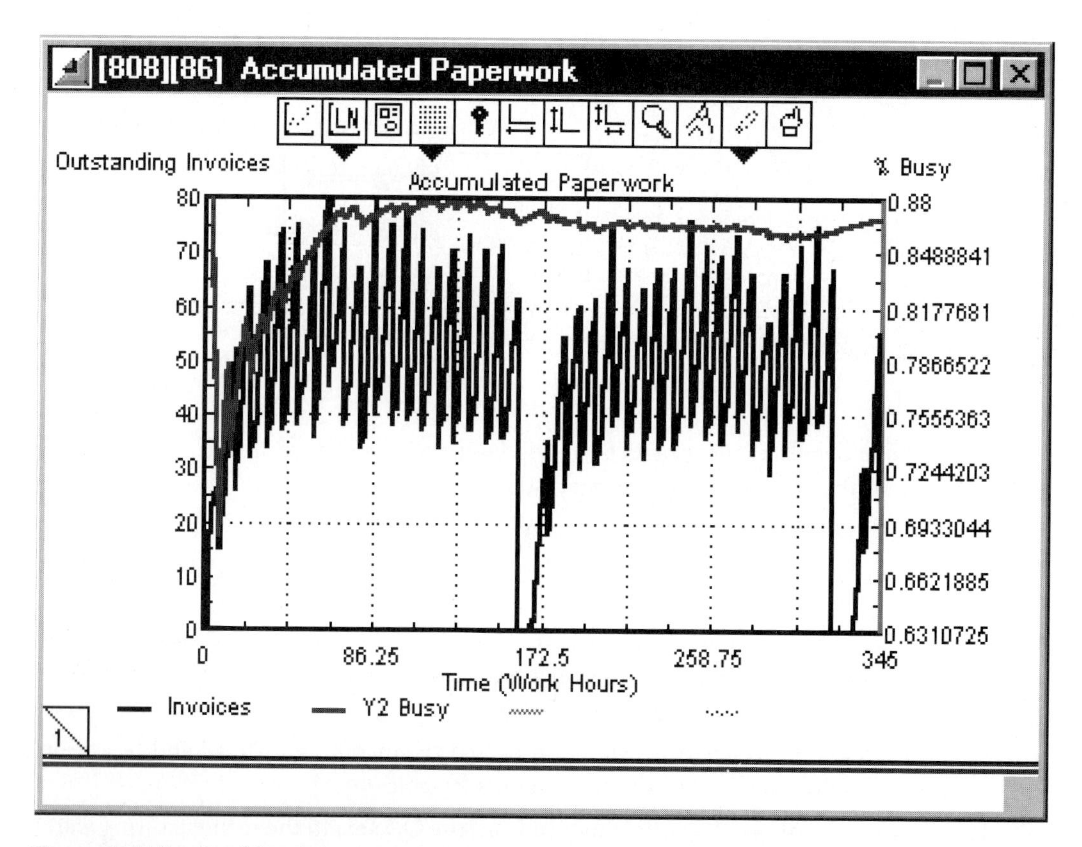

Figure 18.7 Results of Simulation.

This graph shows that the four personnel in Accounts Payable were busy about 88 percent of the time (except for lunch and breaks, 100 percent busy) and the backlog of work, even after month-end processing, remained constant. In other words, month-end processing created a new backlog and was counterproductive.

Second Attempt at Reengineering

The VP of finance decided to investigate a number of changes to the processes.

1. The ability of all personnel to place orders was ended; instead, certain personnel, not necessarily mangers, were given the ability to place orders.
2. Orders that were $5000 or less would be placed using credit cards—there would be no paperwork. All the tracking of orders would be done by the credit card company, a service that was provided at no charge in exchange for the percentage the credit card company would receive from the vendors.
3. Orders would be linked to a computer system accessible by the credit card company and individuals placing orders.
4. The credit card company would issue payments to vendors once per month after receiving notification of delivery.
5. A bar code system would be implemented in the Receiving area. The bar code system interfaced to the computer system and would issue delivery confirmation, update inventory levels, determine who would receive the order, and so on.
6. Vertical partnering arrangements would be made with certain vendors, such as vendors of janitorial supplies, in which those vendors would analyze the stock of material on hand, replenish when necessary, and enter an invoice into the computer system.
7. Purchasing agents would handle only large purchase orders and would be prohibited from trying to solve problems associated with other orders.
8. The QA Department, instead of inspecting all incoming material, would work with vendors who presented quality problems to ensure that shipments were of high quality when they left the vendors' plants.

Some of these changes involved technology, and others involved cultural changes. The VP of finance decided to use simulation to assist in making the cultural changes and to provide a rationale for investing in technology. When the "to-be" model was developed, the same process parameters were measured as in the "as-is" model. For example, after several iterations of simulation, it was determined that, given the dramatic reduction in paperwork, the Accounts Payable Department could be reduced to one individual, and the three remaining individuals could be reassigned to other tasks in other departments. The same graph that was shown in Figure 18.7, representing staff utilization and backlogs, is shown for the "to-be" process in Figure 18.8.

Note that the one remaining person would be busy only about 40 percent of the time, and that the backlog would be reduced from an average of 70 invoices to an average of 16. (Month-end processing still occurred—some habits are hard to break.)

Simulation of the proposed process also revealed that the Purchasing agents would have more than 80 percent of their time available for negotiations, and that the

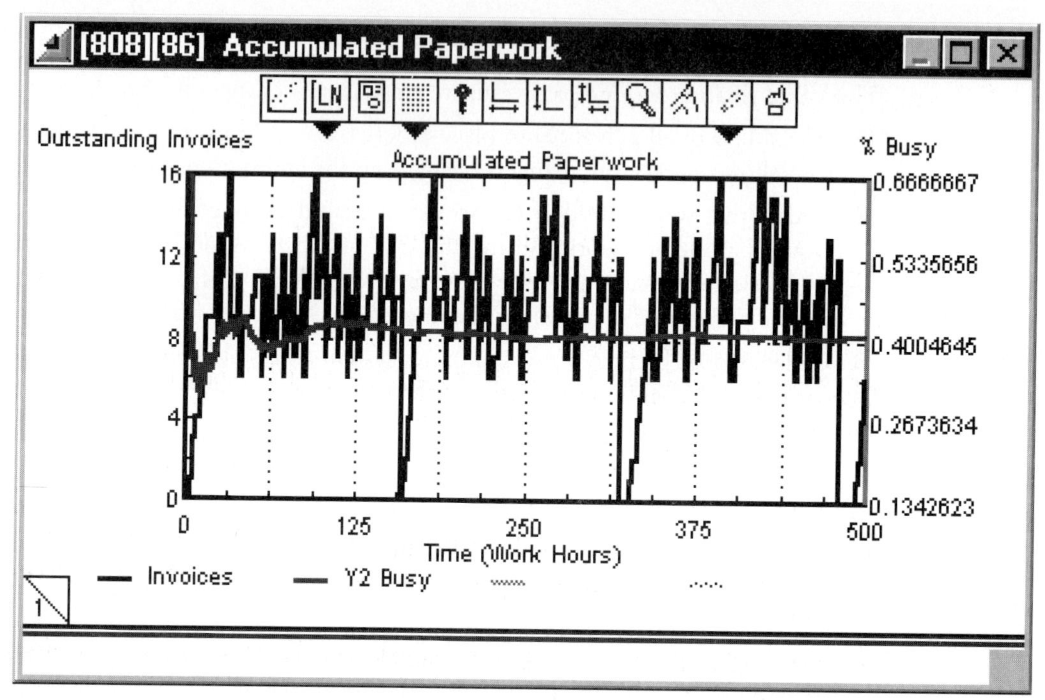

Figure 18.8 Results of Second Simulation.

promise of quick payment would yield price concessions from vendors. Staff could be reduced in a number of functional areas due to the reduction of other types of paperwork and these reductions would help fill needs in other areas of the company.

Before one penny of investment was made, and before new procedures were put into place, company management reviewed the model and simulation with all those who would be affected. Some personnel were wary, suspecting that the effort was really a downsizing effort. Management was prepared, suggesting areas of reassignment for those personnel. Some managers wanted to maintain control and were opposed to the credit card idea, even though the simulation showed that cycle time would be reduced and cost savings could be achieved. In these cases, upper-level management had to insist upon compliance with the plan—cultural change is not always easy and consensus is not always possible.

For the most part, all personnel involved were enthusiastic and pleased with the direction their jobs would be taking. QA personnel, for example, had always thought that working with vendors would eliminate most problems, and now they were being given the chance to see if they were right.

As the computer systems and process changes were implemented, other unanticipated benefits were realized. For example, it was discovered that vertical partnering in many areas reduced the warehousing space required for some material. In addition, most vendors were anxious to participate in vertical partnering opportunities since it meant a guarantee of business.

Summary

This case study makes a number of very important points about reengineering.

- Empowerment is a good concept based on trust—trust of employees and trust of managers. Unfortunately, managers and employees alike do not always act in the trustful manner we would like them to.
- Causal reasoning is a powerful way of thinking about the effects of change; however, what we assume to be true is not always true.
- When changes yield counterproductive results, the changed process is often worse than the original process.
- People relate better to facts than to supposition. Simulation deals with facts and helps reinforce the value of process changes, breaking down cultural barriers.
- On occasion, proven strong management techniques are necessary to affect change.
- Simulation is a powerful mechanism for testing the investment required for change before the investment is made.
- Simulation can help overcome the negative feelings caused by the lack of success of previous reengineering efforts.

19

Financial Analysis Using CAPRE Technology

Introduction

So far this book has concentrated on the application of CAPRE technology to process analysis, focusing on cycle time, throughput, productivity, and cost. These are all very important factors when deciding how best to implement a process or series of processes. Simulation is also a powerful tool for the analysis of the overall financial health of an organization. Financial analysis has traditionally been done using spreadsheets and add-ons to spreadsheets. While these are powerful tools, they are still, for the most part, static tools and at best can supply only limited "what if" scenarios. This chapter will demonstrate how CAPRE technology can be used to determine various types of financial measures.

Return on Investment (ROI) Analysis

Return on Investment (ROI) analysis is the type of analysis that answers the questions

- How long will it take before we realize a profit?
- How much profit will we make in a certain period of time?

The goal of ROI analysis is to determine the shortest time period required to see a positive cash flow and to determine the profit that can be realized within a window of opportunity. ROI can be affected by many factors, such as policy decisions, technological decisions, operational implementation, and so on.

Consider the example of a small manufacturer that wants to introduce a new product line. The manufacturer is trying to determine whether he should fully automate his manufacturing line or partially automate the line. Full automation requires a large upfront investment in technology, while partial automation requires less of an investment but more personnel. In addition, the partially automated lines are less productive than the fully automated lines. Complicating the analysis is that the sales of the product will grow over time, possibly requiring an additional investment in technology or more personnel.

There are a number of steps that must be taken before doing ROI analysis. First, the manufacturer must determine the process flows required to create a finished product. This is done best using CAPRE technology and looking at the inflow of raw material, processing rates, and outflow of finished products. The simulation can be set up to run numerous scenarios, building in machine downtime, variable processing, and so on, so a picture of average processing capacity emerges.

Figure 19.1 shows the model used to perform the ROI analysis. Notice that named connectors are used in the model, such as Management Labor, Cost of Stamper, and so on. These represent some of the parameters that will be entered into the simulation by the user of the simulation.

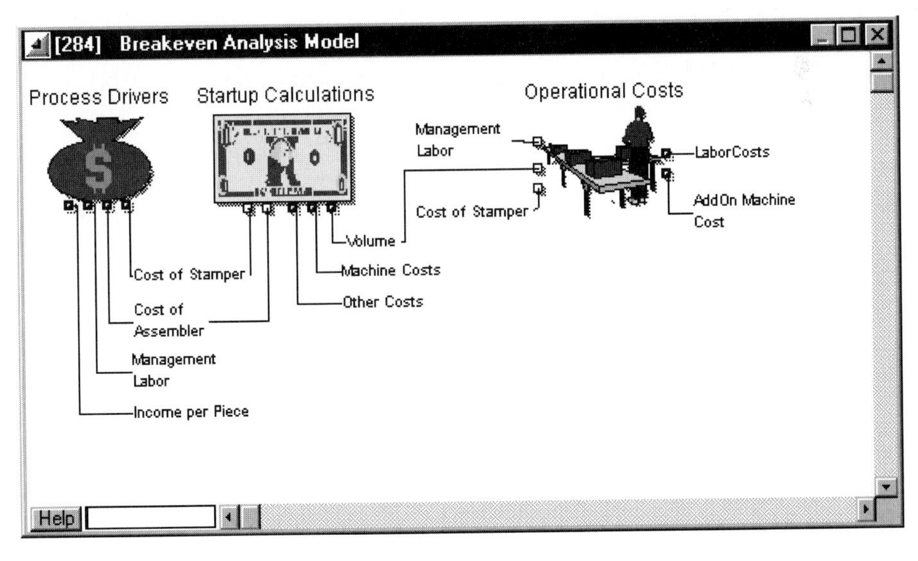

Figure 19.1 Overall Model of Manufacturing Process.

The model shown in Figure 19.1 performs analysis in the following manner.

1 It determines the startup costs for the process. These consist primarily of the cost of the machinery to be used in the process, namely a stamper and an assembler.

2 It determines the daily costs of the management personnel required for the process using burdened labor rates supplied by the user of the simulation.
3 It determines the daily sales volume of the finished products. Sales projections are usually obtained from marketing projections and can be entered in a number of ways, such as in a table or a graph.
4 It determines the cost of raw materials on a daily basis, using the volume information to analyze the number of specific items of raw material required to produce each finished product.
5 It determines the production capacity of the manufacturing lines, using the daily volume information and processing capacity provided by the user of the simulation. For example, capacity could be 1200–1600 finished products per hour for the fully automated process and 800–1200 finished products per hour for the partially automated process.
6 It predicts the daily cost of labor for the process. Labor requirements will change over time as volume increases.
7 It predicts the costs of additional machinery requirements based on volume. For example, if volume increases to the point that one set of machines cannot handle the load, then the model will assume that another set of machines will be purchased.
8 It determines the periodic maintenance costs required for the machinery used in the process.
9 Finally, it determines the revenue that can be realized from the process using the daily volume figures and a revenue-per-product figure input by the user of the simulation.

Figure 19.2 shows the result of the simulation for the highly automated process.

Figure 19.3 shows the results of the simulation for the partially automated process.

A comparison of the two plots shows that the partially automated process provides a positive return on invested capital faster than the highly automated process, and also that the partially automated process provides more of a profit at the end of a four-year period—the ROI analysis period chosen by management.

With this information, and with this model, a manager can begin to perform some "what-if" analyses. For example, the manager can explore less expensive technology that provides less productivity but may ultimately provide more of a profit due to the lower costs; sales projections can be modified to look at best-case, likely-case, and worst-case scenarios, and so on.

The value of simulation over spreadsheets is that the parameters can be changed one at a time or many at a time; in fact, they can be changed automatically. For example, Figure 19.4 shows the results of a simulation that automatically changes the parameters in the simulation to provide ROI analysis of both processes.

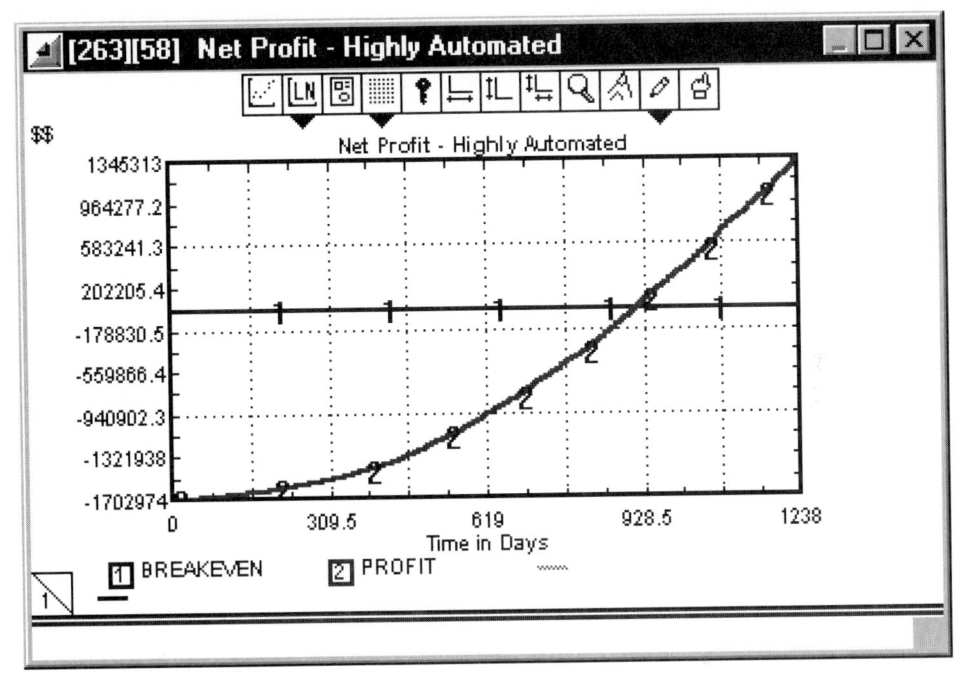

Figure 19.2 Results of Simulating Highly Automated Process.

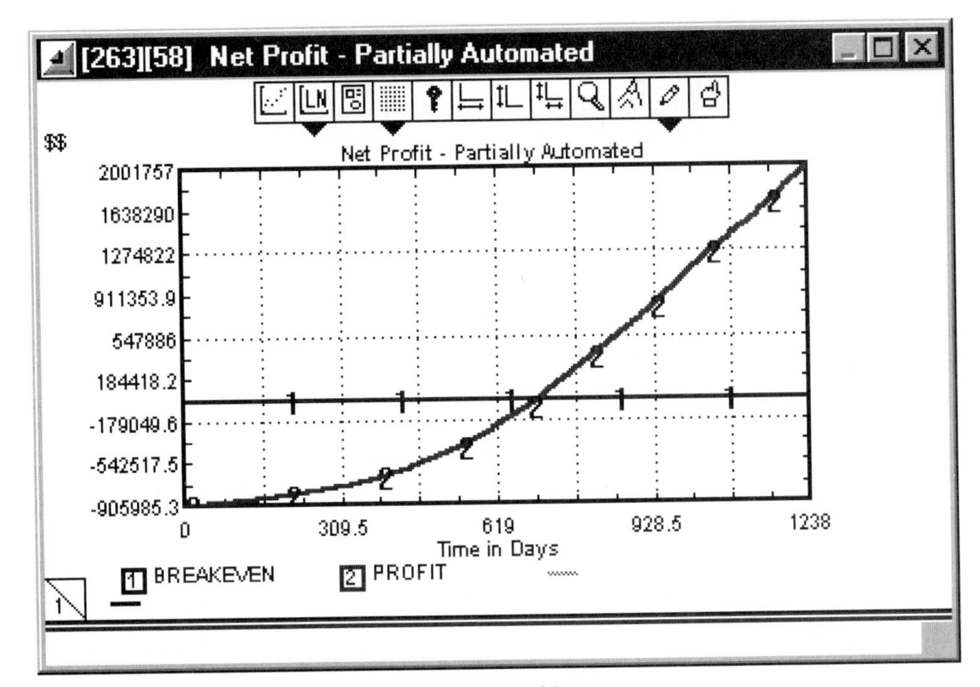

Figure 19.3 Results of Simulating Partially Automated Process.

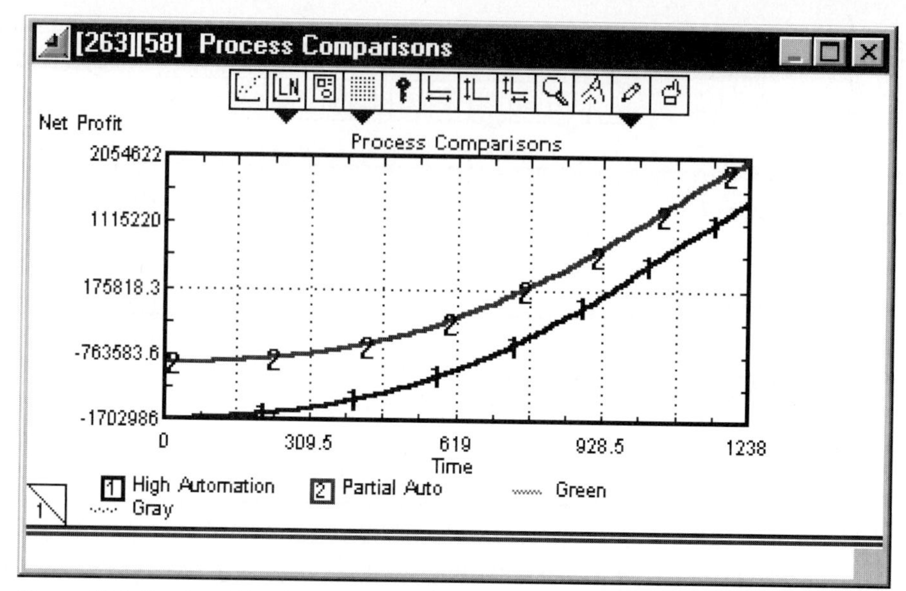

Figure 19.4 Results of Simultaneous Simulations.

This graph shows what we already know. It also shows, however, that the highly automated process is closing the gap between it and the partially automated process. If the product is one that the company decides to continue manufacturing, it might want to compare the two processes for a longer period of time, as shown in Figure 19.5.

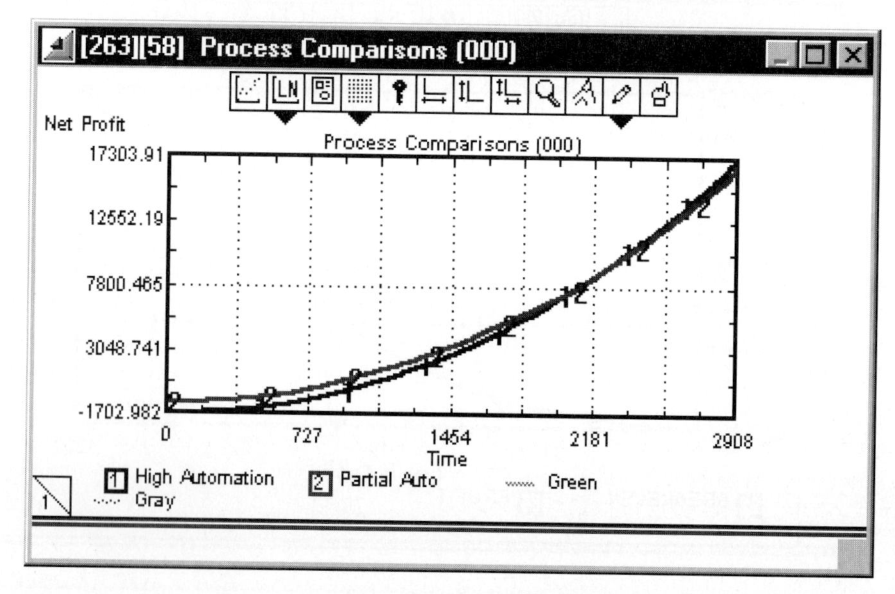

Figure 19.5 Comparison of Processes over a Ten-Year Period.

The graphs show that the two results are very close. By "zooming in" on the upper-right-hand corner of the graph, we can get better detail of the comparison. Figure 19.6 shows a graph of the last few days of the simulation.

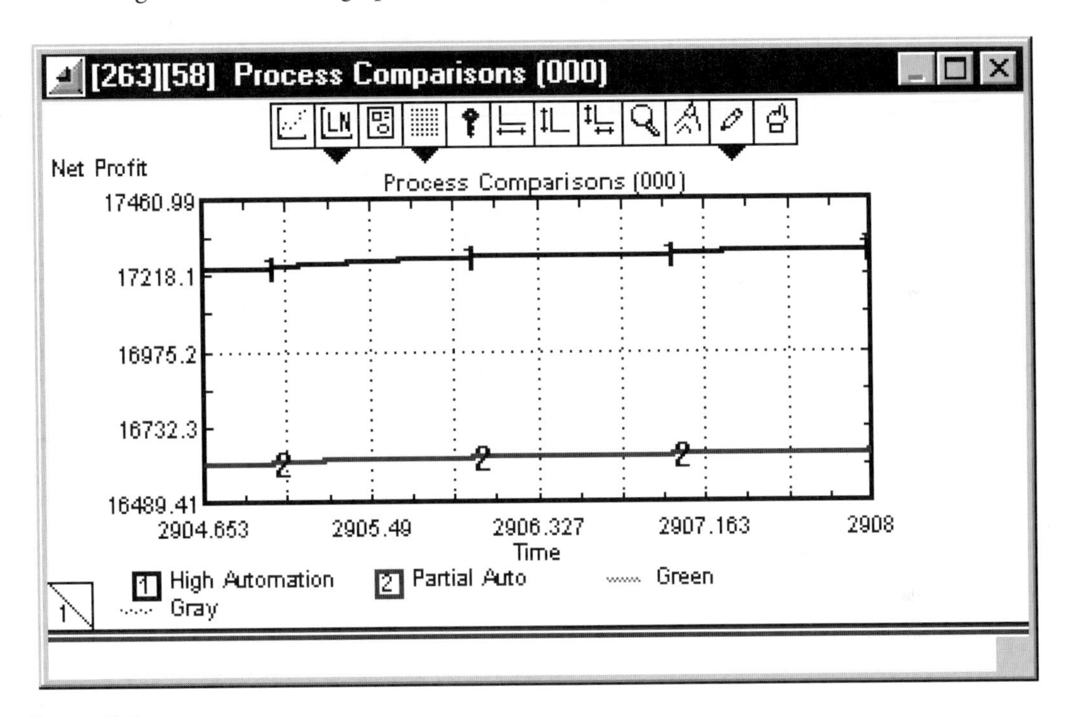

Figure 19.6

This graph reveals that the highly automated process returns more of a profit at the end of ten years than the partially automated process. This information provides management the opportunity to consider whether it wants to maximize profits over a longer period of time or maximize profits over a shorter period of time.

The advantage of simulation over static analysis, such as in spreadsheets, is that simulation is dynamic. Parameters can be changed easily and scenarios tested quickly. Moreover, a simulation is a "living" financial model in that it can use historic data and projected data and, as conditions change, such as the cost of technology, the model can be changed as well. In addition, after ROI analysis has been completed, actual process design and simulation can begin, primarily to test the assumptions made about productivity. If the process simulations determine that the assumptions were incorrect and that, for example, productivity in the partially automated process is less than expected, then the ROI analysis can be repeated.

Upon completing both sets of analysis, company management has been able to analyze the financial viability of a new product offering, as well as test its assumptions about manufacturing flow, all without investing in personnel or equipment.

Operational Performance Measurement

This book promotes a systems analysis approach to Business Process Reengineering. This approach is different from the traditional approaches of TQM and CPI, which were focused only on tasks within hierarchical organizations. When changing the way we look at business, we must also change the way we measure the success of that business. The ROI example is an example of financial analysis using nontraditional financial analysis methods. This concept can be expanded further and applied to a business as a whole, not just a piece of a business. The approach discussed here is called *operational performance measurement.*

Traditional performance measurement systems are based on functional hierarchies, which were created to support "command and control." As companies move away from rigid hierarchy to process or network organization structures, their performance measurements must also change. Because of this, traditional performance measurement systems suffer from the following.

- Traditional performance measurements can be distorted by accounting entries and "nonrecurring" items.
- Current performance systems have weak links to the organizational strategy and are not easily linked to actions and decisions by managers throughout the organization.
- Traditional measures lack predictive power and do not capture key business changes until it is too late.

Larry Procter of Nolan, Norton and Company suggests that "managing a business operation with traditional methods is like steering a car by looking through the rear view mirror." Operational performance measurement, on the other hand, provides a framework of measurement parameters and allows every level in the organization to participate in meeting the strategic objectives. This framework combines strategic analysis, financial analysis and operational analysis and utilizes techniques such as Economic Value Added (EVA™)[1] analysis.

Economic value added is a method of analyzing the financial "health" of an organization by

- Using accounting data to relate operational performance to shareholder value.
- Measuring operational performance on a cash flow basis.
- Expanding business focus by linking the impact of operation and asset management.
- Linking the drivers within an organization that create value to its performance.

Graphically, EVA analysis is shown in Figure 19.7.

[1]EVA is a trademark of Stern Stewart, New York, NY.

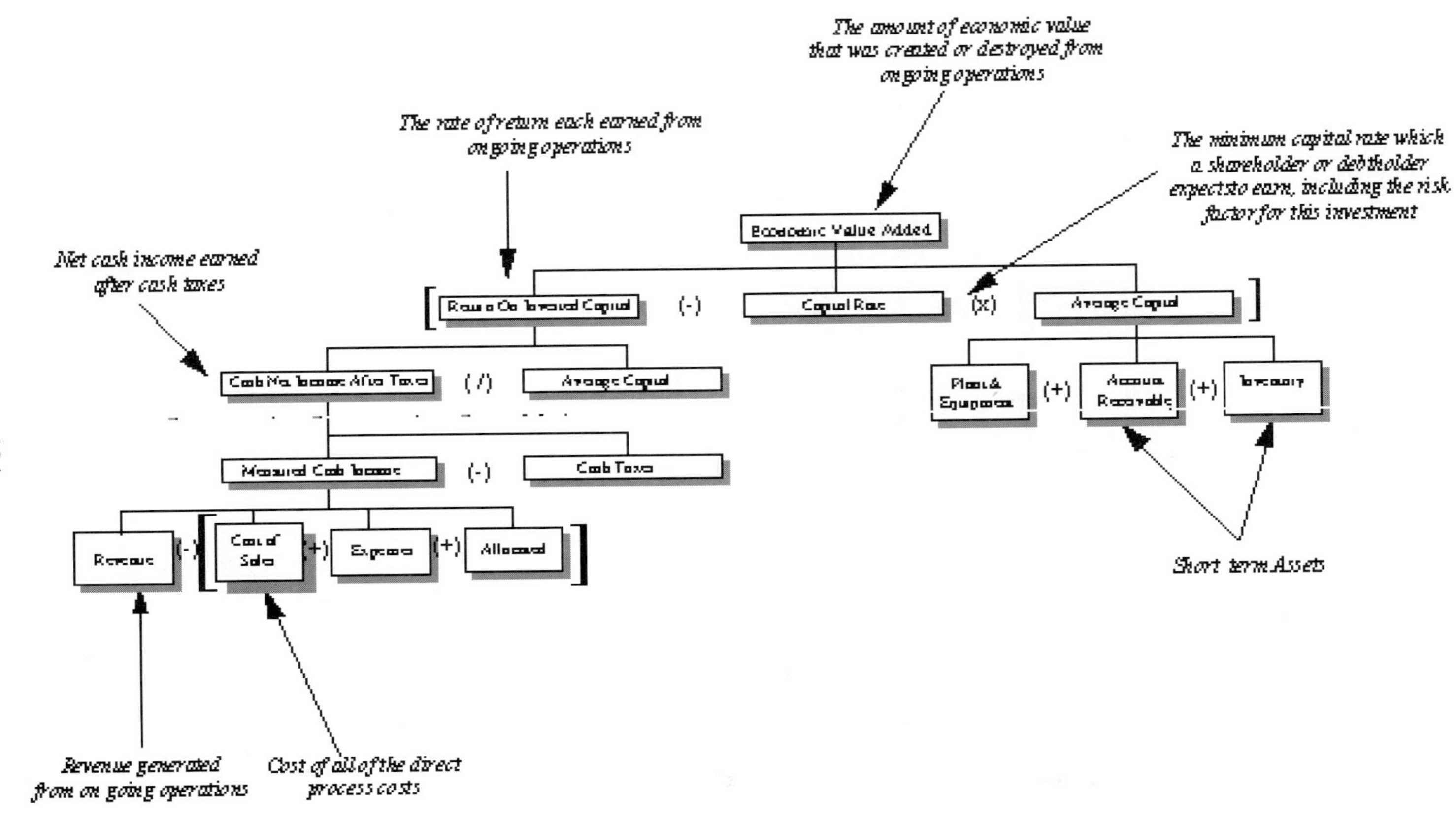

Figure 19.7 Graphical Representation of Economic Value Added Analysis.

This would appear to be a traditional flow of financial information, but there is more to be explained. Operational performance measurement also utilizes the concept of a *balanced scorecard*, which links many apparently disjoint parameters to each other. For example, a balanced scorecard would link employee satisfaction to productivity, which is then linked to cycle time and cost, which is linked to customer satisfaction, which is linked to sales, and so on. Graphically, balanced scorecard analysis is shown in Figure 19.8.

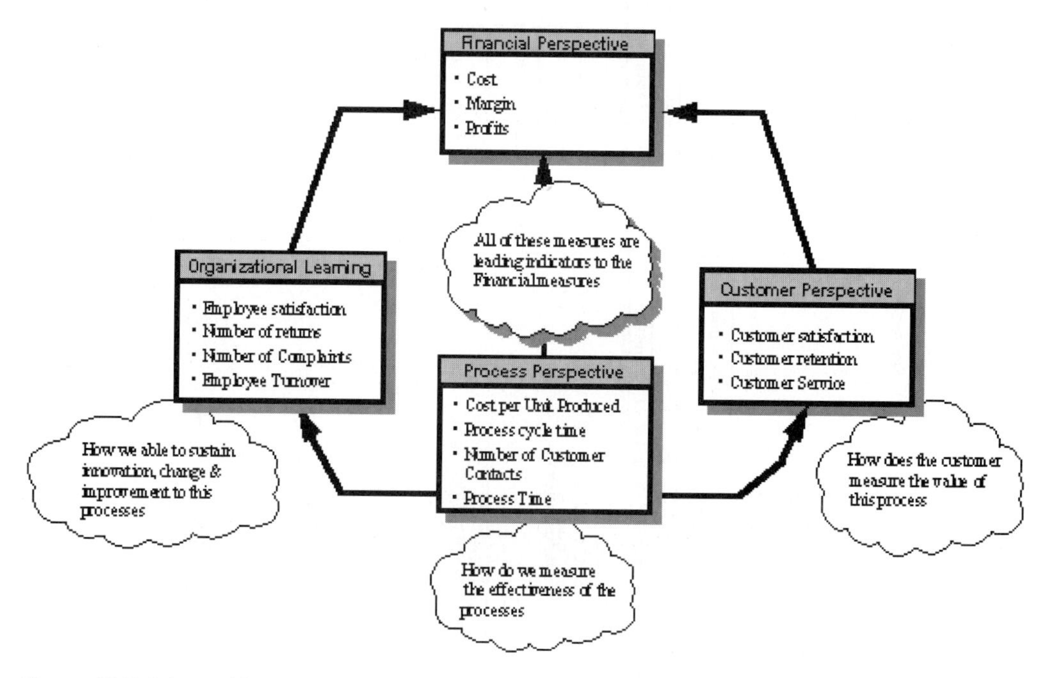

Figure 19.8 Balanced Scorecard.

What is interesting about Figure 19.8 is that it represents *causal analysis*. This figure shows the relationships between business parameters and we can imply that an increase in employee satisfaction will result in an increase in productivity and quality which will result in an increase in sales and so on. What is also interesting about this figure is that "customer satisfaction" is merely one of many parameters—it is part of the analysis of a business perspective.

Now we can discuss how simulation is ideal for analyzing operational performance measurement. Figure 19.9 is a portion of an Extend EVA simulation for a manufacturing client.

You will notice that it looks very much like the diagram shown in Figure 19.7; however, underlying each hierarchical box in the model is simulation logic. For example, under the box labeled Return On Invested Capital is the simulation block shown in Figure 19.10.

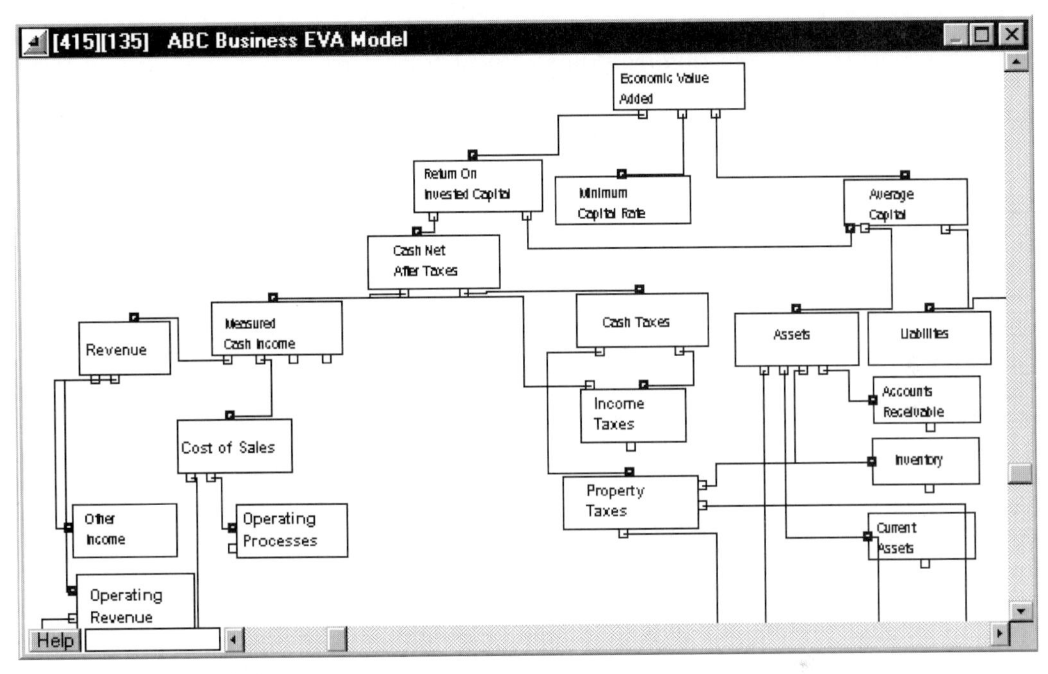

Figure 19.9 EVA Model.

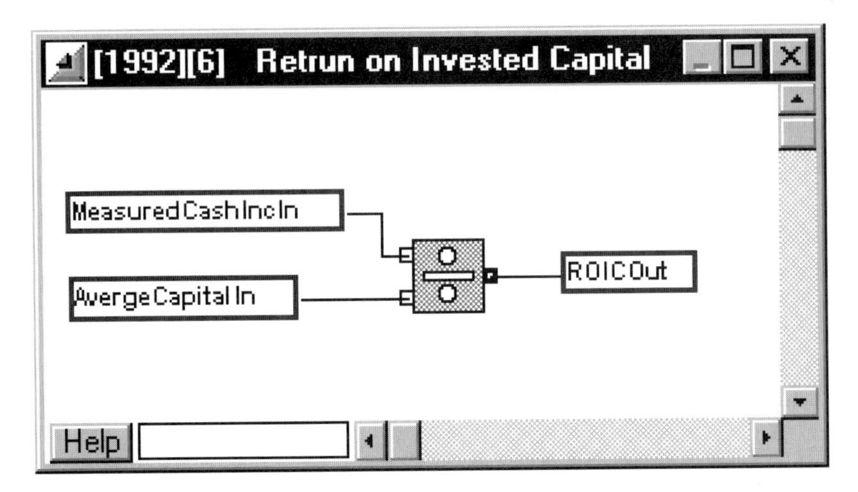

Figure 19.10 Details of Return on Invested Capital Block.

This block is simply a division of Measured Capital by Average Capital, which yields Return on Invested Capital. The MeasuredCashIncIn and AverageCapitalIn connectors are inputting data from other similar formulas in the simulation, and the ROICOut connector is supplying data to another part of the simulation. The simulation can be viewed as one gigantic formula broken into pieces and linked. But there is more to the simulation than simple calculations. First, data must be supplied and, depending upon the calculation, the amount of data varies. For example, consider the block labeled Operating Processes. Figure 19.11 shows this block after it has been expanded.

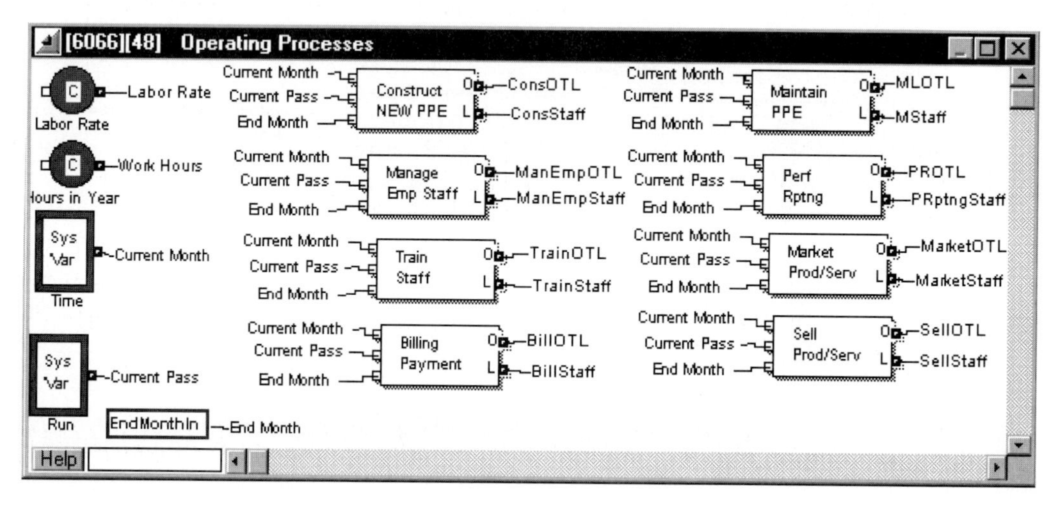

Figure 19.11 Details of Operating Processes Block.

Figure 19.11 shows that the costs of operations consist of the sum of the costs of many processes. Each process is further broken down, as shown in Figure 19.12.

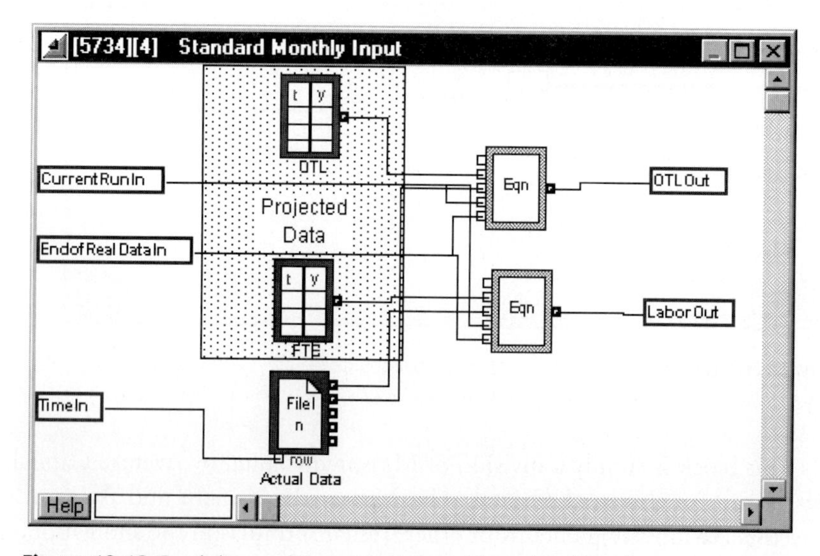

Figure 19.12 Breakdown of Cost Calculations Hierarchical Block.

The data for the cost calculations for the processes are composed of both historic data (input through a file) and projected data (input through tables). The model is set up to execute a number of times, running first with actual data and next with projected data. It can also be run with actual data for some period of time and then cut over to projected data. Processes can be simulated separately, as described earlier when discussing ROI analysis, and the results written to files. These results can then be input into the EVA model and used to determine the impact, for example, of a new product introduction.

In addition, the model allocates costs to all operational processes from other support processes. For example, Human Resources is considered a support process and the cost of that operation is allocated to all revenue-producing processes.

Finally, the model consists of "what-if" drivers. For example, if it is true that employee satisfaction is related (in a causal relationship) to productivity, then there should be some formula that relates the costs of providing employee satisfaction and another that relates employee satisfaction to productivity. Figure 19.13 shows how this might be accomplished in Extend.

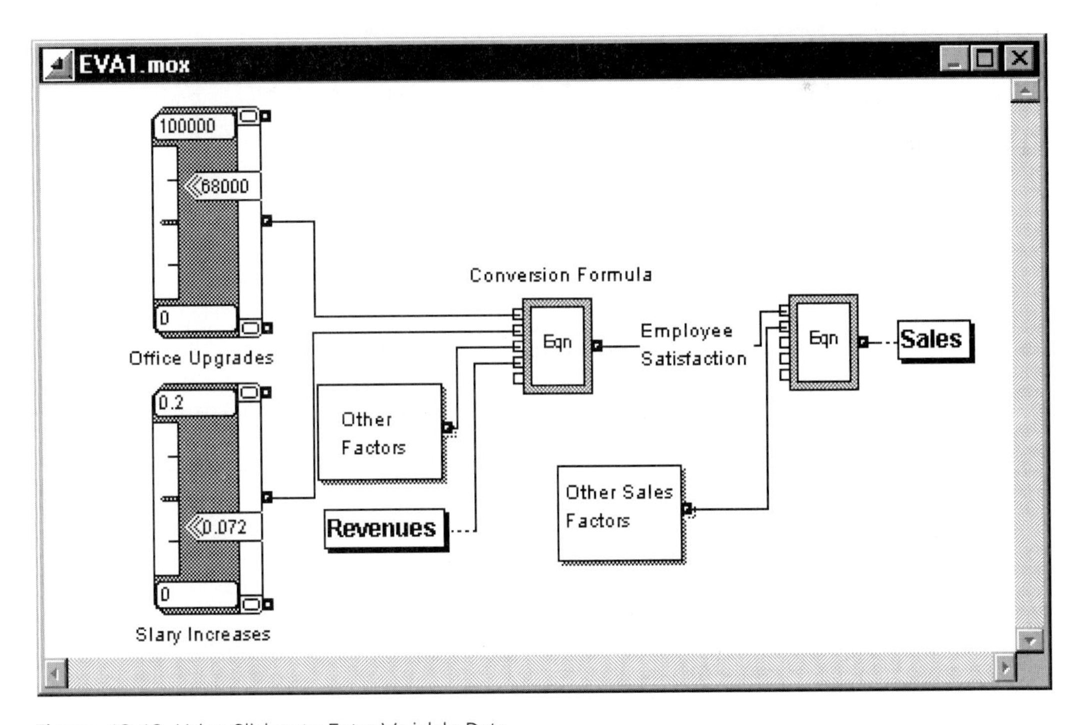

Figure 19.13 Using Sliders to Enter Variable Data.

The *sliders* labeled Office Upgrades and Salary Increases can be adjusted by a user of the model. Assuming that there are other factors leading to employee satisfaction, and that those can be easily changed, then it is easy to play "what-if" gaming scenarios to test the effect of spending on employee satisfaction and ultimately on sales. In addition, there is a relationship between sales and revenues and, since revenues affect the ability to spend money, it can be seen that employee satisfaction not only affects the process, but it is also affected by the process. This is called a *continuous closed loop*.

Finally, after all of the strategic, financial and operational relationships, and formulas have been defined in the model, the simulation can be executed. Figure 19.14 shows the result of the simulation.

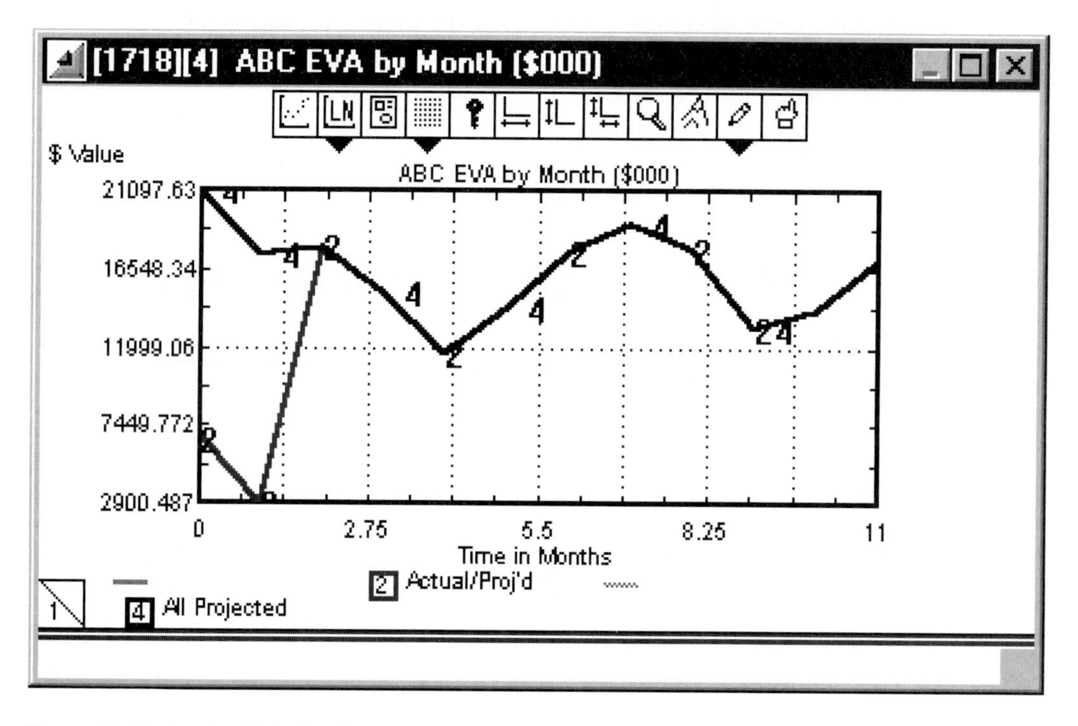

Figure 19.14 Result of EVA Simulation.

The part of the graph labeled with the number "2" represents the simulation using actual data up to a point, in this case, two months, after which it uses projected data. The graph labeled with the number "4" represents the simulation using projected data. This particular model has more than 200 parameters, any or all of which can be changed by a user of the simulation. In addition, it also has many causal equations that use the data supplied by users to calculate other parameters. Although the simulation is extremely complex, it is easily used and modified for dynamic analysis. This is a capability that cannot be supplied with traditional financial modeling tools.

Long-Term Financial Strategic Simulation

The simulation just described deals with the impact of changes to business processes at a somewhat detailed level. There is also a need to determine, at a higher level of abstraction, the impact of complex strategic decisions. Decisions cannot be made in a vacuum; instead, the impact of a decision on one part of a business must be measured against the business as a whole. Using the electric utility as an example, I will demonstrate how simulation can be used to effectively capture the overall impact of any business decision.

Most businesses use a simple formula to project profitability, namely, Revenues – Cost of Sales = Profit. In general, it is also assumed that, as sales increase, revenues increase, but so does the cost of sales, since more inventory is required to meet demand. In the electric utility business, this relationship can be viewed as the causal diagram shown in Figure 19.15.

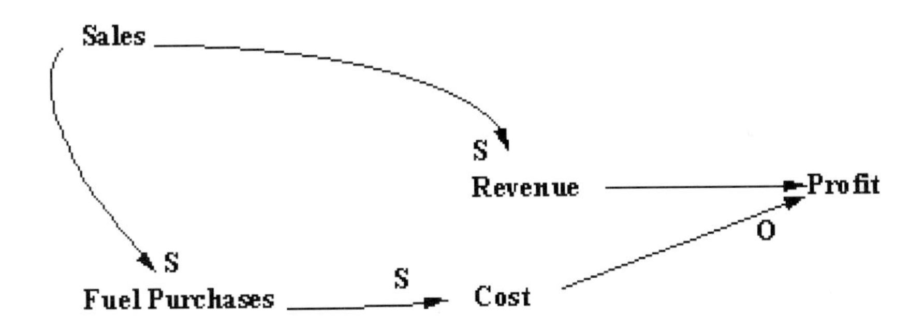

Figure 19.15 Relationship of Sales to Revenue and Cost.

This causal diagram is read: "As sales increase, so do revenues and profit. As sales increase, so do costs, and profits decrease." If sales are greater than costs (an assumption from this point on), then profits increase with sales.

However, utilities are faced with a small dilemma—they can choose to purchase low-cost, low-efficiency fuel, or they can choose to purchase high-cost, high-efficiency, fuel. The impact on profit is shown in Figure 19.16.

This causal diagram shows two cost factors: As the efficiency of the fuel decreases, (increases), the amount of fuel needed increases (decreases) and, as the efficiency goes up or down, the price goes up or down.

To this point, any decision about what type of fuel to purchase can be made with a simple spreadsheet. However, the decision is more complex than it seems. Utilities are being encouraged to burn high-efficiency fuels that give off fewer emissions, and if they do not, the penalties levied by the Environmental Protection Agency increase as the emissions increase. This causal relationship is shown in Figure 19.17.

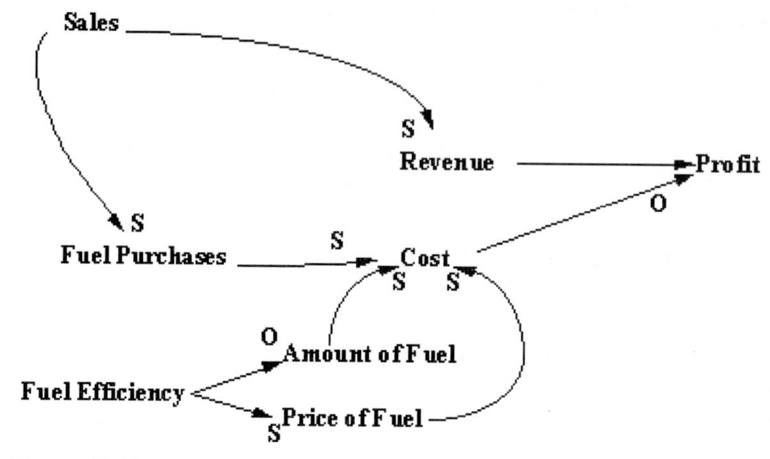

Figure 19.16

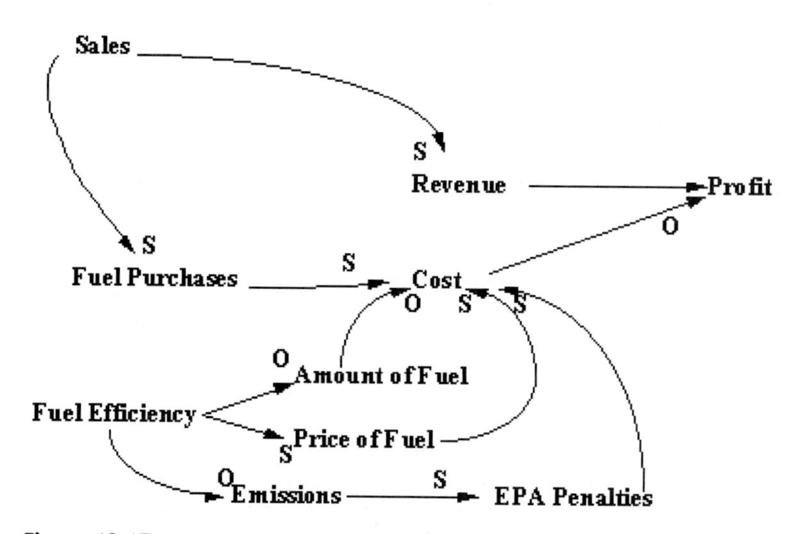

Figure 19.17

This diagram is read: "As efficiency of the fuel decreases (increases), the emissions increase (decrease), as do the EPA penalties levied." At this point a trade-off can be made based on the amount of the fines and the cost of fuel. However, the complexity of the decision does not end with this relationship. Utilities are given an emissions allowance, and if they fall short of the allowance, they can sell the remainder to other utilities. This can be a profitable deal and it encourages the use of higher-cost fuel. The factor of emissions allowance sales is captured in Figure 19.18.

The sale of allowances has added a "wrinkle" to the decision about the type of fuel to buy—higher-priced fuel provides an opportunity to sell allowances, which

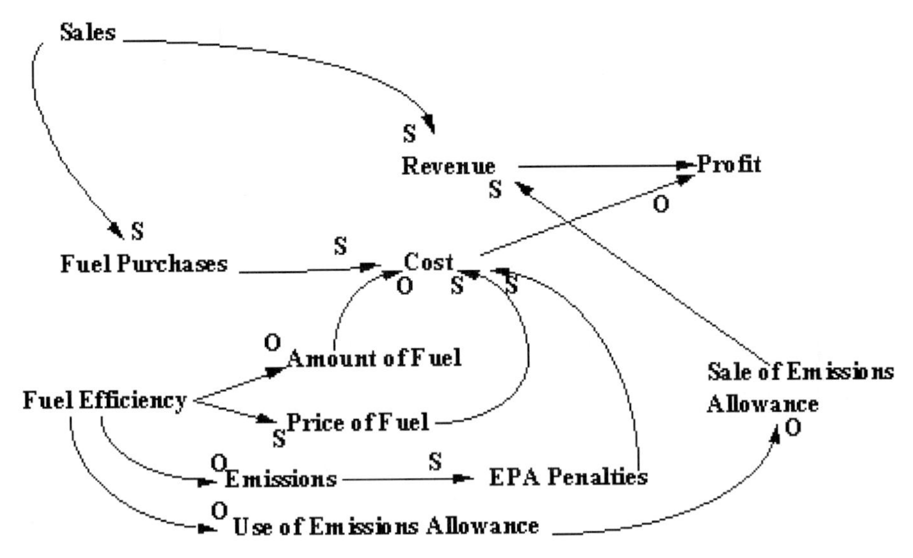

Figure 19.18

affects revenue. However, this assumes two factors: Demand will be low enough so that allowances are left over to sell and the cost of higher-efficiency fuel will not affect the size of the utility's market. Moreover, the price of fuel, the size of the market, and so on, will change over time. This new complexity is shown in Figure 19.19.

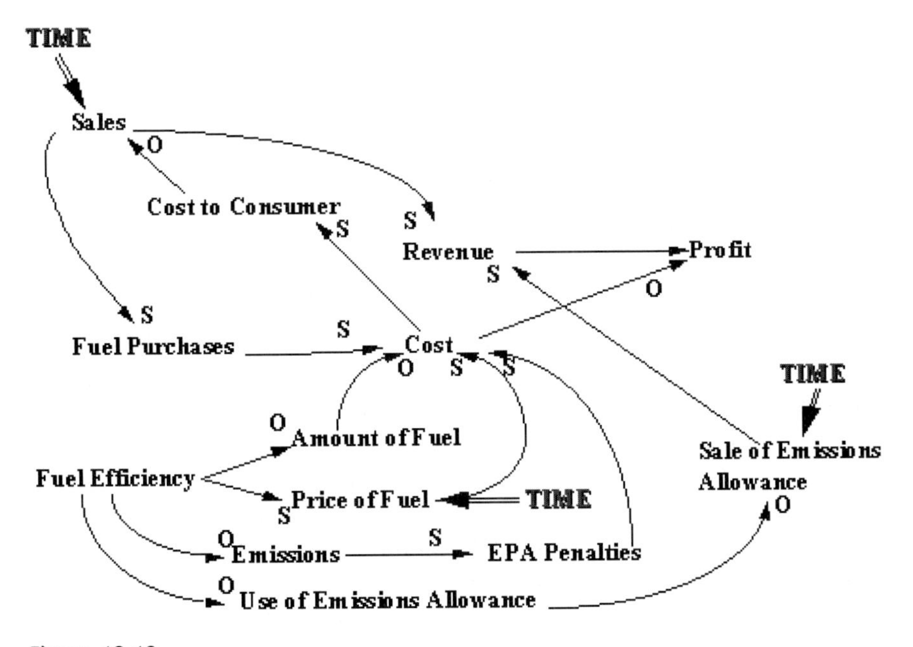

Figure 19.19

The following challenges are faced by any utility.

- The utility must measure the impact of fuel efficiency on price, emissions, potential sales of emissions and so on.
- The utility must determine the impact of corporate profit goals on price charged to consumers and, therefore, market share.
- The utility must consider other factors, such as preventive maintenance, in its analysis. Different types of fuel present different maintenance issues.
- The utility must measure the impact of a changing market over time.

These challenges are simply too complex to be analyzed using spreadsheets and other standard tools. Spreadsheets are not iterative, so they cannot explore multiple scenarios; spreadsheets are not dynamic, so the numbers cannot change; and so on. Software systems are iterative and dynamic but require the assistance of a software expert.

Simulation can be used to explore the strategic decisions just described. Simulation is dynamic, it is iterative, and it is easy to use. Figure 19.20 is an example of an electric utility model developed with the CAPRE tool Extend.

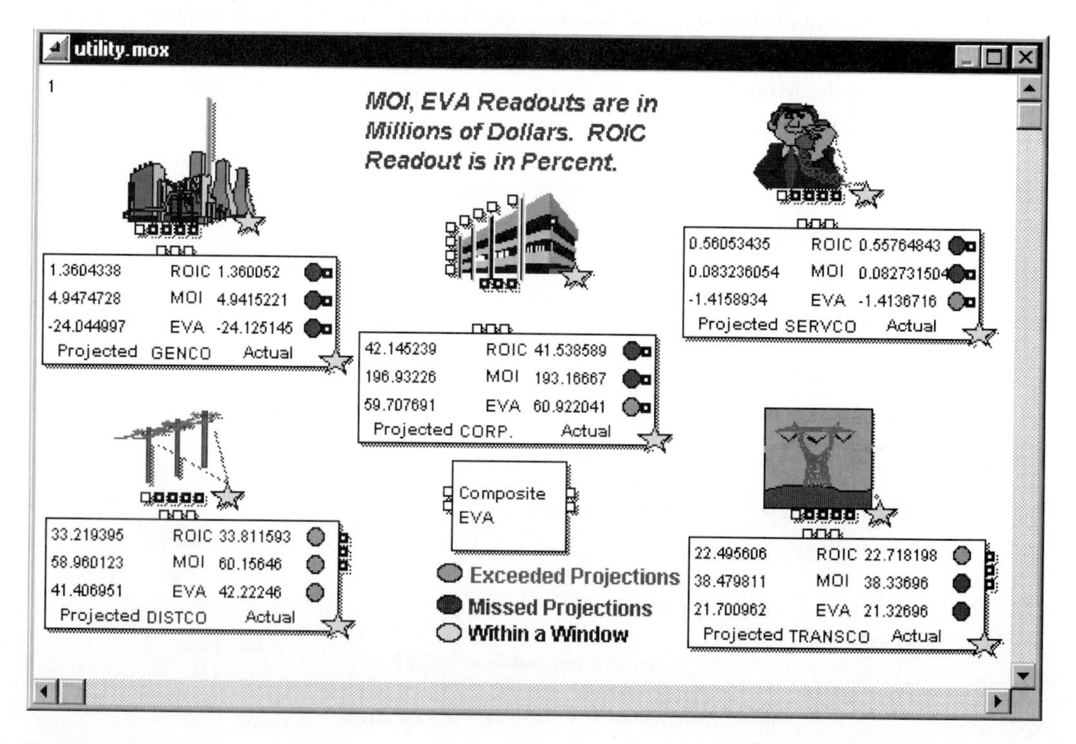

Figure 19.20

This simulation determines Return On Invested Capital (ROIC), Measured Operating Income, and EVA for several of the utility's business units. Each of the graphical representations is a hierarchical block that contains layers of detail. For example, Figure 19.21 shows a portion of the simulation that determines the Cost of Sales. Part of the Cost of Sales is the Cost of Fuel, shown as an input to the hierarchical block labeled Generate Power. Another part of cost of sales is contained in Purchase or Sell Power. This block contains a complex set of equations, some of which explore the relationships described earlier in this chapter. Yet another part is contained in the hierarchical block labeled Construction of PP&E (physical plant and equipment). Each of these is complex and subject to change due to market conditions. The question is: Can a simulation be made user-friendly enough to allow for rapid change of these formulas and/or inputs so that an inexperienced person can use the simulation?

Figure 19.22 shows how Construction of PP&E can be changed easily and quickly.

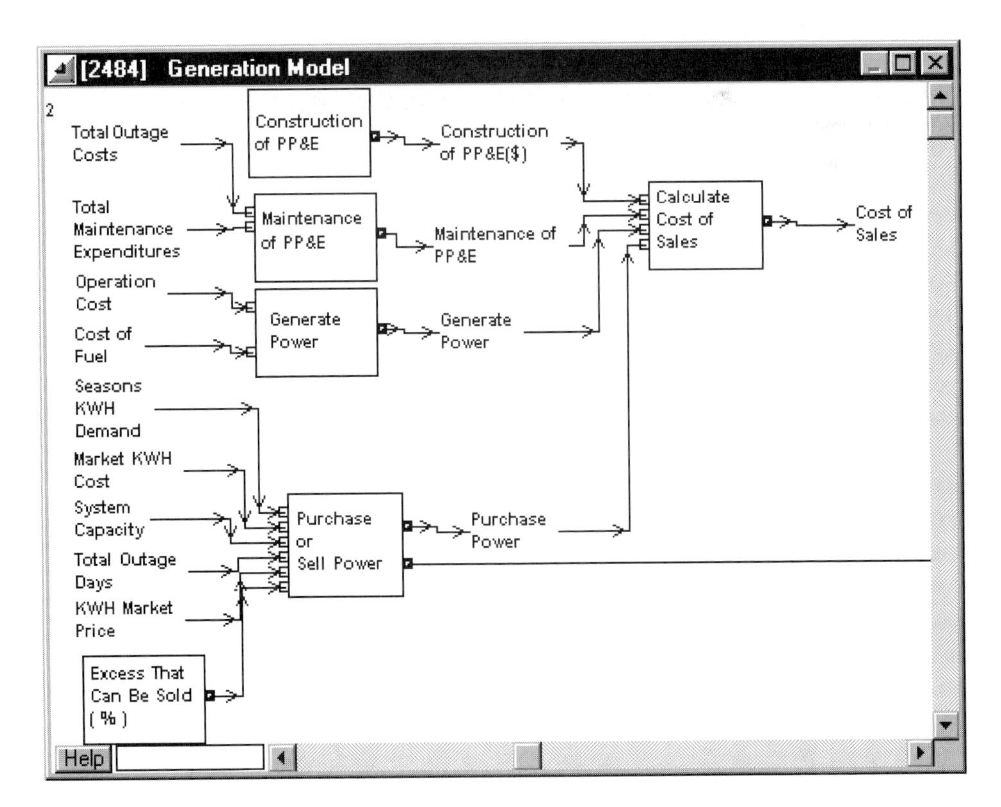

Figure 19.21

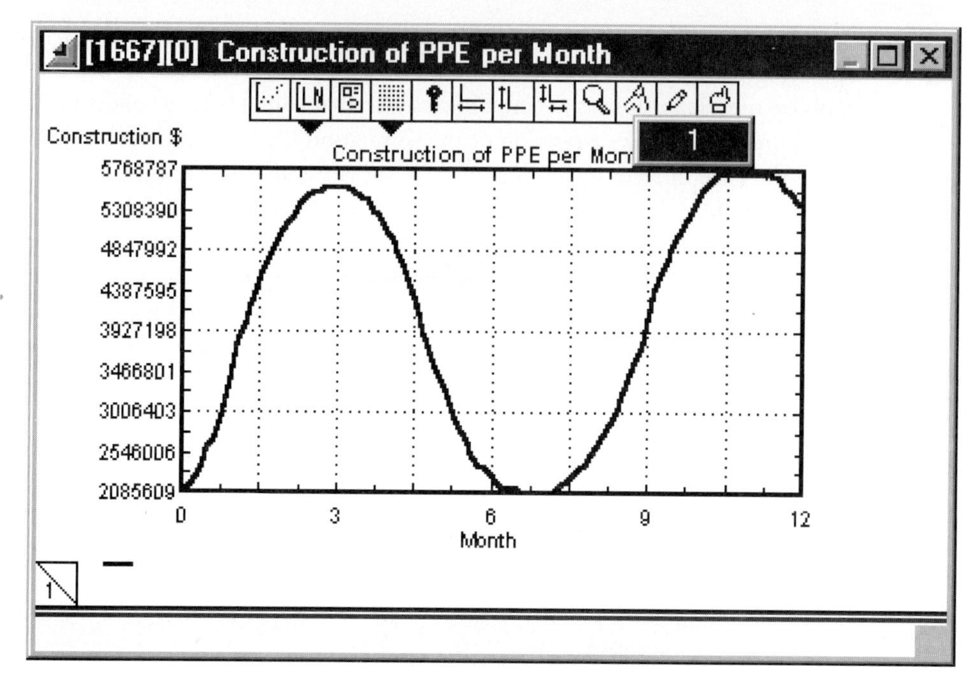

Figure 19.22

This figure shows a graph with projected expenditures plotted over time. If a user of a simulation wanted to change those projections, he or she need only change the plot! This "what-if" scenario change could not be made any easier.

Figure 19.23 shows two methods of determining the cost of fuel. One is a set of costs generated from a table—these are the projections made at the beginning of the year. The second is a formula which projects demand from customers, the amount of fuel needed based on the efficiency of the fuel, and ultimately the cost of the fuel based on demand and efficiency.

The formula is read as "cost is equal to the demand divided by the BTU per MW required to generate power (this yields BTU required)." The BTUs required are translated into tons of fuel required by dividing the BTU by a fuel rating (BTU per ton). This is then multiplied by the cost of the fuel being used. The choice of the cost calculation is made by the user by simply clicking a switch. The tabular method can be used to establish a business baseline, and the formula method can be used to test multiple what-if scenarios, varying the demand for power, the type of fuel used, the cost of fuel, and so on. Each method is powerful and necessary, yet neither requires an understanding of simulation by a user of the simulation.

This simulation can also interrogate remaining emissions allowances and, if they are low, decide that the utility must buy higher-priced fuel and, if they are low, decide to buy lower-priced fuel. It can explore demand, maintenance requirements, and

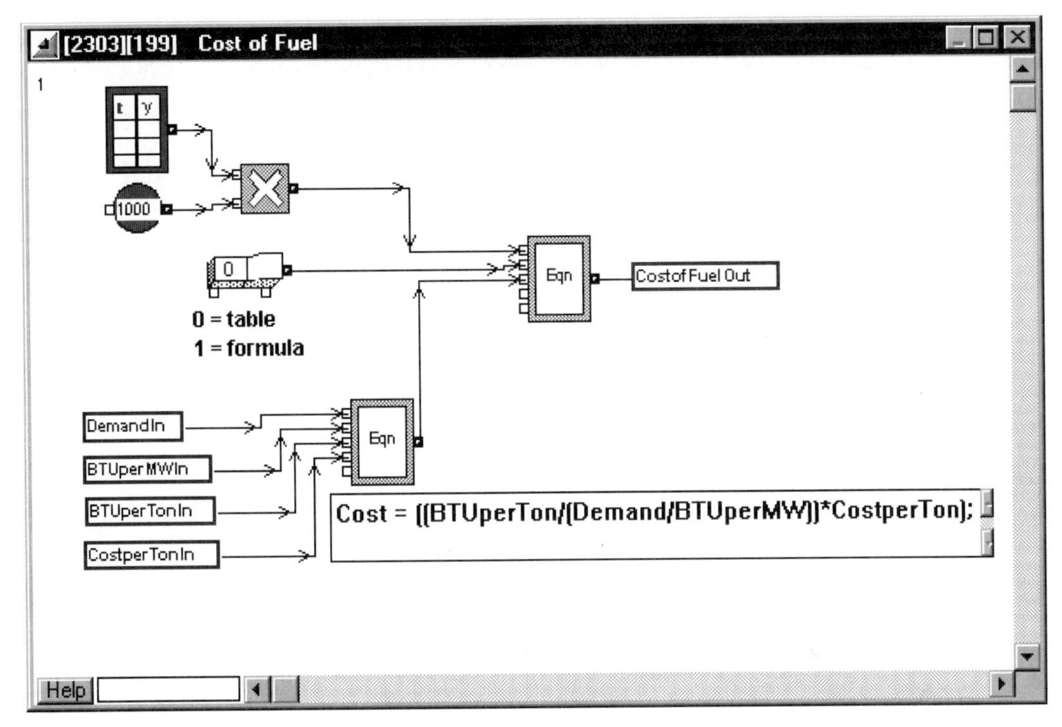

Figure 19.23

many other factors in a similar way and project costs and revenues. This type of dynamic analysis cannot be performed with currently popular tools.

Summary

The types of analysis just described are very complex and ideally suited for simulation. Financial and process data go hand-in-hand. It is often important to analyze the financial viability of a process before implementing it and then analyze the effects of process implementation on the overall business. In addition, it is also important to analyze the effects of overall business performance on the processes themselves. These analyses are all interrelated and can only be accomplished through simulation and CAPRE technology.

20

Getting Started: Suggested Training Approaches

This chapter suggests methods of training that can be used to effectively introduce causal reasoning and computer aided process reengineering to organizations that are committed to TQM and CPI. The goal of each approach is the same: to develop process reengineering facilitators and process engineers. It is the emphasis of each approach that is different.

A process reengineering facilitator is a person who is familiar with the process maturity model and automated process analysis concepts. A process reengineering facilitator is not necessarily familiar with any particular process under investigation but has an ability to interact with process experts to extract important information about detailed aspects of a process.

Why train people as process reengineering facilitators? Experience indicates that experts in processes are likely to skip over information they consider to be unimportant; however, process reengineering facilitators, having no preconceived notions about a process, consider all information important. The use of process reengineering facilitators can be extremely beneficial, since it brings together two sets of experts—the process experts and the process reengineering experts. Therefore, the training suggested here is not limited to understanding the application of modeling and simulation to process reengineering; it also stresses the need to understand the fundamentals of process maturity evolution.

A process engineer does not have to be a process reengineering facilitator. Process engineers are skilled in the use of automated analysis tools, such as CAPRE technology, and can work with process reengineering facilitators to develop comprehensive models of processes under investigation.

The first training approach is methodical in nature and is spread over several sessions. Students have the opportunity to practice what they have learned between each session and apply their skills to business processes. It is this training session that will develop process reengineering facilitators who will also be skilled process engineers.

The second training approach is one in which students begin modeling immediately and learn the Rules of Process Reengineering in a nonsequential manner by learning and applying computer aided process reengineering technology. This session will produce process engineers who are familiar with, but not necessarily expert in, process reengineering facilitation techniques.

A Gradual Method of Learning CAPRE

This method of training is aimed at students at all levels and with a mixed set of skills. Ideally, the students will be familiar with PCs and have some idea of graphical, computer-based software products. The students do not have to be experienced with either modeling or simulation.

Session 1: Introduction of the Maturity Model

Although the process maturity model developed by the Software Engineering Institute is theoretical, it is intuitively correct and usually meets with immediate acceptance. Managers and workers alike seem to accept the fact that processes can be grouped into levels of maturity; moreover, they are immediately able to identify the levels of the process in which they participate. Both the maturity model and the concept of an organized, progressive method of process reengineering are welcomed and appreciated.

In this training session, the instructor will review the maturity model by discussing the characteristics of processes at each level and the tools used to document those processes. Then the instructor will lead the students in interactive discussions about their business processes. In those discussions, the instructor will ask that the students determine the levels of the processes being discussed and discuss their reasons for making those determinations. This session should be relatively unstructured with an emphasis placed on encouraging open dialog.

If the training is spread over several days, the instructor can issue a "homework" assignment. Students will be requested to review one or two business processes that were not discussed in the training session and determine the maturity level of those processes.

Session 2: Introduction of Process Maturity Migration

There are two ways to approach training in process maturity migration. The first method is to invest in some standard CPI/TQM product, such as The Flying Star Ship Factory, and follow the chronology presented in this book. An alternate approach is to simply bring people together to discuss process reengineering efforts that have been successful and some that have failed. The role of the instructor in this training session is to (1) gather as much information about the reasons for the success or failure of the attempts at reengineering, and (2) suggest methods of improving communication, developing documentation, and building process maps.

The instructor will accomplish this by leading the students in discussions on a number of topics, such as,

- How effective was communication between managers and workers? Did management actively seek out suggestions from workers?
- How effective was communication among the workers? Did workers feel free to suggest process changes to workers outside their work area?
- Did workers in one task consider the recipients their work as their "customers"? If so, were they certain they were providing their customers exactly what they wanted or needed? If not, why not?
- What tools were used to develop schedules, assign tasks, determine level of effort required, and so on? How effective were those tools? Were the results of using the tool accurate or inaccurate?
- What was the quality of the products produced? How much rework was required, or how many products were scrapped? What were the reasons for the rework?
- How accurately does process documentation reflect reality? Are details left out or glossed over? If diagrams are included, are they understandable? How often is the documentation updated to reflect process changes?
- If process flow diagrams have been developed, how accurate are they in presenting an accurate view of a process? Do the flowcharts show parallelism, AND conditions, transactions, priorities, and so on? Are the flowcharts ever referenced or were they developed simply to satisfy some company requirement?

After these discussions, the instructor should introduce methods and mechanisms for increasing the effectiveness of verbal communication. Since the purpose of this training session is to discuss process maturity migration, this discussion will be relatively brief; it is not intended to be an interpersonal skills development session. The instructor will emphasize that there is no single *best* way to communicate verbally, but there are many good ways. The instructor will discuss methods such as

- *Coaching versus Judging*—a technique that stresses avoidance of expressing opinions about someone else's comments.
- *Instant Feedback*—a technique in which a process facilitator will validate someone's statement by saying "So what you are saying is...."
- *Brainstorming*—getting groups of people together to discuss a process.

Next, the instructor will discuss methods of developing good written documentation. This may seem like a fundamental skill that everyone has, but this is simply not the case. The instructor will discuss the need to

- Assume that the reader of documentation has no previous knowledge of a process or process task.
- Define all the elements (inventory, reports, computer software, and so on) that must be present before a task can begin.
- Make a description of a task complete by including details that might seem unnecessary.
- Be repetitious. Repetition in documentation for the purpose of reinforcement is valuable.
- List all the conditions that can end a task and the actions that are taken when such a condition occurs.
- Define the product that is created by a task.
- Define quality measures for the task, if any.

After this discussion, the instructor will present methods of developing flowcharts and process maps. Once again, there is no one best method of developing flowcharts, but there are several techniques which can be combined which, when applied, will result in more informative diagrams. Some of the techniques to be discussed are described next.

1. Describe AND conditions in flowcharts graphically by introducing an additional symbol. For example, consider the diagram shown in Figure 20.1. The addition of a simple symbol, as shown in Figure 20.1, can explicitly define a condition in which two tasks must both be completed before a third can be started, thereby making the flowchart more informative.
2. Add symbols to represent multiple-path options based on conditions (Figure 20.2). The addition of a symbol like this allows conditions to be more explicitly presented in a flowchart and enhances the information presented in a flowchart.
3. Invent a symbol to represent a condition that brings a process, or part of a process, to a halt (Figure 20.3).

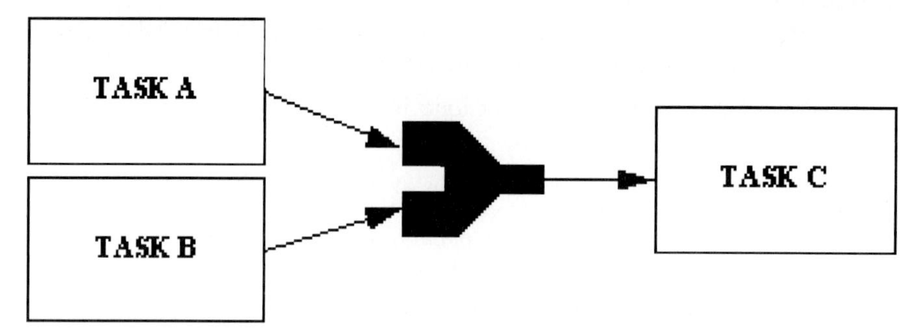

Figure 20.1 AND Condition Depicted in a Flowchart.

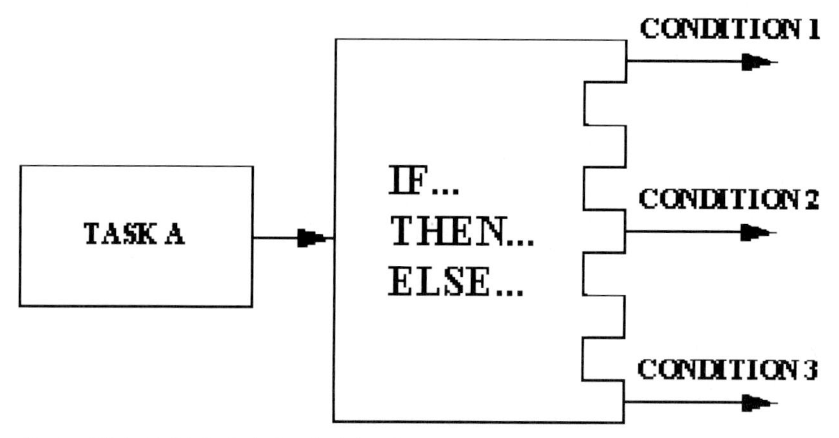

Figure 20.2 Multiple-Path Symbol for Flowcharts.

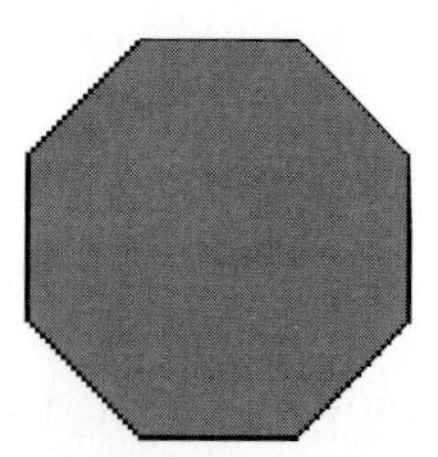

Figure 20.3 Stop Sign Symbol.

These are simple techniques that not only increase the value of process maps and flowcharts but also begin to introduce students to the concept of behavioral analysis. Figures 20.1, 20.2, and 20.3 all represent the depiction of a condition that defines the behavior of a process. Increasing process mapping skills will increase the ability of students to learn computer aided process reengineering techniques.

After this training session, the students will have learned Process Reengineering Rules 1, 2, and 3, and the knowledge of those rules is essential to be able to move on to modeling and simulation. In addition, students will be given some time to put into practice what they have learned.

As a homework assignment, each will be requested to develop a map of some process with which they are familiar and be prepared to use that map when being trained in Process Reengineering Rules 4 and 5.

Session 3: Introduction of Causal Reasoning

It is appropriate to introduce causal reasoning after the discussions of Process Reengineering Rules 1, 2, and 3. In those discussions, the instructor and students discussed attempts at process reengineering, and the students learned

- Effective communication concepts.
- Effective documentation concepts.
- Methods of depicting conditional behavior in process maps.

It is the third point that is most critical for understanding causal reasoning. A causal relation can be defined as a relation of one of the two types shown in Figure 20.4.

Figure 20.4 Types of Causal Relations.

The former relation is the type of relation that is typically described in process maps and flowcharts; the latter is the type of relation that is captured in enhanced process maps and models. Causal reasoning cannot be done effectively without a knowledge of both types of relations.

In this session, the instructor will return to the discussions of process reengineering attempts presented in the previous session. The instructor and students will theorize why certain actions were implemented and what the reasoning was for taking those actions. In the case of failures, students can relate why expectations were not met. If The Flying Star Ship Factory or some other Continuous Process Improvement product is used, the students can discuss what improvements they expected when they implemented process changes, and if those improvements were realized.

Another way to introduce causal reasoning is to discuss some current events and the causal loops that have caused those events. Before conducting the class, the instructor will suggest that students scan the news for examples and bring them into the class.

Regardless of the method used, the instructor will develop a few causal loops to teach the technique, after which the students develop the loops themselves. This should be done in two ways: by the students in pairs during 10- or 15-minute sessions and interactively with the whole audience. The instructor will concentrate on examples that result in three types of archetypes presented by Senge in *The Fifth Discipline*: vicious loops, virtuous loops, and fixes that backfire. These are the most prevalent archetypes in industry and are sufficient for the level of training recommended here.

Session 4: Introduction of Process Reengineering Rules 4 and 5

After students have had some time to work with the Process Reengineering Rules 1, 2, and 3 and with causal reasoning, it is appropriate to introduce the concept of measuring processes and modeling. These are the elements of process reengineering that provide the highest payback; therefore, it is important that they be taught gradually and thoroughly.

In session 2, students were requested to create maps or flowcharts of processes with which they were familiar. In this training session, the instructor will use those maps to begin applying process parameters, such as time required for a step, time to move an item from step to step, number of people required to perform a task, and so on. The instructor will point out deficiencies in the maps when they are discovered and recommend the addition of detail when it is required. When decisions are included in the maps, the instructor will help the students develop IF...THEN...ELSE statements that reflect the logic associated with those decision blocks. At this point, the students will begin to increase their understanding of behavioral modeling.

The next part of this session will concentrate on the actual development of enactable models. The instructor will present Extend+BPR and discuss how each iconic block is used in modeling. The instructor will also present simulation techniques, such as the use of random number generators, event generators, and so on. Finally, the instructor will run through some models to demonstrate the use of Extend+BPR and simulation techniques.

From this point on, the students will develop models of the processes they have been discussing. This activity should be done in pairs, since the initiation into modeling can be intimidating. This is the most important aspect of the training, and it should not be rushed. At the end of this sessions, students will be well educated in the process maturity level, the methods used to ascend through levels, process engineering, and computer aided process reengineering. These students can now work with other employees to assist in process reengineering activities; that is, they can now act as process facilitators.

Suggested Course Outline for the Gradual Approach

The following is a course outline, with suggestions for the length of time required for each session. Sessions can be run consecutively, or with some time between each session (1 day–1 week) to allow students to practice what they have learned.

Session 1	Introduction of the Process Maturity Model
Session 2	Introduction of Process Reengineering Rules 1 through 3
Session 3	Introduction of causal Reasoning
Session 4 (Part 1)	Introduciton of Process Reengineering Rules 4 and 5
Session 4 (Part 2)	Introduction of Computer Aided Process Reengineering (CAPRE) Technology
Session 4 (Part 3)	Introduction of Advanced Use of CAPRE Tools

The Accelerated Method

The accelerated method of introducing computer aided process reengineering follows the same outline as the gradual approach, but the emphasis is on modeling and simulation. Organizations that utilize this method will already have invested in TQM, CPI, and interpersonal training, and several employees will already be working as process reengineering facilitators.

Selection of students for accelerated training is very important. Not only should the prospective students be trained as discussed earlier, but they should also be familiar with graphical, computer-based tools, whether on PCs or workstations. They should also have a firm understanding of behavioral analysis, knowledge of modeling, and an appreciation of simulation. Experience indicates that software engineers are well suited for this training, since much of software is composed of IF...THEN...ELSE reasoning, or behavioral modeling. The knowledge of behavioral modeling is easily transferred to human processes.

Session 1: Introduction of the Rules of Process Reengineering

In this training session, the process maturity level and the Process Reengineering Rules will be introduced and discussed. However, the first three process maturity levels and process reengineering rules will be only briefly addressed, and there will be little or no interactive discussion regarding actual case histories.

Instead, discussions will center on the development of models as process maps, behavioral analysis, and the use of computer aided process reengineering technology. After the discussion of the first three process maturity levels and the associated tools and technologies, the instructor will introduce the subject of process parameters. Using a predefined map of a process selected for the training exercise, the instructor will solicit opinions from the students about

- What process parameters can be associated with the map.
- How those parameters could be used in simulations.
- What parameters could be targets for improvement.

The instructor will then ask the students to develop possible IF...THEN...ELSE rules that could be used to define possible conditions of the diagrammed process. When these have been developed, the process map will be changed to reflect the alternative paths suggested by the conditions and rules.

When this exercise has been completed, the students will have developed a model of the training exercise process and will use this model in a later session to develop a simulation of the process.

Session 2: Introduction of Causal Reasoning

At this time, the instructor will introduce causal reasoning. The emphasis in the accelerated method will not be simply on the development of causal diagrams, but also on how causal relations can be captured in a model. The instructor and students will work with the model developed in the first session, look at the conditions and rules that were defined, and develop causal diagrams to reflect those rules.

When the causal diagramming effort has been completed, the instructors will determine the probabilities associated with each causal relation. This information will be included in the simulation of the process that is addressed in the next session.

Session 3: Introduction of Computer Aided Process Reengineering (CAPRE) Technology

In this session, the instructor will introduce Extend+BPR, review the use of the toolset blocks, and present some examples. This will take place as an interactive discussion rather than in lecture format, and the students will be encouraged to ask questions. The instructor will then spend some time discussing object-oriented concepts and the use of those concepts in process modeling and simulation.

The students will then build an enactable model of the training exercise process, working in teams of two. The role of the instructor will be the same as described in Session 4 of the gradual approach, but the instructor will focus more on providing information. The goal of the instructor is to encourage students to work as independently as possible, taking advantage of the on-line Help and documentation features of Extend+BPR. In addition, students will be required to develop on-line documentation to explain the model being created.

After the modeling exercise has been completed, the students will review and critique each other's models. The focus will be on completeness of the model, richness of scenario development, and the use of documentation features to explain the model. The purpose of the critique is to instill in the students' minds the importance of making a model useful to a viewer as documentation of a process.

At this point, all of the Process Reengineering Rules have been invoked in the following manner:

1. Rule 3 (map the process), Rule 4 (measure the process), and Rule 5 (simulate the process) were all invoked during the development of the model.
2. Rule 2 (document the process) was invoked at some time during model development when students were instructed to use on-line documentation and Help facilities.
3. Rule 1 (talk about the process) was invoked during the critiquing session.

Once the critiquing session has been completed, students will continue to enhance the model and gain experience with advanced modeling techniques.

Suggested Course Outline for the Accelerated Approach

Session 1 Introduction of the Process Maturity Model and Process Reengineering Rules

Session 2 Introduction of Causal Reasoning

Session 3 Introduction of Computer Aided Process Reengineering (CAPRE) Technology

Results of Training

Depending on the preexisting skills of the students who are trained, both the gradual approach and the accelerated approach will provide an organization with individuals who can work independently to improve processes, or work as part of a process reengineering facilitation team to affect change. The approach chosen relates to an organization's investment in TQM and CPI training and its investment in technology. The recommended approach is the gradual approach, since students will be given adequate time to develop both sets of skills. The accelerated approach is suitable for organizations that feel ready to move beyond traditional TQM and into computer aided process reengineering.

21

Some Final Thoughts

What Computer Aided Process Reengineering Cannot Do

People in the business of predicting changes to processes, the economy, world stability, and so on, often say, "I don't know what I don't know." This simply means that there are certain events or conditions that cannot be anticipated so, therefore, there is no way to capture them in a model. For that matter, so many events go on every day, you would not *want* to capture them in a model.

In process reengineering efforts, the more detail that goes into a model, the greater the benefits. The company that chooses to model detailed contingency plans, for example, will be in a better position than the company that does not. Some managers may look at a modeling effort as a potential waste, given the extremely low odds of certain events; other managers may look at a modeling effort as a wise investment, given the potential payback. Adding detail to a model is a matter of choice that is facilitated by CAPRE technology.

> Unfortunately, many organizations want to avoid the hard work that is associated with creating change.

I spoke at length with one client, for example, about the five Rules of Process Reengineering, and the effort associated with the migration between maturity levels.

In addition, I explained that before a modeling effort can begin, a great deal of information must be collected about the process. After we discussed his particular needs, the client decided not to use simulation because (*this is the truth*), "Your method will make us have to think too much."

As amazing as that may seem, it made me realize something very important: Management will, in general, take the easy approach to reengineering, no matter how minimal the benefits.

I like to say that as a rule of thumb, if it is easy, management will do it, no matter what the payback is; if it is hard, management will not do it, no matter what the payback is. Following the philosophies of TQM and CPI is easy; simulation is not. The payback from TQM and CPI is minimal, whereas the payback from simulation can be tremendous.

Computer aided process reengineering is not foolproof—it is possible to build bad models. Just as adding detail to a model requires time, so does building a good model. Process reengineering is (or should be) an iterative process that requires collaboration of managers and workers alike. The more iterations and scenarios that are considered, the better the resulting model. The application of computer aided process reengineering, just as with TQM and CPI, requires a commitment from management and an investment in training and equipment. Organizations that commit investments in TQM, CPI, and computer aided process reengineering will be rewarded many times over.

More on Causal Reasoning

Visit any casino and you will hear dealers urging people to place bets by yelling "The more you bet, the more you win!" That *could* be true, but it could also be true that the more you bet, the more you lose.

One problem with the common implementation of causal loops is that they imply a certainty. In other words, the diagram shown in Figure 21.1 reads: "An increase in A implies an increase in B."

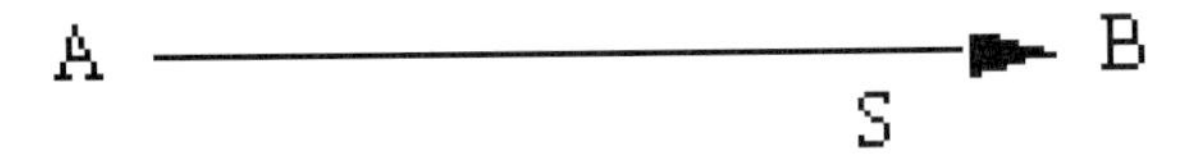

Figure 21.1 Causal Relation.

There is an implied certainty in the way this is interpreted. In reality, when using causal diagrams, there are many occasions when one might want to depict the condition shown in Figure 21.2.

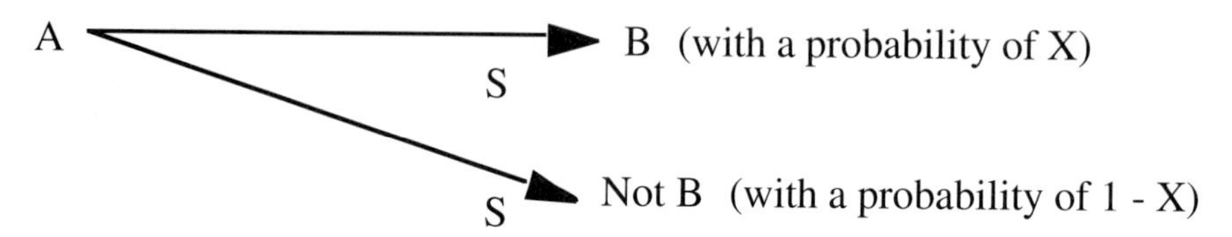

Figure 21.2 Causal Relation Depicting Two Possible Results.

This is the actual case when considering an option. There is a probability of event A causing event B; there is also a probability of event A causing the opposite of event B (or Not B), no matter how small it might be. When considering a course of action, it is important to consider what negatives as well as what positives may result.

This type of reasoning can be modeled. Take "the more you bet, the more you win" scenario. Figure 21.3 is a model of a roulette game in which, with every spin of the wheel, one dollar is bet on each of the same three numbers. Using a random number generator as the wheel, this model will calculate the winnings and losses of the bettor. Each win pays $36, and each loss results in a loss of $3. The simulation reveals that there is a small loss at the end of the simulation.

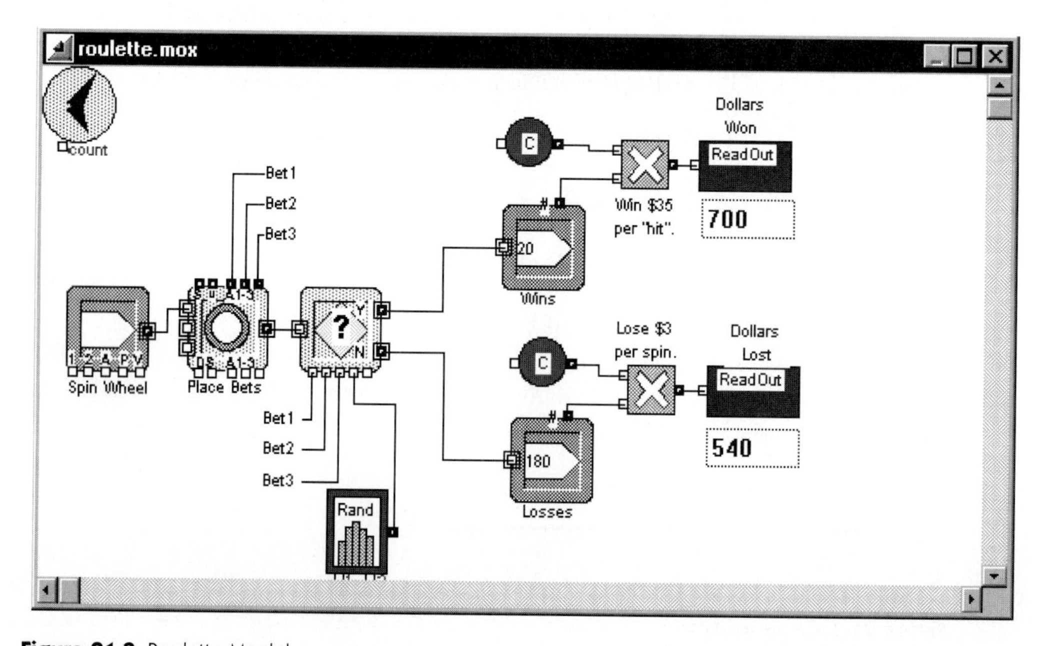

Figure 21.3 Roulette Model.

This simulation shows that there are two possible results of a bet—a win and a loss. As is most likely in a casino, the losses in roulette will be larger than the winnings.

In making business decisions, one must weigh the potential positives against the potential negatives. The "fixes that backfire" archetype indicates that there can be negative side effects or results from a decision. When using causal reasoning in decision analysis, all possible results should be considered. The risk of fixes that backfire, or at least the surprises associated with them, will be reduced. When considering process reengineering changes, all possibilities should be modeled to determine potential paybacks and potential losses. Again, this reinforces the iterative nature of process reengineering.

Spheres of Influence

Spheres of influence is a theory of why things sometimes happen the way they do. It deserves more discussion, since it is an important consideration when considering process or organizational changes.

No matter how much an organization may want to change, the ability to change is governed by influential factors that have various weights. These factors are the spheres of influence. Figure 21.4 captures the essence of spheres of influence very nicely, in this case the influence of varying layers of management.

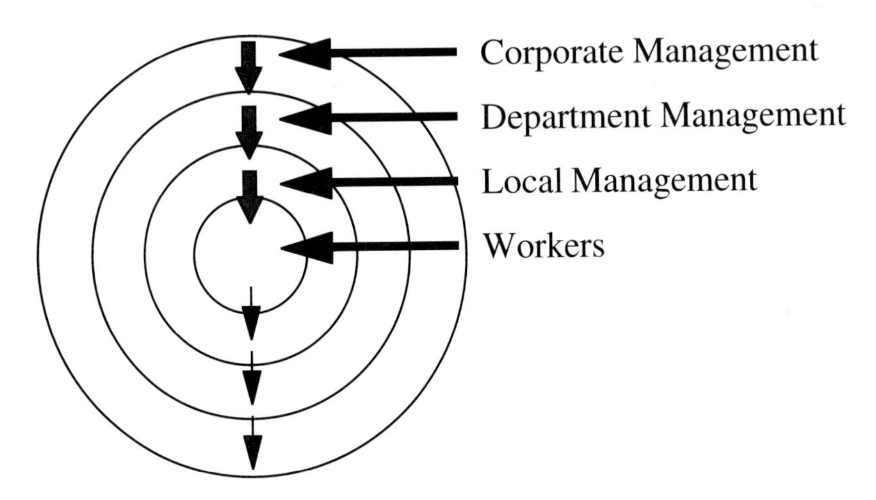

Figure 21.4 Spheres of Management Influence.

Figure 21.4 shows individuals or organizational groups lower in the "pecking order" have less influence on those higher in the order, and vice versa. When discussing CPI or TQM, the influence (or lack of influence) of workers on process change is overwhelmed by the influence of management. It is this type of sphere that Deming is referring to when he says, "Eliminate boundaries between management and workers." In a sphere of influence, the factors on the outermost layers have more influence than those on the inner layers.

This presents an interesting situation. It is the workers, after all, who make the products or services sold by a company. Intuitively, the knowledge of the workers should be highly regarded. Unfortunately, it is often "numbers" that drive a business and, when workers are let go or ignored, their knowledge is lost or buried.

One example of how spheres of influence affected a program is the space shuttle. The space shuttle program was intended to be a mechanism to deploy space-based objectives in a more cost-effective manner. For most of the program, the astronauts in capsules and on the shuttle were test pilots or military personnel, and, as part of their job, they knew there was a risk and accepted that risk.

The opinions of the astronauts were highly regarded, and their recommendations were always given serious consideration by National Aeronautics and Space Administratation (NASA). Similarly, the opinions of NASA personnel were highly regarded by the astronauts. Therefore, the spheres of influence were equal, representing a cooperative agreement between workers (astronauts) and management (NASA). This is reflected in Figure 21.5.

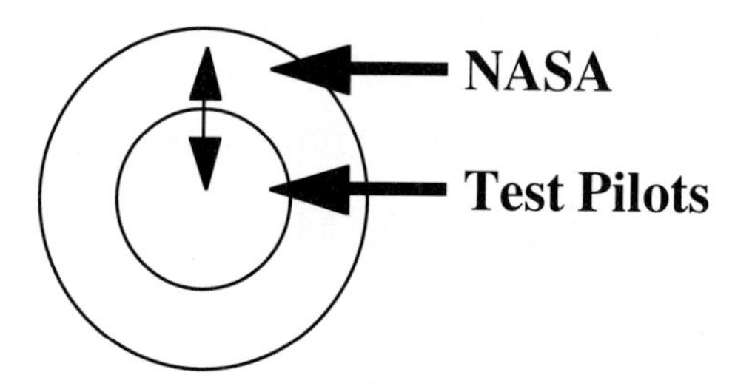

Figure 21.5 Cooperative Spheres of Influence.

The public, however, saw one successful flight after another and perceived there was minimal risk in launching a shuttle. Pressure began to build on NASA to make the program more open. Politicians, ever aware of public opinion, decided to make a political gain from the shuttle program by putting "everyday citizens" into space. In fact, government politicians began to "help" the public with the formation of its opinions regarding the space shuttle. Thus, the sphere of influence for the space shuttle program went from an equal, cooperative partnership between NASA and the astronauts to the relationships shown in Figure 21.6.

In Figure 21.6, we see that politicians helped form public opinion, public opinion forced NASA to make changes to the program, and the opinions of the test pilots

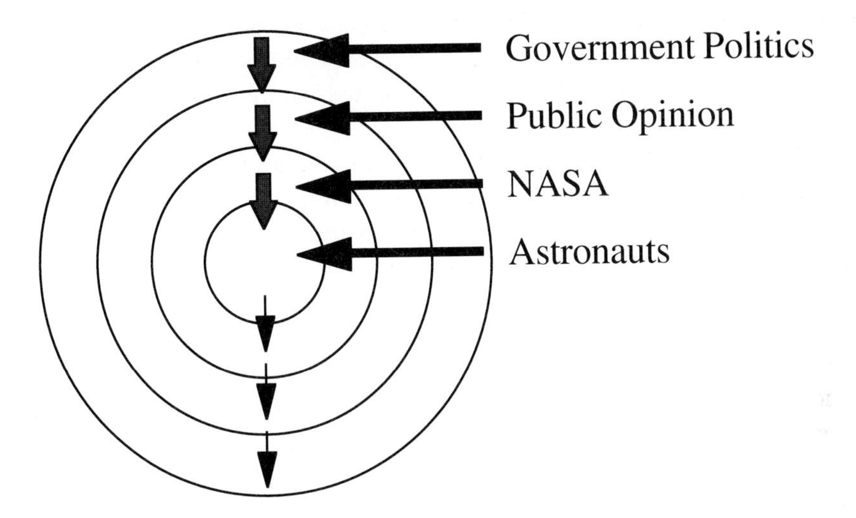

Figure 21.6 Space Shuttle Program Spheres of Influence.

were lost somewhere. Unfortunately, just as the program became the most political, several events happened.

- A contest was held to select the first teacher to go into space.
- The teacher was selected and underwent more than six months of training.
- While the teacher was training, she was the subject of tremendous media exposure.
- The launch of the shuttle was televised.
- The shuttle exploded.

Had the shuttle been manned by military personnel and not a teacher and had the launch not been so highly "advertised" and televised, the program may have been able to absorb the loss more easily. The public and the politicians would most likely have accepted the explosion as an accident, and NASA would have taken what it considered to be necessary actions to avoid a similar problem To this day, however, when the shuttle "tragedy" is discussed, it is almost always discussed in terms of the teacher who died.

This example shows how spheres of influence changed the whole nature of a process. The shuttle program was a strictly scientific program with scientific processes that was changed into a political program with political processes.

Peter Quennell of the Change Institute has been researching the "Russian doll" (doll within a doll within a doll...) relationships of systems within larger systems within still larger systems (a process, within a corporation, within an industry, within a country, and so on). He has found powerful evidence that, in a well-functioning system, many process issues, including those of empowerment and boundaries, are neatly taken care of, and that attempts to change the processes that make up those systems can prove to be distracting and even dysfunctional.

When the NASA space shuttle program began, no one had to tell the astronauts that they were empowered, or that they could communicate directly with management. This was a well-functioning system in which change was self-directed. The space shuttle program was, however, a system within a system. The outer "system" was the political system that consisted of Congress, the Senate, the press, and so on. It was only when other branches of government began acting as strong managers of NASA did open communications begin to close down. Shuttle astronauts became "workers," rather than members of a management team. If any good has come out of the *Challenger* disaster, it is that the spheres of influence that crept into the shuttle program have been eliminated.

Process reengineering facilitators have to deal with spheres of influence on an ongoing basis. The willingness to implement carefully considered changes over time as opposed to quick fixes is inversely proportional to the economy. As the economy turns downward, the impulse to implement quick fixes and abandon iterative process reengineering increases. It is important that process reengineering be addressed utilizing both top-down and bottom-up approaches, in which the needs of management and the ideas of workers are considered. For this to happen, spheres of influence must be eliminated.

CAPRE technology can help. As demonstrated in the purchase order process example, modeling and simulation were used to prove a point to management. Management will almost always argue with opinions but will likely listen to and deal with facts. CAPRE technology provides workers (and managers) with the tools they need to determine the facts of proposed process reengineering changes. This helps eliminate the spheres of influence, thereby facilitating change.

Deming and Computer Aided Process Reengineering

Deming's philosophies have been referenced in this book when the migration of processes through the Software Engineering Institute maturity model was discussed, particularly when discussing Level 1 processes. The reason Deming was referenced is simple—he is the most widely quoted Continuous Improvement advocate in the world. His 14 Points for Management are standard educational material in virtually all TQM and CPI training exercises. A careful examination of Deming's 14 points shows the following:

- Most apply to processes at Level 1 or Level 2.
- They reinforce the need for computer aided process reengineering.
- Application of the 14 points alone will not improve industrial and governmental processes.

Two statements by Deming particularly support the concepts put forth in this book. These statements are quoted from Deming's book *Out of the Crisis.*

About Point 12 (Remove Barriers That Rob People of Pride of Workmanship)

Deming relates this story: "A production worker told me that instructions for every job... are printed and visible, but... by the time she is halfway through she is... so confused she is afraid to go on."[1] This story clearly reinforces Process Reengineering rule 2: document the process so it can be understood and repeated.

About Point 14 (Take Action to Accomplish the Transformation)

Deming says: "Every activity, every job is a part of a process. ... Planning requires prediction.... The effect of a change may be studied... by *simulation*... avoiding experimentation."[2] This statement embodies the philosophy of this book. Changes must be tested before they are implemented. It is this point that summarizes the need to look at processes from the perspective of maturity levels to determine how to migrate through those levels, to establish process parameters, and to simulate changes before implementing them.

Conclusion

We live in the age of technology. Simulation is used in every aspect of our lives. Planes are built using simulation, pilots are trained using simulation, economic decisions are made using simulation. When it comes to making business decisions, we somehow think we can find answers to questions by talking about them.

If change and improvement were so easy, the emergence of new philosophies would cease. But this is not the case. The current popular theory is Kaizen; before that it was Theory Z; before that, quality circles, and so on. Soon Kaizen will be out of date and a new philosophy will appear.

Modeling and simulation are not philosophical. They are not emotional. They are *factual.*

Through this book, I hope that business leaders are convinced to look beyond their current methods of implementing change and into the power of computer aided process reengineering.

[1]Deming, *Crisis* (see chap. 3, n. 5), p. 78.

[2]Ibid., pp. 87–88.

Appendix A

The Flying Star Ship Factory Process

The Flying Star Ship Factory is a product of Block Petrella Weisbord, Inc., 1009 Park Avenue, Plainfield, NJ. 07061. The following is a brief description of the process.

This Continuous Process Improvement training exercise is a model of an assembly-line process. At a minimum, there are two paper cutters, one or two folders for each color paper (yellow and white), two assemblers, one or two painters, one inspector, one material handler, and one manager. The number of people performing any task depends on the size of the class.

Workers performing the same task are grouped at tables, which are usually placed in a sequential order (Figure A.1).

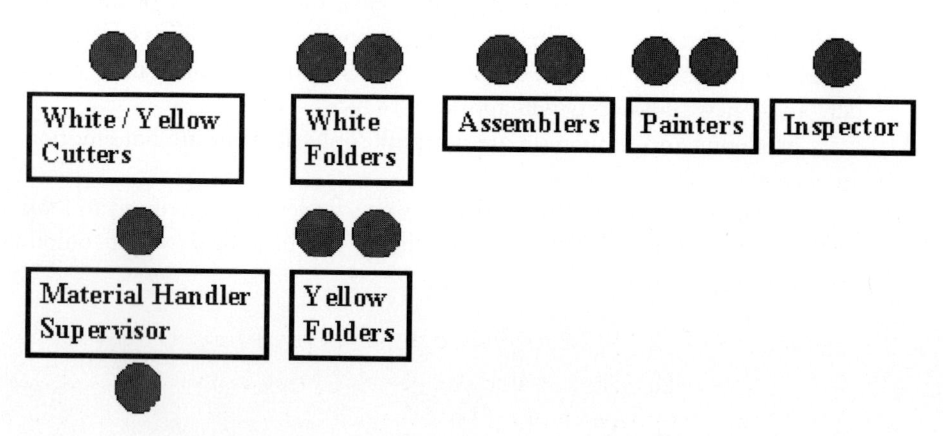

Figure A.1 Table Arrangement in The Flying Star Ship Factory Process.

The process proceeds as follows.

1 Pieces of yellow and white paper are cut to specified sizes.
2 The cut pieces are placed in a stack to be moved to the folders by the material handler.
3 The yellow and white pieces of paper are folded to specified shapes. Figure A.2 shows what a folded yellow piece of paper would look like. Similarly, Figure A.3 shows what a folded white piece of paper would look like. For each diagram, the arrows indicate additional folds.

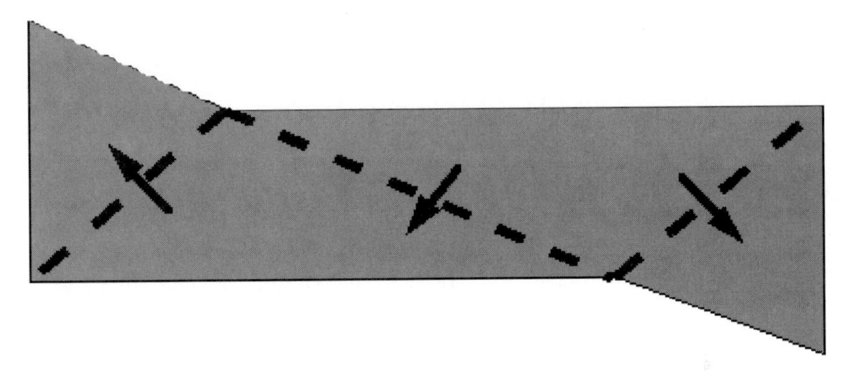

Figure A.2 Folded Yellow Paper.

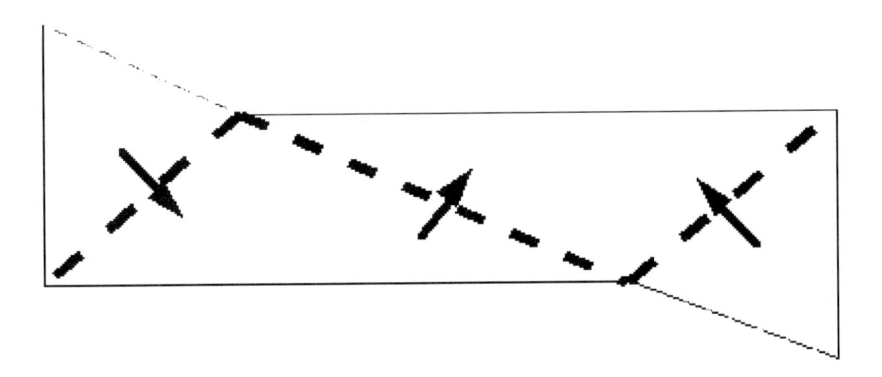

Figure A.3 Folded White Paper.

4 The folded pieces are moved to the assemblers by the material handler.
5 The yellow and white folded pieces of paper are assembled into a four-corner, star-shaped product by tucking folds of each type of paper into folds of the other. Figure A.4 shows the assembled product.

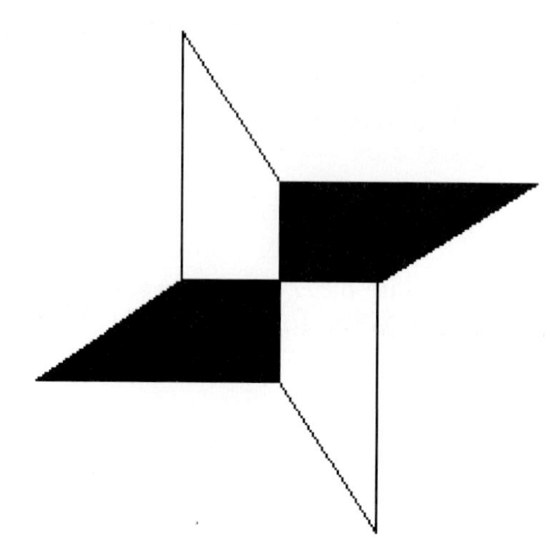

Figure A.4 Assembled Origami Product.

6 The assembled product is moved to the painter's table by the material handler.

7 The front of the white portion of the star-shaped product is painted various colors. Figure A.5 depicts the final product.

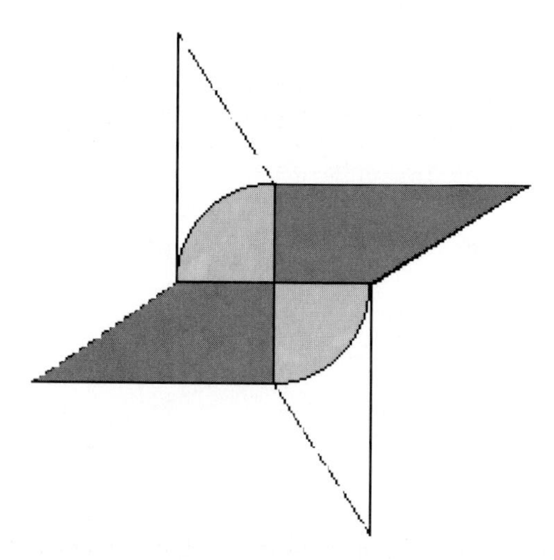

Figure A.5 Final Painted Flying Star Ship Factory Product.

8 The painted (final) product is moved to the inspector's table by the material handler, where it undergoes an acceptance inspection.

9 The inspector delivers the "quality" product to the customer for acceptance.

This is the basic process. Of course the tasks occur in parallel, so it is not necessarily as sequential as the description just provided.

The process operates under the restrictions described earlier in the book and then attempts are made at process improvement. These attempts are successful both in cycle time reduction and quality improvement.

Experimenting with CAPRE Using the Flying Star Ship Factory Process

There is no guarantee that the improvements suggested during the training exercise are the best improvements of the process; therefore, I recommend modeling the process and simulating the effects of various changes. In each case, estimate what the improvement would be and then test your assumptions.

1 Assign the material handler to one of the tasks and have the workers move their products themselves. Using Extend+BPR, this is accomplished very easily (Figure A.6).

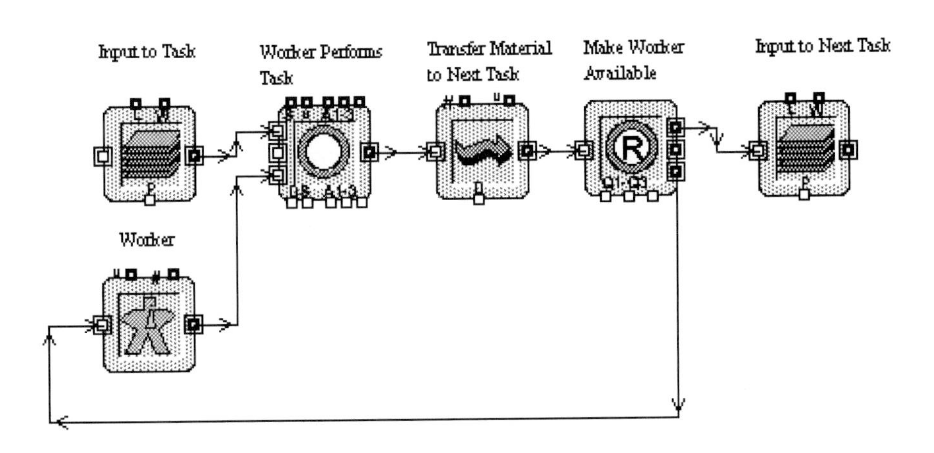

Figure A.6 Workers "Pushing" Products.

In this model, a worker gets an input item, performs a task, moves the finished product to the next storage area, and then repeats his task.

2 Allow the material handler to move among tasks based on the amount of raw material at each work area. This decision of where the material handler should go next would be modeled using Extend+BPR (Figure A.7).

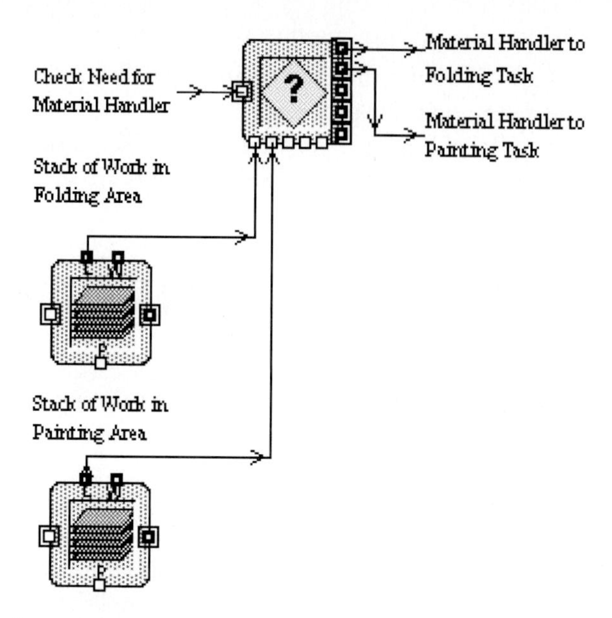

Figure A.7 Determining the Need for a Material Handler.

First, the inventory level of each task in the process would be measured and compared. Using IF...THEN...ELSE logic, the comparisons would be used to determine what task the material handler would perform next.

The addition of the material handler as a temporary worker in any task in the process would be accomplished as shown in Figure A.8.

In Figure A.8, the material handler would perform a task and then exit back to the Decision block.

3 Add inspections at each stage of the process by creating a certain percentage of rejects using random number generators. Reroute rejected products back into the model the first time they pass through the simulation and allow them to exit the model the second time through. The effect of rework on productivity can then be measured.

Determining how many times an object is passed through a model is accomplished with Extend+BPR as shown in Figure A.9.

In this model, a Measurement block is used to read the values of a Time Through attribute and an Accept/Reject attribute which is set earlier in the process.

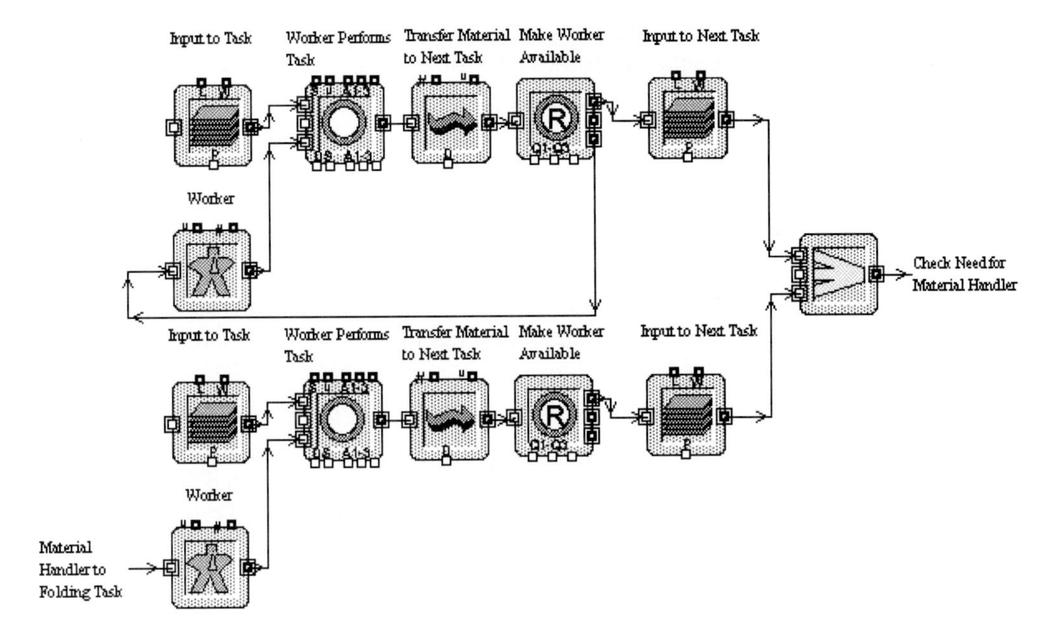

Figure A.8 Depicting a Temporary Worker.

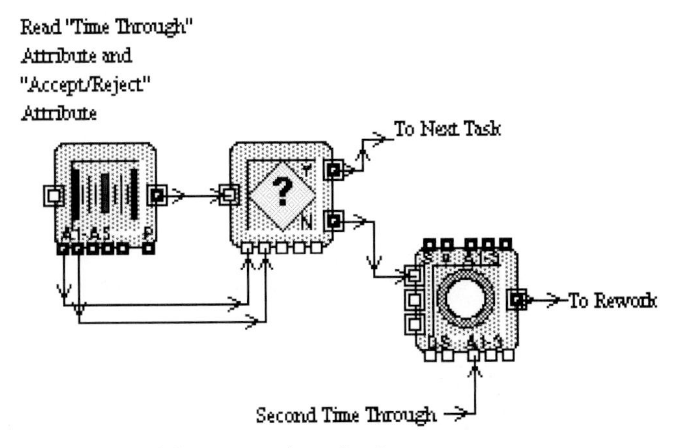

Figure A.9 Modeling Passes through a Process.

The rule in the Decision block will (1) allow any item that has already been through the process once to take the Yes path, or (2) determine if the item will pass through or go back for rework based on the value of the Accept/Reject attribute.

When an item is rejected, the Time Through attribute is set to reflect the fact that it has been through the process once.

4 Simulate the insertion of machinery in one or more tasks. For example, replace the cutters with an automated cutting machine and assign those personnel to other tasks. The only change required to the model of the process is a change to the processing time associated with the task that has had a machine inserted.

There are many such examples of process reengineering changes that can be simulated. These simulation exercises will greatly increase the model developer's knowledge of simulation techniques and demonstrate the utility of CAPRE technology. The Flying Star Ship Factory is not only an excellent TQM training tool; it is also an excellent exercise for learning how to model and simulate organizational processes.

Index

LICENSE AGREEMENT AND LIMITED WARRANTY

READ THE FOLLOWING TERMS AND CONDITIONS CAREFULLY BEFORE OPENING THIS SOFTWARE PACKAGE. THIS LEGAL DOCUMENT IS AN AGREEMENT BETWEEN YOU AND PRENTICE-HALL, INC. (THE "COMPANY"). BY OPENING THIS SEALED SOFTWARE PACKAGE, YOU ARE AGREEING TO BE BOUND BY THESE TERMS AND CONDITIONS. IF YOU DO NOT AGREE WITH THESE TERMS AND CONDITIONS, DO NOT OPEN THE SOFTWARE PACKAGE. PROMPTLY RETURN THE UNOPENED SOFTWARE PACKAGE AND ALL ACCOMPANYING ITEMS TO THE PLACE YOU OBTAINED THEM FOR A FULL REFUND OF ANY SUMS YOU HAVE PAID.

1. **GRANT OF LICENSE:** In consideration of your payment of the license fee, which is part of the price you paid for this product, and your agreement to abide by the terms and conditions of this Agreement, the Company grants to you a nonexclusive right to use and display the copy of the enclosed software program (hereinafter the "SOFTWARE") on a single computer (i.e., with a single CPU) at a single location so long as you comply with the terms of this Agreement. The Company reserves all rights not expressly granted to you under this Agreement.

2. **OWNERSHIP OF SOFTWARE:** You own only the magnetic or physical media (the enclosed disks) on which the SOFTWARE is recorded or fixed, but the Company retains all the rights, title, and ownership to the SOFTWARE recorded on the original disk copy(ies) and all subsequent copies of the SOFTWARE, regardless of the form or media on which the original or other copies may exist. This license is not a sale of the original SOFTWARE or any copy to you.

3. **COPY RESTRICTIONS:** This SOFTWARE and the accompanying printed materials and user manual (the "Documentation") are the subject of copyright. You may not copy the Documentation or the SOFTWARE, except that you may make a single copy of the SOFTWARE for backup or archival purposes only. You may be held legally responsible for any copying or copyright infringement which is caused or encouraged by your failure to abide by the terms of this restriction.

4. **USE RESTRICTIONS:** You may not network the SOFTWARE or otherwise use it on more than one computer or computer terminal at the same time. You may physically transfer the SOFTWARE from one computer to another provided that the SOFTWARE is used on only one computer at a time. You may not distribute copies of the SOFTWARE or Documentation to others. You may not reverse engineer, disassemble, decompile, modify, adapt, translate, or create derivative works based on the SOFTWARE or the Documentation without the prior written consent of the Company.

5. **TRANSFER RESTRICTIONS:** The enclosed SOFTWARE is licensed only to you and may not be transferred to any one else without the prior written consent of the Company. Any unauthorized transfer of the SOFTWARE shall result in the immediate termination of this Agreement.

6. **TERMINATION:** This license is effective until terminated. This license will terminate automatically without notice from the Company and become null and void if you fail to comply with any provisions or limitations of this license. Upon termination, you shall destroy the Documentation and all copies of the SOFTWARE. All provisions of this Agreement as to warranties, limitation of liability, remedies or damages, and our ownership rights shall survive termination.

7. **MISCELLANEOUS:** This Agreement shall be construed in accordance with the laws of the United States of America and the State of New York and shall benefit the Company, its affiliates, and assignees.

8. **LIMITED WARRANTY AND DISCLAIMER OF WARRANTY:** The Company warrants that the SOFTWARE, when properly used in accordance with the Documentation, will operate in substantial conformity with the description of the SOFTWARE set forth in the Documentation. The

Company does not warrant that the SOFTWARE will meet your requirements or that the operation of the SOFTWARE will be uninterrupted or error-free. The Company warrants that the media on which the SOFTWARE is delivered shall be free from defects in materials and workmanship under normal use for a period of thirty (30) days from the date of your purchase. Your only remedy and the Company's only obligation under these limited warranties is, at the Company's option, return of the warranted item for a refund of any amounts paid by you or replacement of the item. Any replacement of SOFTWARE or media under the warranties shall not extend the original warranty period. The limited warranty set forth above shall not apply to any SOFTWARE which the Company determines in good faith has been subject to misuse, neglect, improper installation, repair, alteration, or damage by you. EXCEPT FOR THE EXPRESSED WARRANTIES SET FORTH ABOVE, THE COMPANY DISCLAIMS ALL WARRANTIES, EXPRESS OR IMPLIED, INCLUDING WITHOUT LIMITATION, THE IMPLIED WARRANTIES OF MERCHANTABILITY AND FITNESS FOR A PARTICULAR PURPOSE. EXCEPT FOR THE EXPRESS WARRANTY SET FORTH ABOVE, THE COMPANY DOES NOT WARRANT, GUARANTEE, OR MAKE ANY REPRESENTATION REGARDING THE USE OR THE RESULTS OF THE USE OF THE SOFTWARE IN TERMS OF ITS CORRECTNESS, ACCURACY, RELIABILITY, CURRENTNESS, OR OTHERWISE.

IN NO EVENT, SHALL THE COMPANY OR ITS EMPLOYEES, AGENTS, SUPPLIERS, OR CONTRACTORS BE LIABLE FOR ANY INCIDENTAL, INDIRECT, SPECIAL, OR CONSEQUENTIAL DAMAGES ARISING OUT OF OR IN CONNECTION WITH THE LICENSE GRANTED UNDER THIS AGREEMENT, OR FOR LOSS OF USE, LOSS OF DATA, LOSS OF INCOME OR PROFIT, OR OTHER LOSSES, SUSTAINED AS A RESULT OF INJURY TO ANY PERSON, OR LOSS OF OR DAMAGE TO PROPERTY, OR CLAIMS OF THIRD PARTIES, EVEN IF THE COMPANY OR AN AUTHORIZED REPRESENTATIVE OF THE COMPANY HAS BEEN ADVISED OF THE POSSIBILITY OF SUCH DAMAGES. IN NO EVENT SHALL LIABILITY OF THE COMPANY FOR DAMAGES WITH RESPECT TO THE SOFTWARE EXCEED THE AMOUNTS ACTUALLY PAID BY YOU, IF ANY, FOR THE SOFTWARE.

SOME JURISDICTIONS DO NOT ALLOW THE LIMITATION OF IMPLIED WARRANTIES OR LIABILITY FOR INCIDENTAL, INDIRECT, SPECIAL, OR CONSEQUENTIAL DAMAGES, SO THE ABOVE LIMITATIONS MAY NOT ALWAYS APPLY. THE WARRANTIES IN THIS AGREEMENT GIVE YOU SPECIFIC LEGAL RIGHTS AND YOU MAY ALSO HAVE OTHER RIGHTS WHICH VARY IN ACCORDANCE WITH LOCAL LAW.

ACKNOWLEDGMENT

YOU ACKNOWLEDGE THAT YOU HAVE READ THIS AGREEMENT, UNDERSTAND IT, AND AGREE TO BE BOUND BY ITS TERMS AND CONDITIONS. YOU ALSO AGREE THAT THIS AGREEMENT IS THE COMPLETE AND EXCLUSIVE STATEMENT OF THE AGREEMENT BETWEEN YOU AND THE COMPANY AND SUPERSEDES ALL PROPOSALS OR PRIOR AGREEMENTS, ORAL, OR WRITTEN, AND ANY OTHER COMMUNICATIONS BETWEEN YOU AND THE COMPANY OR ANY REPRESENTATIVE OF THE COMPANY RELATING TO THE SUBJECT MATTER OF THIS AGREEMENT.

Should you have any questions concerning this Agreement or if you wish to contact the Company for any reason, please contact in writing at the address below.

Robin Short
Prentice Hall PTR
One Lake Street
Upper Saddle River, New Jersey 07458